Modeling Farm Decisions
for Policy Analysis

Also of Interest

*Agricultural Economics: Principles and Policy, Christopher Ritson

†Science, Agriculture, and the Politics of Research, Lawrence Busch and William B. Lacy

The Public Role in the Dairy Economy: How and Why Governments Intervene in the Dairy Business, Alden C. Manchester

Farm Management in Peasant Agriculture, M. P. Collinson

Developing Strategies for Rangeland Management, National Research Council

†Planning and Conducting Applied Agricultural Research, Chris O. Andrew and Peter E. Hildebrand

Readings in Farming Systems Research and Development, edited by W. W. Shaner, P. F. Philipp, and W. R. Schmehl

Energy Analysis and Agriculture: An Application to U.S. Corn Production, Vaclav Smil, Paul Nachman, and Thomas V. Long

Industrialization of U.S. Agriculture: An Interpretive Atlas, Howard F. Gregor

Agriculture as a Producer and Consumer of Energy, edited by William Lockeretz

Farming Systems in the Nigerian Savanna: Research and Strategies for Development, David W. Norman, Emmy B. Simmons, and Henry M. Hays

The Political Economy of Mechanization in Agriculture, Barry L. Price

International Agricultural Trade: Advanced Readings in Price Formation, Market Structure, and Price Instability, edited by Alexander Sarris and Andrew Schmitz

†Available in hardcover and paperback.
*Available only in paperback.

Westview Special Studies in Agriculture Science and Policy

Modeling Farm Decisions for Policy Analysis
edited by Kenneth H. Baum and Lyle P. Schertz

Policies and programs aimed at the agricultural production subsector affect both directly and indirectly the economic conditions in the sector and the economic viability of farms. Agricultural policymakers and economists thus are continually challenged to provide the information necessary to project and evaluate the likely effects of alternative policies and programs. This book, the outgrowth of a conference sponsored jointly by the Economic Research Service, U.S. Department of Agriculture, and the Farm Foundation, addresses that challenge.

The contributors focus on the potential of economic and behavioral models at the micro level—the individual farm being the micro unit—to depict the complex nature of growth and adjustment decisions by farm firms in response to various policy and regulatory environments. Although recognizing that, because the microeconomic aspects of dynamic change are so complex, micromodeling cannot provide a complete correspondence to actual behavior, they demonstrate the usefulness of farm growth and adjustment models for evaluating the costs and benefits of governmental policies and their impacts on the performance and efficiency of the farming sector and individual farms. They also identify ways to specify, quantify, and evaluate efficiency, productivity, and performance at the farm level, and examine ways to improve the quality of farm growth and adjustment modeling research.

Dr. Baum and **Dr. Schertz** are agricultural economists with the Economic Research Service of the U.S. Department of Agriculture.

Published in cooperation with the

Farm Foundation

and the

Economic Research Service, U.S. Department of Agriculture

Modeling Farm Decisions for Policy Analysis

edited by Kenneth H. Baum
and Lyle P. Schertz

Routledge
Taylor & Francis Group

LONDON AND NEW YORK

First published 1983 by Westview Press

Published 2018 by Routledge
52 Vanderbilt Avenue, New York, NY 10017
2 Park Square, Milton Park, Abingdon, Oxon OX14 4RN

Routledge is an imprint of the Taylor & Francis Group, an informa business

Library of Congress Cataloging in Publication Data
Main entry under title:
Modeling farm decisions for policy analysis.
  (Westview special studies in agriculture science and policy)
  Papers presented at the Micromodeling Conference held Nov. 18–20, 1981, at Airlie House, Va., and sponsored by the U.S. Dept. of Agriculture's Economic Research Service and the Farm Foundation.
  Includes index.
  1. Agriculture—Economic aspects—United States—Mathematical models-Congresses. 2. Agriculture—Economic aspects—United States—Decision making—Mathematical models—Congresses. I. Baum, Kenneth H. II. Schertz, Lyle P. III. United States. Dept. of Agriculture. Economic Research Service. IV. Farm Foundation (Chicago, Ill.) V. Micromodeling Conference (1981: Airlie House) VI. Series.
HD1755.M62   1983   338.1'873'0724   83-6869
ISBN 0-86531-589-2

ISBN 13: 978-0-367-01986-0 (hbk)
ISBN 13: 978-0-367-16973-2 (pbk)

Contents

Foreword xiii

Preface xv

PART 1
THE HISTORICAL AND THEORETICAL SETTING

1. A Perspective on the Evolution and Status of
   Micromodeling in Agricultural Economics,
   John E. Lee, Jr. 1

   DISCUSSION OF CHAPTER 1

   Farm Modeling: An Overview, *Eldon E. Weeks* 10

   Current Perspectives of a Former ERS Model Builder,
   Neill Schaller 15

2. Farm Decisions, Adaptive Economics, and Complex
   Behavior in Agriculture, *Richard H. Day* 18

   DISCUSSION OF CHAPTER 2

   Some Thoughts on Adaptive Economic Policy and
   the Behavioral Science of Modeling the Farm Firm/
   Household, *Kenneth H. Baum* 50

   A Proper Perspective for Considering Adaptive
   Economics, *David Zilberman and Gordon C. Rausser* 56

   The Challenge of Chaos: A Plea for Parsimony in
   Theoretical Constructs, *Barry Richmond* 61

PART 2
MACRO-MICRO RELATIONSHIPS

3. Complementarities Between Micro- and Macro-Systems
   Simulation and Analysis, *C. Robert Taylor*.................. 63

4. Incorporating Aggregate Forecasts into Farm Firm Decision
   Models, *Donald O. Mitchell and J. Roy Black*............... 73

   DISCUSSION OF PART 2

   Complementarities Between Micro- and Macro-
   Simulation Models, *James W. Richardson*.................. 88

   Understanding and Human Capital as Outputs of the
   Modeling Process, *Thomas A. Miller*..................... 91

   Modeling and Evaluating Policy and Institutional
   Impacts on Farm Firms, *Marvin Duncan*.................. 96

PART 3
NATIONAL POLICY PERSPECTIVES ON MODELING
FARM BEHAVIOR

5. Implications of National Agricultural Policies in Modeling
   Behavior of Farm Firms, *Kenneth R. Farrell*............... 101

PART 4
INSTITUTIONS AND IMPLICATIONS FOR MICROMODELS

6. Financing Growth and Adjustment of Farm Firms Under
   Risk and Inflation: Implications for Micromodeling,
   Peter J. Barry... 111

7. Tax and Other Legal Considerations in the Organization
   of the Farm Firm, *J. W. Looney*......................... 135

   DISCUSSION OF PART 4

   Dealing with Institutional and Legal Information in
   Models of the Farm Firm, *Steve A. Halbrook*............ 146

The Role of Financial and Legal Concepts in Micromodel Building, *Duane G. Harris*....................... 149

Institutional Aspects of Firm-Level Models,
George D. Irwin 153

PART 5
RISK MANAGEMENT IN MODELS OF THE FARM

8. Cash Flow, Price Risk, and Production Uncertainty
   Considerations, *Vernon R. Eidman* 159

9. Hedging Procedures and Other Marketing Considerations,
   Wayne D. Purcell....................................... 181

   DISCUSSION OF PART 5

   Modeling Farm Decisionmaking to Account for Risk,
   Glenn A. Helmers...................................... 195

   Thoughts on Modeling Risk Management on the Farm,
   Steven T. Sonka....................................... 200

   The Need for Prescriptive Risk-Management Models:
   A Suggested Methodological Approach, *James N. Trapp*... 204

PART 6
SIMULATION MODELS

10. FLIPCOM: Farm-Level Continuous Optimization Models
    for Integrated Policy Analysis, *Kenneth H. Baum and
    James W. Richardson*.................................... 211

11. FLOSSIM: A Multiple-Farm Opportunity Set Simulation
    Model, *Jerry R. Skees*................................. 235

12. FAHM 1981: A System Dynamics Model of Financial
    Behavior of Farming-Related Households, *Lyle P.
    Schertz, Linda Calvin, and Lois Schertz Willett*.............. 261

    DISCUSSION OF PART 6

    Multiple-Use Farm Models, *David W. Parvin, Jr.*.......... 279

Positive Farm Models: A Comment on Aesthetic Versus
Functional Value, *Katherine H. Reichelderfer*.............. 281

Micromodeling: A Systems Approach at the National
Level, *John A. Miranowski*.............................. 286

PART 7
OPTIMIZING MODELS

13. A Polyperiodic Firm Model of Swine Enterprises, *Wesley
    N. Musser, Neil R. Martin, Jr., and Donald W. Reid*.......... 289

14. A Recursive Goal Programming Model of
    Intergenerational Transfers by Farm Firms,
    Craig L. Dobbins and Harry P. Mapp, Jr.................... 310

15. Generalized Risk-Efficient Monte Carlo Programming:
    A Technique for Farm Planning Under Uncertainty,
    Robert P. King and George E. Oamek 332

    DISCUSSION OF PART 7

    A Comparison of Optimizing Modeling Methodologies,
    Robert House 351

    Optimizing Models: Some Observations and Opinions
    on the State of Progress, *Garnett L. Bradford*............. 358

    Optimizing Farm Models: Some Concerns,
    Verner G. Hurt 364

PART 8
USE OF FARM MODELS IN EXTENSION ACTIVITIES

16. Specification and Use of Farm Growth and Planning
    Models in Farm Management Extension Activities,
    Ted R. Nelson... 367

17. Specification and Use of Farm Growth and Planning
    Models in Policy Extension Activities,
    Otto C. Doering, III 375

DISCUSSION OF PART 8

A Conceptual Micromodel for Extension Farm
Management and Policy, *Harold W. Walker* 385

Computer Software Needs in Farm Management
Extension Activities, *Gerald Perkins*..................... 389

Use of Farm Models in Extension Activities: An Old
Concept with New Techniques, *B. H. Robinson*........... 393

PART 9
NEEDS OF THE FUTURE

18. Micromodeling in Perspective, *B. R. Eddleman*.............. 399

19. The Use of Micromodels, *R. J. Hildreth*................... 403

Index 409

Foreword

A principal purpose of the Farm Foundation is to improve the level of knowledge available regarding agricultural and rural problems. For that reason, we were pleased to cooperate with the Economic Research Service, U.S. Department of Agriculture, in sponsoring a conference in November 1981 related to farm level- or micro-modeling and policy analysis, and then to support the publication of the proceedings.

The agricultural economics profession is continually challenged to enhance its contribution to society—through dialogue among members and interaction with policymakers—with the objective of developing more effective methods of anticipating the effects of alternative policies without advocating specific policies. This book continues this dialogue and contributes to an increased capability by agricultural economists to respond to agricultural problems and policy issues.

R. J. Hildreth
Managing Director
Farm Foundation

Preface

Microeconomic modeling has been an important tool for agricultural economists for several decades and promises to be important for addressing the research problems of the 1980s as well. This volume explores the possibilities for using micromodeling to analyze how individual farm businesses react to and are affected by farm policies. Although this purpose represents only one potential use of micromodeling, effective modeling for policy analysis necessitates a broad look from several historical, analytical, and institutional perspectives. The Micromodeling Conference held November 18–20, 1981, at Airlie House, Virginia, under the auspices of the U.S. Department of Agriculture's Economic Research Service and the Farm Foundation reflected these concerns.

The objectives of the conference were:

- To develop a better understanding of the "state of the art" of micromodel building, and
- To identify how this state of the art can contribute to the building of micromodels for policy analysis and their experimental use in policy-related research.

Those who participated in the conference substantially realized the first objective of developing a better understanding of the state of the art involved in micromodeling. We hope that the availability of this book will help others to share in this understanding and will stimulate all of us to advance the art of micromodeling. Just as importantly, though, the conference points up the need for greater attention to the second objective. It is evident that much thought and work is needed to better answer "so what" questions. For example, which micromodeling approaches should be used for specific policy issues, and how should aggregate and micromodels be linked? Many insights into this kind of question are scattered throughout the book. However, more time than was available during the conference and much experimentation will be needed to adequately develop the potential of micromodeling for policy analysis.

The proceedings of the conference, presented here, are thus likely to satisfy several needs. A number of papers address questions of theory and describe current models. Balancing this technical discussion are papers that examine the potential of micromodeling for policy analysis and the economic, institutional, technical, and professional forces that are likely to shape micromodeling work by agricultural economists and extension specialists.

In sequence, the book focuses on a historical perspective of ERS micromodeling efforts in the 1960s; changes in the institutional and policy setting; theoretical contrasts important to micromodeling, such as credit arrangements and tax rules that affect behavior of micro units in farming; behavior in response to risk; examples of simulation and optimizing models; use of micromodels in farm management extension and policy extension work; and a perspective on the conference as a whole, including a look toward the future.

Policymakers are challenged to selectively discontinue some farm programs and continue to modify and initiate yet others. Their primary challenge is to develop a set of policies and regulations that accomplishes legislative objectives while avoiding unexpected effects and costs. In turn, agricultural economists are challenged to provide insights into the direct and indirect costs and benefits of alternative farm-related programs and regulations. Responses to policies, and therefore the effects of policies, differ widely among people involved with farming. Consequently, there is an increasing realization that wise selection among alternative policy options requires us to anticipate how policies may differently affect people involved in farming as well as to anticipate the potential aggregate effects of policies on broad measures of well-being, such as U.S. farm income.

Members of our profession pride themselves on being able to relate effectively to complex policy issues. Yet, in spite of extensive research, model building, and policy analysis by agricultural economists, we are left with substantial uncertainties about the likely effects of many government policies and regulations. The rapid changes occurring in farming, uncertainties as to whether they are a result of or in spite of government policies and regulations, and the perceived inability of economists to convincingly discuss the effects of alternative policies and programs contribute to these doubts.

Sector and other aggregate models already provide analyses and related estimates useful in policy selection. We expect that policymakers will continue to rely on these models. In contrast, however, formal microeconomic models have been utilized in only a limited manner to support decisions on agriculturally related policies. These circumstances lead us to ask whether increased use of micromodels would contribute to wiser policy decisions. The substitution of micromodels for aggregate models is not at issue, for the use of micromodels would not be effective without aggregate models that incorporate markets for factors such as land, labor, and credit, as well as for products.

An examination of the potential contribution of micromodels to policymaking is appropriate for all policy and program situations. Policymakers want to anticipate the varied effects of policies among individuals, especially if some of the effects would be politically undesirable. It is possible, of course, that some persons pressing for particular policies would prefer to limit the awareness of distribution and equity effects. However, this position is inconsistent with democratic principles of government and an informed electorate.

Agricultural economists have constructed micromodels for farm management to help individual farmers improve their financial welfare in response to policies, program provisions, and the general economic environment. Although these models are not intended primarily to support policy analysis, they represent a rich knowledge base of analytical techniques and research information. In addition, the economists who developed these models are an important reservoir of expertise on micromodeling and possess unique understandings of the behavior of individuals involved in farming.

In conclusion, we would like to express our appreciation and remember those who have given their time and support in preparing this book for publication. Preparation of the text could not have occurred without the diligence and patience of Sylvia Bryant, Sandy Swingle, and Marionetta Holmes. The editorial suggestions and expertise supplied by Ray Bridge and Judith Latham have been invaluable. The strong support for the conference and publication of these proceedings by Jim Hildreth, Kenneth Farrell, John Lee, Gene Mathia, and David Harrington has never wavered. Finally, we would like to thank the many participants who worked hard and who made the conference a remembered event.

Lyle P. Schertz
Kenneth H. Baum
Washington, D.C.

Modeling Farm Decisions
for Policy Analysis

The Historical and Theoretical Setting

1. A Perspective on the Evolution and Status of Micromodeling in Agricultural Economics

John E. Lee, Jr.

Abstract During the 1950s and 1960s, agricultural economists in government used models of firm behavior to help farmers cut costs and improve efficiency. Later, the focus shifted to longer term growth and financial management, then to the study of aggregate supply response. Macromodeling largely supplanted micromodeling in the Economic Research Service during the 1970s, reflecting the new concern for national and international agricultural issues. Micromodeling is again proving useful for 3 specific needs: understanding likely responses of firms to specific economic conditions and policy provisions, understanding the likely distributive effects, and providing additional detail and likely behavioral responses not well specified in macromodels.

Keywords Distributive effects, Economic Research Service, farm growth model, micromodel, policy analysis

Micromodeling activity and interest in the Economic Research Service (ERS) parallel the evolution of the ERS program itself, and, to a large extent, the evolution of agricultural economics.

In the 1920s, 1930s, and 1940s—even the 1950s—much of the profession's attention focused on problems consistent with a commonly held perception of the farm sector. The farm sector was perceived as

The author is the administrator of the Economic Research Service, U.S. Department of Agriculture.

a large number of small, inefficient farms whose operators were caught in the trap of trying to continuously adopt new technology to survive, and by so doing, kept both productivity and total output rising faster than demand. Hence, all the attendant feedback problems of chronic overproduction, depressed prices, and depressed incomes also occurred. In that setting, the profession gave high priority to economic analysis designed to help farmers improve their management skills by reducing costs and increasing efficiency. Farm accounting and budgeting techniques became the predominant tools of farm management and production economics during the twenties and thirties.

By the 1940s and 1950s, more attention was given to firm growth as the adjustment the farm sector needed to take advantage of new technology and the resulting low prices. Regional projects were established to determine the quantity and mix of resources and the mix of output that individual farms needed to assure adequate levels of family income (Colyer and Irwin; Day and Sparling; Northeast Dairy Adjustments Study Committee; Sundquist et al.). Studies of firm growth strategy became a major professional activity (Irwin).

Although the use of farm accounts and budgeting permitted a form of firm modeling, micromodeling as we know it today became important in the fifties and sixties with the rapid emergence of linear programming (LP) techniques as the key analytical tool of production economists. This development, facilitated by newly available computer capacity, greatly increased analytical flexibility to study cost minimization, profit maximization, minimum resource levels needed for specific income levels, and optimum management adjustments to changes in market conditions and agricultural policies.

The convergence of computer availability, rapid improvements in computational speed, LP, developments in related software programs, and the strong interest in problems of farm firm management and farm-sector adjustments led to a virtual explosion of micromodeling activity.[1] One wit commented that the decade and a half from the mid-1950s to the late 1960s was largely devoted to empiricizing and computerizing the graphic displays in Earl Heady's textbook! In fact, it is hard to overestimate the importance of Professor Heady's and other economists' clear expositions of input-input, input-output, and output-output relationships to the explosion of micromodeling activity that trained a majority of a whole generation of production economists. Parallel developments were also occurring in marketing economics. Many of these studies focused on optimal size and efficiency strategies for individual marketing firms, especially the optimal processing plants. Successful LP models optimizing the decision process in a single decision period then led to simultaneously optimizing over multiple or sequential decision periods (Day). This methodological extension encouraged study of growth strategies and longer term financial management strategies. The introduction of simulation (basically a logical display of sequential decision processes and decision criteria) and recursive programming techniques aided these temporal optimization studies.

The other logical extension of these new analytical tools was for spatial analysis of the more aggregate commodity supply behavior. For policy reasons, it is important to know the shape of the aggregate supply response function for major commodities with price supports. Large surpluses and burdensome storage costs caused policy analysts to devote more attention to how to modify programs and their implementation to reduce federal budget cuts. This research required improved understanding of how farmers' group behavior was likely to respond to various economic and policy incentives.

The most visible institutional response to the multiple interests in farm-level micromodeling in the late 1950s and 1960s was the establishment of a set of cooperative USDA and land-grant regional adjustment studies (S-42, NC-54, W-54, and others). Using LP as the basic tool, these studies focused massive research resources on firm-level studies (firm growth, optimum response to policies and prices, minimum resource requirements for specified family incomes, firm cost minimizations, and so forth), studies of more aggregate state and regional supply functions, and optimum regional use of resources and mix of output.

The approaches to use of microeconomic theory and micromodels to deal with aggregate or macro issues varied. First, the data requirements for realistic LP models were awesome. Agricultural economists probably collected more firm-level data on resources and production practices during this period of increased funding of regional adjustment studies than in any period before or since. Because of the large data requirements, estimating aggregate supply functions by simply aggregating supply functions for every individual firm became impractical. Alternative approaches included aggregating representative farms or treating a particular land base such as a river basin, county, state or multiple-state region as a large farm. Consequently, great effort was expended on defining representative farms within homogeneous regions. Ultimately, all these efforts confronted the same basic problem of aggregation error, which in turn was caused by an inadequate data base.

Eliminating or reducing aggregation error became a major research preoccupation of quite a few economists in the 1960s. A substantial body of literature was written on this topic because the problem was seen as the major obstacle to the development of credible aggregate supply functions. Although the theoretical conditions for completely eliminating aggregation error were eventually determined (Lee), the necessary conditions made impractical the aggregation of individual or representative farms into realistic, short-term estimates of likely supply response.

In summary, ERS was fully involved, often in a leadership role, in these micromodeling activities. In the 1940s and 1950s, many poor farmers still had inadequate resources to be efficient or to earn incomes competitive with the nonagricultural sectors. Improving their well-being was a priority of public policy and hence a legitimate activity of ERS economists. The 1960s brought increased interest in, and relative priority of, a more complete understanding of aggregate supply behavior for use in formulating and analyzing policy. Micromodeling, especially in the regional adjustment studies, appeared to be a promising approach to this policy research. Thus, at the end of the 1960s, ERS had developed a long history of studying firm-level production and marketing problems.

Shifts in the 1970s: From Micromodeling to Macromodeling

In the 1970s, ERS made a significant resource shift away from firm models to macroeconometric models. This shift was a rational response to changing needs and partly reflected ERS's unique role and responsibilities. The continuing frustration with data and methodological problems, a growing concern with regional and national commodity policy issues, the rapidly growing interdependence of U.S. agriculture with world markets, and the changing organization of the farm sector were all contributing factors.

The market shocks surrounding the Soviet grain sale and the world food crisis of the early 1970s also caused a major internal reevaluation of ERS programs and priorities. ERS researchers realized that the future of U.S. agriculture and the well-being of farmers were increasingly tied to international market developments. This situation implied considerably more uncertainty and volatility in demand and prices than we had become accustomed to during the relatively stable 1950s and 1960s. The information system of USDA needed major improvement if farmers and policymakers were to be kept adequately informed of likely international and domestic developments and their consequences. Several investigations were made as to why ERS had failed to forecast the events of the early 1970s. The most important investigation was commissioned by the Council of Economic Advisors and conducted by Karl Fox. The Fox report noted several deficiencies. A critical deficiency was the lack of a comprehensive analytical framework or model for assuring consistency among the commodity components of the ERS forecasting program and for linking commodity developments to aggregate farm-sector indicators and general economic conditions. A new emphasis on econometric models for both forecasting and policy analysis emerged.

The growing concern with national policy issues, personnel ceilings (beginning in 1972), and declining budgets in real dollars reemphasized the ERS resource shift away from farm-level and regional-level adjustment problems. Evidence began to accumulate that showed that the farming sector had been undergoing a gradual change toward a bimodal distribution of farms into two groups: a relatively small number of farms with enough resources to be efficient and to generate returns to resources and incomes competitive with the nonagricultural sectors, and a large number of farm households that typically did not depend primarily on farming for a living, but whose total incomes from all sources compared favorably with that of nonfarm families. This evidence, together with growing concern about possible concentration of market power by fewer and larger farmers, greatly reduced the urgency of ERS micromodeling for advising farm operators, particularly about growth strategies. Moreover, availability of similar private advisory services had been increasing.

Finally, new areas of policy concern required further redirection of resources from traditional farm management and production economics research and even from firm-level marketing research. These new areas included: environmental problems; structure and performance of the food delivery system; food distribution programs; regulatory impact studies; international supply, demand, and trade; transportation; energy; improved understanding of demand shifts; and renewed interest in study of agricultural policy alternatives and implementation procedures.

Gradually by the early to mid-1970s, ERS had come to view itself as the economic intelligence arm of the national agricultural policy establishment. With limited resources ERS focused on providing economic understanding and information about agricultural issues of broad national concern. In addition, ERS maintained primary responsibility for providing current public supply, demand, and price information to increase the efficiency of price discovery and resource allocation in private markets.

The ERS reorganization of 1973 represented almost a total shift away from microlevel analyses toward national analyses. The Commodity Economics Division was responsible for aggregate supply, demand, and price analysis and structure and with performance information for the major commodity subsectors—for study of the macro- or aggregate economics of the commodity subsectors. The National Economic Analysis Division was responsible for research on cross-commodity and farm-sector level economic issues from an aggregate perspective.

The typical farm research, started by Sherman Johnson and carried out for many years under Wylie Goodsell's leadership, was phased out.

After 1973, the agency did not designate a staff or unit specifically identified to do farm management research or microanalysis of marketing firms. The shift away from microanalysis was virtually complete except for some continued, large-scale, regionalized LF work, mainly with the various models at Iowa State University, partially funded by the Natural Resource Economics Division.

The ERS emphasis on issues of broad national and international policy significance as compared with local or firm issues is not likely to be reversed. With further declines in resource availability, we may have to refocus our program even more closely on the fundamental responsibilities of USDA. However, we are also becoming aware that we are missing some key economic information and analytic capability that require understanding of decision processes and behavior at the firm level. The need for this information increases as past experience and data become less relevant to the behavior of markets in agriculture today. Concern with the differential impacts of policies and market phenomena on different types and sizes of farms and an appreciation of the diversity within the farm sector in terms of well-being, performance, and other criteria also have grown.

For example, when we set out to examine and explain the impact of federal policies on the structure of agriculture, we had to understand how these policies were translated into decision behavior and resource use at the firm level. Moreover, we had to understand how a range of financial, physical, and managerial characteristics of a firm interacted with these policies to generate different kinds of farm growth and adjustment behavior. Our primary research objective was not to study individual farm behavior but rather to use the derived behavioral insights to more adequately understand and anticipate changes and their implications in the economic and institutional character of the farm sector. Policymakers needed to understand not only the aggregate economic affects, but also the differential or distributional effects. For some policy purposes, these concerns may likely be the most important to policymakers. In other words, our econometric and statistical models for sectors and subsectors do not provide the necessary insight or detail either to analyze questions related to whose behavior and well-being are affected and how (by specific tax policy and regulatory provisions, credit policies, commodity programs, and so forth) or to assess the implications of their continuing effects.

A type of aggregation error in our macroeconomic indicators has also created some misleading information that is of concern to researchers and policymakers. For example, considerable weight is given to changes and direction of aggregate net farm income. But in years when net farm income rose significantly for the nation, as many as half of the nation's farm operators still suffered losses. In 1979, a year of relatively high overall net farm income, a noticeable contingent of farmers protested on the mall in Washington, D.C. Increasingly, policymakers wanted to know who is well off, who is not, and why. To provide this information, ERS had to examine variations in farm characteristics, financial structure, and other market circumstances. We needed farm models that permitted simulation by researchers of various firm and household characteristics in order to understand their effects on the financial situation of farmers.

Another example relates firm response behavior to farm program provisions. If the proposed provisions have no precedent, the researcher has to be able to understand something about conditions and logical behavior at the farm level to feel confident about deriving reliable forecasts of aggregate farm-sector response. Knowing who is likely to participate and how the benefits are likely to be distributed may be important in developing farm programs. The aggregate supply response to two alternative program provisions may be similar, but the distribution of response may differ greatly.

Recently, we have begun to reexamine our need for firm-level analytical capability and are in the process of reestablishing the nucleus of a small in-house research group responsible for micromodeling activity. It will likely remain relatively small in terms of resource commitments. But I anticipate that microanalysis will play an increasingly important complementary and supplementary role to the agency's macroeconomic research program.

Micromodeling Needs for the 1980s

Farming has changed. Indeed, the entire economic setting for the food and fiber system has changed. Thus, the priorities of publicly employed agricultural economists, including those of us in ERS, have changed. These changed priorities require both new analytical capabilities and new uses of existing tools including increases in micromodeling and analysis of firm behavior. But the major change from the fifties and sixties, in addition to some significant methodological, software, and hardware improvements, is that the objective of the microeconomic inquiry has shifted from a farm management orientation, that is, providing assistance to individual farmers, towards a more complete understanding of aggregate sector behavior and of probable farmers' responses and their distributive implications.

Universities will continue to need to train students in microanalysis because they are training people for farm management and analysis in the private sector as well as for future public research. In fact, we might expect that the preponderance of these training activities will be for the private sector because the provision of microanalytic and management services to farm firms is a rapidly growing private-sector activity. Training students for that part of the private sector is an appropriate role for educational institutions. Further, the growth of such private services should not be viewed by public institutions as threatening competition. In fact, trying to compete with private firms in providing firm-level analytical and management services to today's commercial farmers may not be a high priority for publicly employed agricultural economists.

The needs of ERS for micromodels and related analytical capacity have been well identified. In summary, these needs are:

- To understand likely responses of farms in various regional, commodity, size, financial, and other situations to various market conditions and policy provisions in order to qualitatively understand, but not necessarily quantify, likely aggregate responses.
- To understand likely distributive effects and farmer responses to various policy and market situations; and
- To use micromodels with macro- and econometric models to help provide additional detailed information and likely behavioral responses not well specified in the macromodels.

In my view, it would be a mistake to renew attempts to use micromodels to estimate aggregate behavior because large data requirements and aggregation problems remain. Today's econometric models are far more efficient for estimating aggregate supply, demand, and price behavior. Nevertheless, the uses of micromodels for the purposes I have identified define a legitimate and important subset of activities for economists in the years ahead. These activities are sufficiently important to justify our investment in this conference and to improve the usefulness of firm/household behavioral models.

First, I want to express my support for a renewed cooperative effort between ERS and university economists both in advancing micromodeling analytical methodology and in applying the resulting analytical capability to important economic issues. Second, I want to emphasize that renewed ERS interest in micromodeling is not a return to the objectives of the 1950s and 1960s. Rather, our interest is to reach back and pick up an important analytical perspective to use with our larger tool kit for helping to evaluate the priority agricultural issues of the 1980s.

Notes

1. Many of the developments covered in the early part of this paper are reviewed more thoroughly in Day and Sparling.

References

Colyer, D., and G. D. Irwin. *Beef, Pork and Feedgrains in the Cornbelt: Supply Response and Resource Adjustments.* Missouri Agr. Expt. Sta. Research Bulletin 921. 1967.

Day, R. H. *Recursive Programming and Production Response.* Amsterdam: North-Holland Pub. Co., 1963.

————, and E. Sparling. "Optimization Models in Agricultural and Resource Economics," *A Survey of Agricultural Economics Literature* (ed. G. G. Judge, R. H. Day, S. R. Johnson, G. C. Rausser, and L. R. Martin). Vol. 2. Minneapolis: Univ. of Minnesota Press, 1977, pp. 93–127.

Fox, Karl. "An Appraisal of Deficiencies in Food Price Forecasting for 1973, and Recommendations for Improvement." Unpublished report prepared for President's Council of Economic Advisors, 1974.

Heady, Earl O. *Economics of Agricultural Production and Resource Use.* Englewood Cliffs, N.J.: Prentice Hall, 1952.

Irwin, G. D. "A Comparative Review of Some Firm Growth Models," *Agricultural Economics Research,* Vol. 20, 1968, pp. 82–100.

Lee, John E. Jr., "Improving Aggregation Validity," in *Implications of Structural and Market Changes on Farm Management and Marketing Research.* CAED Report No. 29. Center for Agricultural and Economic Adjustments, Iowa State Univ. Ames, 1967.

Northeast Dairy Adjustments Study Committee. *Dairy Adjustments in the Northeast.* Univ. of New Hampshire Expt. Sta. Technical Bulletin 498. 1963.

Sundquist, W. B., J. T. Bonnen, D. E. McKee, C. B. Baker, and L. M. Day, *Equilibrium Analysis of Income-Improving Adjustments on Farms in the Lake State Dairy Region.* Univ. of Minnesota Agr. Expt. Sta. Technical Bulletin 246. 1963.

Discussion of Chapter 1

Farm Modeling: An Overview

Eldon E. Weeks

ERS Micromodeling Objectives

I am glad that Dr. Lee observes a renewed interest in ERS in micromodeling focused on the 2 objectives: improved understanding of likely aggregate responses to new economic and institutional developments through better understanding of decision behavior at the firm level and improved insight into the distributional effects of new developments. These are important objectives. Also, micromodeling is an appropriate approach for addressing these objectives.

Objective 2 has considerable importance and should command considerable attention. It offers, in addition to the prospect of more new information on distributions of impacts, an opportunity for researchers to participate in the early development of new, advanced modeling activities. Although the availability of data may be somewhat limiting, ERS enjoys a comparative advantage to influence any new large data-gathering activities for distributional studies.

Dr. Lee's concern with his first objective is understandable. It would, however, be distressing to a number of economic scientists interested in studying the behavior of farmers if behavioral analyses were not considered to be the initial procedural necessity for improved understanding of aggregate (farmer) responses to new economic and institutional developments. We need to know more about farmer behavior even though—and because—policy objectives may change for many reasons.

Whence Farm Modeling

I have a vivid recollection of the introductory lecture in a class I took that was to use farm budgeting procedures in the conduct of analyses. The professor recounted how, in a similar class in his graduate training, his professor suggested that the German monks of the 14th century had used many of the critical concepts encompassed by budgeting

The author is a program analyst with the Program Analysis and Evaluation Staff, Cooperative State Research Service, U.S. Department of Agriculture.

procedures in planning for agricultural operations on monastery lands and in arriving at the amounts of rent owed by the operators. I do not know whether this is true, but it sounds plausible.

Certainly the case farm study procedures initiated by W. J. Spillman in his search to find out why some farms prospered more than others took on the flavor of modeling efforts. A big stride forward was made when the categories of costs and returns were defined to reflect neo-classical economic distinctions of factors of production. Another stride was made when budgeting procedures were adapted for investigations relating to policy impacts, economies of size, and capital planning.

Methods and procedures for microanalysis and resulting models of the farm have developed over a long period. The scope of farm-level analyses has expanded because of concern about farmers as impacted by and as participants in social issues as well as the increasing capability of computers.

Interest in Micromodeling

A number of agricultural scientists and a few economists began the serious study of farm management in USDA and the land-grant universities early in this century. Farm incomes dropped in the early 1920s and stayed low until World War II. During this period, farm management studies of case or exemplary farms were conducted with increasing frequency as the Depression spread and deepened and as more people became interested in trying to develop advice for farmers that would increase their incomes. These studies included the most important aspects of farming—selection of enterprises and technical practices, insurance, finance and credit, soil and water management, and size of farm. A review of these experiment station publications, journals, and farm magazines shows, however, that these early studies often lacked a strong adherence to a standardized and rigorous use of economic theory.

During the 1930s and 1940s, farm budget studies became increasingly related to theoretical cost and return curves and to the principles of statistics. Applications and development of economic and statistical theory were also simultaneously experiencing a period of rapid development. This interdependence became particularly noticeable after the end of World War II when the pace of farm adjustments increased. Farm supplies and equipment became more available, market prices were relatively high, and many farmers could afford to purchase new technologies. Agricultural economists conducting farm management studies remained close to farmers as sources of primary data. As a result, full farm budgets were frequently constructed, and then various aspects of finance, practices, enterprise selections, and alternatives were examined with partial budgeting analyses using desk calculators.

The farm analyses of the 1930s and 1940s extended into the 1950s and 1960s as more complex problems were analyzed with the development of computers and deterministic mathematical programming, a technique well suited to the study of farms. In addition to using these models and computational aids for studying farm adjustment problems, agricultural policy impacts and other issues, we saw a significant number of studies with the additional objective of modeling methodology development. The development of recursive and dynamic forms of mathematical programming, somewhat paralleled by the later development of simulation techniques, changed the focus from analyses of adjustment strategies of farms to marketing and growth problems. More recently, micromodeling activities using stochastic (probabilistic) techniques have been employed to analyze risk and uncertainty as another facet of farm decision behavior.

It is conceivable that a next step in microsimulation modeling of farms will follow some of the technologies and methodologies presently used in distributional analyses of household consumption and policy impacts by analysis of stochastically merged large data bases.

Choosing Among Micromodels

Many of us are likely to carry around in our heads synthesized cognitive models of individual farms representing types and sizes with which we are most familiar. We may also have various synthesized and interrelated models of the farm and of the farmer. One or a combination of these informal models is used to interpret farm impacts from environmental events or to interpret managerial behavior-related impacts in a given environment.

Agricultural analysts, however, frequently develop more formal models of farms when processing data to generate results appropriate for scientific conclusions. The informal models then serve in one or several roles. First, they may encourage the analyst to chose one family of formal model types or methodology and may guide actual model specification and data selection. Second, these informal models may play crucial roles as referents when the analyst judges the results obtained from formal modeling as acceptable or not. Over time this conventional/ referent wisdom is influenced by cumulative results of formally modeled circumstances and situations, as well as by changes in the real world.

Credible microeconomic farm modeling thus depends upon sharp perceptions and good judgment. Consequently, the results of any modeling exercise are judged according to acceptability criteria that may not be completely formal and objective. Indeed, there are dangers of inadvertently using the same informal models to judge the results as were used to develop the formal model.

Perhaps the critical criterion is that policymakers concerned with real world circumstances and issues will use informal micromodels of organizations and institutions to make important judgments, whether or not there are inputs from economic analysts using more formal models. If this is true, abstinence from formal micromodeling activities is, effectively, a failure to deliver state-of-the-art descriptions, analyses, and predictions of farmer adjustment behavior to changing environmental circumstances and a failure to provide continuing tests of the validity and consistency of both the underlying assumptions and the internal structure of the relevant disciplinary theory.

What Is a Farm Firm?

About 98 percent of the farms including about 89 percent of the land in farms with sales of $2,500 or over were operated by unincorporated farming institutions, with proprietorships strongly dominant, according to the 1974 Census of Agriculture. Both the Standard Industrial Classification (SIC) Manual and the 1974 Census of Agriculture define farms as business establishments. Consistency with the National Income and Product Accounting System would place farm business establishments in the Business Sector of the economy and unincorporated farming households in the Households and Institutions Sector.

This inconsistency leads to a quixotic point with the terminology often employed in reference to farm micromodeling. A farm incorporated as a single establishment firm is a single farm firm for most modeling purposes that seek to portray representative situations or circumstances. A farm business establishment operated by an unincorporated operator is not a firm. Thus, by definition in most instances, financial management, investment, and resource use decisionmaking cannot be interpreted as self-contained managerial functions within either farm business establishments or farm firms. An additional complication is that roughly two-thirds of unincorporated farm operator families receive incomes from off-farm sources.

The Issue of Relevant Perspective

In most cases, the objectives guiding farm investments and activities are those of unincorporated households rather than those of farm firms. Most of the managerial choices made in farming are calculated to further the goals of the operator family within its planning horizon, the role(s) of farming in its preference structure, and the range of farm and nonfarm activity alternatives that are practical for consideration. Few farm micro-modeling efforts can accommodate all the decisions that farm operators must face from when they decide to enter farming until they decide to leave it. Nevertheless, recent formal farm-micromodeling efforts are oriented to more complex theoretical representation of selected aspects of effects from managerial adjustments such as risk, farm growth, finance, and fixed factor utilization.

Present activities in modeling farming operations can provide a large amount of relevant and important information that relates policy to the full perspectives of farm decisionmakers over recent decades, and perhaps the future, given a few parameters or restraints. Modeling of farming behavior is rapidly becoming more complex as farm operator families become more intricately related in nonfarming ways to the nonfarm economy. Nevertheless, the tendency is to concentrate more powerful modeling techniques on more specialized aspects of farming and farm operator decisionmaking rather than on the integrative analyses actually needed. Who will integrate these efforts? From the gist of much of my reading and from my own research, I am beginning to think seriously that the theory of the firm and its application to agriculture is more directly relevant to proprietary farm-operator families than to farm business establishments. Consequently, much of the literature has limited relevance to the large majority of farms.

Discussion of Chapter 1

Current Perspectives of a Former ERS Model Builder

Neill Schaller

Dr. Lee has given us an excellent historical overview and stage-setting paper. Others at this conference may be better equipped than I to comment on it from a technical standpoint. After serving as an ERS model builder of sorts during the 1960s, I left the agency to pursue other professional challenges for more than a decade. I share my reactions with that qualification.

My first comment relates to terminology—or semantic housekeeping. The term "macro" is relative. One person's "macro" model may be another person's "micro" model. Lee refers to regional linear programming models as micromodels. Nevertheless, during the 1960s when working on the ERS national model of production response, we thought we were developing a macromodel. Actually, neither label is wrong. The national model was "micro" in the sense that it was designed to capture important differences among farms in production subregions. The model was also macro-oriented in the sense that our purpose was to generate useful estimates of aggregate crop production response. Obviously, we will understand each other better when using the terms "micro" and "macro" if we remember to distinguish between the purpose of a model and the starting unit of analysis.

Secondly, Dr. Lee tells us that micromodeling approaches to the estimation of aggregates, such as total acreage and production response, face the same basic problem of aggregation error. He was at the forefront of research in the 1960s that taught us how to reduce, if not eliminate, aggregation error, given certain conditions. The solution, delineating a sufficiently large number of representative farms, is straightforward but very costly. But what are the alternatives? Does shifting from a micro- to a macro-approach merely replace aggregation error (about which we now know a great deal) with specification error (about which we may know less)?

The author is assistant director of the Natural Resource Economics Division, Economic Research Service, U.S. Department of Agriculture.

Four reasons are cited for ERS' past involvement in micromodeling. They were: to do farm management research, to be able to analyze the effects of public policies and changes in macro variables on individual farms and regions, to learn about actual farm behavior and decision criteria to construct more reliable macromodels, and to develop aggregate estimates through summation. Dr. Lee says, rightfully so in my judgment, that the first reason is no longer a valid objective for ERS. The fourth reason is also questionable today. We simply cannot afford the detail and the high cost of estimating aggregate responses by summing representative farms.

Apparently, the second and third reasons are still important. If so, other questions are raised. First, if ERS perceives a need for micromodels to analyze policy effects on individual representative farms, how much farm-level detail is actually needed? Second, what will it cost to generate enough representative farm information for reliable policy analysis? The third question depends partly on the second question: To what extent will these analyses be complementary or substitutable for macromodel results? It is fairly safe to assume that the number of representative farms needed to do these things will be far less than the number needed to obtain macro estimates with minimum aggregation error. Nevertheless, my hunch is that we really do not know for sure how much symmetry is needed in micro-macro relationships to be able to disaggregate aggregate information.

The ERS national model team of research economists had a dream of being able to work analytically up and down the aggregation ladder. We dreamed, not of building a bigger and better model, but of building a capability to link together different kinds of models, formal and informal. The exact linkage was to be determined by the policy questions to be addressed. We also dreamed of this capability as an effort to help bring together staff economists and researchers, with each learning from the other, thereby making a more useful contribution to the identification and analysis of policy alternatives.

Did we dream the wrong dreams? Or did we simply run out of theory and—lacking the time, appetite or skills to fill that void—abandon the dream? Now, as we chart a new course for the eighties and beyond, it might help to identify the knowledge that the national model and other similar policy research ventures have actually given us.

Due to his limited time and space, Dr. Lee chose to focus on the relationships between production economics and farm-level models. So let me urge you to extend his challenge into other areas of research, such as natural resource economics, for a simple reason. The decision of what micromodels are needed to examine the effects of alternative food, farm, or macroeconomic policies on different farms will not automatically tell us what is needed to address the natural resource issues now or soon to be at the policy forefront. I am thinking, for example, of issues like the depletion of underground water supplies in certain regions, the relationship between ownership of farmland and the effectiveness of different resource conservation policies, or the effects of those policies on future ownership patterns. I suspect there are other questions none of us has thought much about concerning the micro-modeling required to do useful research on consumption, demand, and marketing issues.

I take two risks in raising these questions. One is that they may expose my prolonged absence from the modeling field. The other and more serious risk is that you will interpret my questions as a veiled critique of macromodeling or as a nostalgic plea for a return to the micromodeling emphasis in ERS prior to the 1960s. In fact, I make no such critique or plea. My only concern is that we guide, rather than react to, future directions for ERS. Moreover we should try to move deliberately and with imagination to the next issue agenda: What kinds and qualities of information and intelligence do we in ERS really want from models? And for whom? What do we really mean when we say that we must be able to analyze policy effects on different kinds of farms? What relevant methodologies are available to us, including, but not limited to, programming and other traditional tools? Finally, the inevitable but critical bottom-line question: What can be said about the cost-effectiveness of each micro- (farm) modeling approach and other related analytical tools?

As a final thought I cannot resist citing the need to communicate effectively with others who are not well versed in the language of models and modeling. Last summer, users of ERS research and SRS data presented their views of our work at a 2-day hearing in Washington. I heard a few say that if ERS funding has to be reduced, the agency should first eliminate some of its model building. The comment is not new, but in today's climate of austerity, the mutual benefits of helping our clientele understand, from their perspective, what we are trying to do in this conference, and why, could be substantial.

2. Farm Decisions, Adaptive Economics, and Complex Behavior in Agriculture

Richard H. Day

Abstract Change is pervasive in agriculture. A more adequate understanding of change and of improved government policy can be achieved with theories and models that are dynamic and that incorporate salient features of micro-economic behavior. Adaptive economics provides some of the needed theoretical concepts. The next generation of microdynamic simulation models should incorporate strategic details of economic reality. In this way they will facilitate the design of policy for improving the performance of our economic system. Preliminary dynamic simulation studies already provide useful insights about the complex dynamics of economic change.

Keywords Adaptive economics, agricultural change, dynamic theories and models, microeconomic behavior

Understanding Agricultural Change

Agricultural Change and the Need for Strategic Detail

Production technology, associated farm practices, and related urban-industrial activities have gone through continual transformation as once abundant resources became scarce and as new ideas led to practical innovation and adoption by farms and industrial enterprises.

Within this basic picture of continual change are societies and geographical areas where food production practices may have continued more or less unchanged for long periods. Even vestiges of prehistoric hunting and food-collecting cultures have survived the onrush of the agro-industrial complex until the last decade. Nonetheless, the predom-

---

The author is a professor of economics, University of Southern California.

inate picture is not one of approaching long-run equilibrium or even one of a succession of more or less stable regimes or stages, but rather one of overlapping waves of technology, as the rapid rise of a given mode of production replaces a previously prevalent way of life.

Changes in economic epochs related to agriculture need not take place gradually over the long run. Experience in this century teaches a different lesson. The transition from one technology to another often takes place within a generation or two while significant shifts in new directions can take place within a decade. Moreover, market forces are often cyclical in nature with relatively short-run oscillations in farm prices, production, and incomes playing on the surface of the underlying dynamics of technology and farm practice.

The broad forces of technological change and the shorter run interactions of supply and demand all work themselves out through the activities of individual production and marketing enterprises. Because surrounding economic and environmental conditions vary greatly from one geographical area to another, and as previous developments endow individual enterprises with varying resources and capabilities, the further development and economic performance of individual enterprises may vary greatly from one firm to another. Moreover, the way firms respond to economic conditions influences the further development of the system as a whole.

Changes in economic policy, in effect, involve changes in parameters that impinge on the alternatives, constraints, costs, and returns in the decisionmaking process of individual enterprises. These policies have varied effects among institutions with different histories and different economic and geographical settings.

For all these reasons, if we are to understand the dynamics of agricultural change and to improve our ability to estimate the repercussions of government policies such as implementing or removing price supports and production controls; regulating or deregulating production, banking, and marketing activities; modifying systems of taxation or studying the effects of national fiscal and monetary policy on the performance of the private sector; we need to consider strategic details of microeconomic technology and behavior.

Inessential detail only clutters up problem identification and inevitably complicates analysis. Strategic details are those that matter. Part of the function of sound research is to discover what details matter. In micromodeling, as with other research, if we do not pursue analyses that allow for the incorporation of a great variety of details, we can never investigate and discover which ones are indeed strategic.

Problems of Theory and Method

Economists are and have been concerned fundamentally and primarily with change: implicitly—at least—with economic dynamics. Adam Smith's great classic, which posed and to some degree answered a great many of the economic questions that have been the profession's concern ever since, was drawn from a dynamic perspective. In particular, Smith asked why some economies developed and others did not and what devices of government would be most efficacious in fostering economic growth and the common wealth.

Neoclassical economists such as Jevons, Marshall, von Thunen, and Walras focused on the core of the classical concepts, namely, the pursuit of self-interest, the coordinating and equilibrating effect of competition, and the fostering of the common good through self-interest and competition.

The pursuit of self-interest was formalized in the utility or preference-maximizing household and in the profit-maximizing firm. The coordinating effect of competition was formalized in the concept of a competitive equilibrium price system, tatônement (Walras), and market-groping or partial adjustment (Marshall). The fostering of the common good subsequently became formalized in the concept of Pareto optimality or social efficiency and in the idea that the socially efficient resource allocations could be supported by a competitive price system.

The individual optimality, market equilibrium paradigm was never really intended to describe actual day-to-day economic behavior. Instead, it is argued, it can be used operationally to define the situation that will prevail in the long run when short-term adjustments are allowed to work themselves out. This point of view assumes that the actual behavioral system is stable and will converge to a competitive equilibrium.

These fundamental ideas were in due course augmented with comparative statics, the study of how individual optima and competitive equilibria change when the underlying parameters of preference, technology, and government policy are changed. This development, which was given its definitive modern form by Samuelson, transformed the neoclassical theory from a passive rationale for the market system and laissez-faire to an active instrument for the design of public policy.

Although explicit behavioral content is lacking in the neoclassical system, it has generally been given a behavioral interpretation. This interpretation has been stated most cogently and persuasively by Milton Friedman in his famous essay on positive economics. This view essentially boils down to the proposition that (1) agents behave as if they optimize, and (2) markets work as if they generated equilibria. These are the implicit, but fundamental, assumptions of the neoclassical paradigm.

But, suppose these implicit assumptions are incorrect or that they are not always good approximations. Indeed, suppose that, at least under some conditions of technology, resource availability, and demand, agents do not optimize and markets do not converge to competitive equilibria even approximately.

One answer is, "Such cases are unimportant so those possibilities can be ignored." Another might be, "It does not matter because comparative statics gives the initial direction of policy impact." Still another might be, "Yes, the theory may be wrong, but it gives the right policy answers."

None of these answers is satisfactory. Nonetheless, my own respect for the neoclassical paradigm has grown with experience. Just when I am about to reject it completely, I come across another clever application that illuminates a real problem of current interest or I see a useful new extension of the underlying approach to an unexplored domain. But although my respect has grown, my confidence that it is a fundamentally inadequate theory has also grown. More and more I believe that explicit dynamic models of how economies work are necessary, both to satisfy the desire for understanding and to avoid serious errors in policy formulation that can be made when economic implications are rigorously drawn from wholly implausible behavioral assumptions.

The remainder of this paper explores one approach to dynamic economics at the micro- and macro-aggregative level. It is a coherent body of concepts and related models that makes possible more realistic representations of economic structure and new insights into how economies work. Preliminary results lead to seeming paradoxes and raise profound questions about both methodology and policy.

Adaptive Economics: Toward a Theory of Economic Behavior

Equilibrium, Bounded Rationality, and Disequilibrium

The classical-neoclassical economics paradigm involves 3 kinds of equilibria.

Individual equilibrium means that an individual person, household, or firm is doing the best it can according to its preferences, existing opportunities, prevailing prices, and constraints. There is no incentive for the individual to change.

Market equilibrium means demand does not exceed supply, all plans can be executed without modification, and expectations are realized. There is no need for individuals to change.

Social equilibrium means that the only way any producer or consumer can reach a preferred state is for some one to be made worse off. Someone always objects to change from this equilibrium.

The rationale for using this paradigm as a theory of real economics is as follows:

1. If rational individuals have preferred choices, they will exercise them, changing the situation in the direction of individual equilibrium.
2. If markets do not clear at prevailing prices, prices will adjust so that preferred production and consumption plans will move toward consistency with each other.
3. If a person or group can reach a preferred state without making someone worse off, then no one will be object to change through modified plans and changing market prices.

In reality individuals and groups are imperfect decisionmakers. Even when preferences are well defined, a best choice among available alternatives is difficult and poses a problem. Moreover, perception and cognition are imperfect. In these phenomena we have the basis for Simon's phrase, "bounded rationality." Because of it individuals, even if their external situation is unchanged, will attempt to improve their decisionmaking performance.

Since individuals are boundedly rational and change their behavior even when prices are constant, they are unlikely to choose plans that satisfy the conditions of market equilibrium. How much more likely it is that their plans will be inconsistent as prices change in response to excess demands or supplies. Expectations may not be realized, and regret may follow on choices based on mistaken forecasts.

Because of the implied inconsistency of boundedly rational behavior, market equilibrium is not achieved. As a result, mechanisms must exist that mediate disequilibria, which will enable individuals and firms to produce and consume when their best-laid plans cannot be executed, and which facilitate exchange when too much or too little is offered for sale or purchase. How these mechanisms work will influence the path of quantity and price adjustments. Whether these paths lead to equilibria is a question for theoretical analysis and empirical testing; its answer cannot be taken for granted.

Since agents are imperfect decisionmakers, plans may be inconsistent and disequilibrium mechanisms are at work to keep the system moving. It cannot be assumed that all movements avoid making anyone worse off. More likely both improvements and disimprovements will take place. Objections to change will therefore be the rule.

To summarize: Economizers are boundedly rational; their plans are inconsistent; their fortunes fluctuate in divergent paths; the economy is a disequilibrium system.

Planning, Control, and Disequilibrium Mechanisms

A starting point for a theory incorporating these salient features of reality is the adaptive process, a dynamic system in which a behavioral unit of interest (the agent) responds to its own internal conditions and to prevailing circumstances in its environment; the latter likewise is modified by its conditions and by the actions of the behavioral unit in question. Both agents and environment are described by states and by changes in states. Since agents may influence their own behavior, they experience autofeedback. Since the environment influences itself, there is also autofeedback in the environment. Since agents and environment influence each other, interactive feedback is experienced.

An economy is a system of agents who interact with one another and with their overall environment. From each agent's point of view, other agents are part of the environment. The economy may then be thought of as being made up of a set of interacting adaptive processes, i.e., as an adapting system.

The agents cannot foresee all the consequences of their actions, nor can they perfectly predict the behavior of the other members of the interacting system. They must observe what is going on, process appropriate information upon which plans and actions can be based, determine desired behavior, undertake feasible actions, and observe once again what is going on. Behavior thus breaks down into a sequence of components and feedback effects.

To be feasible, action must be consistent with internal states, but since the plans of a given agent may be inconsistent with those of other agents in the environment, it may be impossible to execute them as planned. Therefore, actions must generally be based, not just on plans, but on internal states which must have the effect of insulating the agent from inconsistencies with the outside. Therefore, behavior results from 2 separate functions: planning, which produces desired actions, and controlling (managing or administering), which produces feasible actions. Since planning takes time, uses resources, and is costly, and since control must be continuously exercised, the planning and controlling functions are distinct, usually with separate rules, procedures, and algorithms. Consequently, people have distinct faculties for reflection or thoughtfulness and for controlling movement in response to varying external conditions (for example, reflexes). Analogously, complex organizations have separate staffs for planning and administration.

The function of the control system is to guide action toward desirable goals in a viable manner, in accordance with prevailing external states. The function of planning is to formulate desired goals that would produce preferred internal states taking into account expected external conditions. Since what one is doing is feasible (or one would not be doing it), control involves the adjustment or adaptation of current behavior in the direction of desired action as it is determined by the planning function.

Actions bring about material, financial, informational, and energy flows which modify internal and external stocks of these several items. The flows among various agents, based on individual adaptive processes, are in general imperfectly coordinated—thus out of equilibrium. The resulting imbalances in flows are mediated by stocks which make it possible for flows into a given agent to be unequal to flows out from the given agent.

The existence of such stocks is the fundamental means by which economic activity can persist when the producers' and consumers' plans and actions are imperfectly balanced. The function of control is, therefore, to adjust actions and thereby rates of flow so as to be compatible with existing stocks.

Special agents or institutions have come into being whose primary function is to coordinate flows by regulating stocks. For example, stores are inventories on display which make it possible for consumers to purchase goods and producers to supply them without either knowing the plans or actions of the other. Banks and other financial intermediaries regulate the flow of purchasing power from uncoordinated savers to investors. Their ability to create credit provides a means of coordinating activities at different points in time.

Ordering and order backlogs, adjustments in filling orders and delivering goods, adjustments in production and construction rates, and inventory fluctuations all provide a flow of information that facilitates adjustment to imbalances in supply and demand. Insurance and other transfer schemes such as unemployment compensation and income supplements provide a flow of goods for agents who would possess no admissible action without them.

Adaptive Economizing: A General Framework

Economizing involves the allocation of scarce resources according to a criterion such as profit, sales, utility, or more generally, preferences. It involves doing the best one can in production or consumption or both, where "best" is clearly defined. By adaptive economizing we mean boundedly rational economizing in response to feedback. Bounded rationality implies imperfect perception and knowledge of the environment. Consequently, existing states are perceived imperfectly and the feedback structure that determines how the environment of a given agent works is only partly understood. Bounded rationality also involves limited memory or data storage and limited powers of ratiocination. These imply limits on the ability to solve complex decision problems. Finally, individual people possess limited capacities for interpersonal communication and cooperation. Organizations tend to amplify the frailties of the human mind.

Because the selection of an action through planning and control poses problems of great complexity, economic agents generally break down their overall problem of achieving objectives into a set of simpler decisions solved more or less sequentially. Thus, the decisions to consume, save, and produce involve hundreds, perhaps thousands, of individual choices that are divided into groups.

Economic activities planned at the beginning of a given period include production, sales, investment, and financial and household activities to be undertaken immediately or in future periods. Some activities may be chosen; others may not be pursued at all. Some may be pursued intensively and have large activity levels; others may be pursued less intensively and have small activity levels. The choice among alternative activities and the determination of activity levels are constrained by technology—the set of available and perceived opportunities (the state of the art)—and by resource availabilities and financial constraints. At any given time these constraining factors and the degree of their limitedness circumscribe the individual's opportunities within a feasible region of activity levels.

Associated with each constraining factor or influence is a constraint function that determines how the various activities are restricted by that factor and a limitation coefficient that describes how limiting the particular constraint is. The latter are inherited from the past and are fixed for the time being. Their effective limitedness, however, can be reduced by the choice of certain activities. Machine capacities inherited from the past can be augmented by the purchase of new machines. The supply of money available for investment can be augmented by borrowing. However, investment in a given year is limited by reinvestment funds, and borrowing is limited by lenders' credit-rationing rules. Hence choices that reduce the effective restriction on other activities in any one year are in their turn bounded by limitations of their own. Adaptation in response to feedback from past decisions and the external environment has the effect of modifying the feasible region from one period to the next. For a given period it is fixed.

The true feasible region may be much more complicated and irregular than the one agents perceive in their own decision process. For example, the perceived feasible region may be well represented by a relatively simple, quite regular polyhedron. The actual feasible region, however, may be a nonconvex, irregular body. For purposes of pure theory as well as for the viability of real economies it is important that the perceived possibilities be included among the actual ones. If they are not, then agents may choose a plan that cannot be implemented. They would be forced to choose a feasible plan as close as possible to the plan or, failing that, to seek new information, reestimate their feasible region, or at the extreme "negotiate the environment"; that is, they would have to seek an accommodation with their environment that would modify the true feasible region so that their viability is restored. While this latter type of behavior exists, it may safely be assumed to be costly. It is avoided as much as possible by adaptive tactics designed to prevent its need from arising.

Boundedly rational choice implies uncertainty (as distinct from risk). It is characteristic that, faced with growing uncertainty as they contemplate change from existing patterns of behavior, many decisionmakers limit a consideration of alternatives to those in a neighborhood of current practice. This has been called local search. The willingness to depart from current practice, that is, the extent of the region searched may depend on experience and on the behavior of other agents outside the firm. Thus, adaptation to current economic opportunity may be more or less flexible. The set of alternatives that may be considered at a given time has been called the zone of flexible response (ZFR). Such zones depend on experience and imitation. This dependence means that economizing is more or less cautious. We shall go into this in more detail below. In contemporary economic theory it is conventional to assume the existence of a complete preference structure over all alternatives or a utility function as a fundamental axiom. We shall also allow for this aspect of rationality. A realistic special case of this concept that seems highly plausible and useful for simulation studies is target planning.

Target planning is based on 2 elements: (1) a desired choice or set of choices, a target, and (2) a concept of distance from the target that defines or describes how close to the target a given choice is thought to be. This distance becomes a utility function to be minimized, or alternatively, its negative becomes a utility function to be maximized. If the target is feasible or contains a feasible subset, then an element from this subset is chosen; if not, a feasible choice closest to the target is selected. Thus, target planning is formally analogous to constrained economic optimizing with utility functions, that is, to conventional neoclassical rationality, when satiation (hitting the target) is allowed.

The measure of distance that describes how close a given choice is to the target set defines a family of indifference curves. Choices are equidistant from the target within a given indifference curve. Consequently, they are equally good according to the target criterion. The target may consist of a single unique point called a bliss point. The indifference curves in this case are concentric, elliptically shaped curves. The concept of distance need not be Euclidean, which means that the ellipses need not be circles. For this reason, target planning can be nearly as flexible as general utility-maximizing behavior.

Adaptation in response to feedback has the effect of modifying the location and shape of the target set, but we suppose (for purposes of formulating the current plan) that the target is given by some adaptive rule based on information about the past.

The formal analysis of target planning involves optimizing behavior, which may be objectionable on behavioral grounds as too strong an assumption. Behavioral theorists have suggested that decisionmakers, when they can, choose solutions that are good enough. Only when the decisionmakers cannot achieve a satisfactory level do they make an all-out effort to maximize. This important hypothesis of satisficing behavior can be readily accommodated into the target-planning framework by positing a "close-enough distance," "aspiration level," or "satisficing level" that defines an acceptable choice in relation to the target. This concept, however, adds nothing to the formal model already described because the distance function describes the set of all choices close enough to the target. The hypothesis of satisficing is thus equivalent to specifying a larger target.

Faced with extreme uncertainty about the physical and market environment, real world decisionmakers may be described as behaving cautiously. One approach to modeling cautious behavior is to follow the expected utility, portfolio approach. Decisionmakers are assumed to behave like sophisticated gamblers on the basis of well-defined odds or probabilities.

An alternative approach uses the concept of target planning just defined. Here adaptive agents form, for purposes of the current planning exercise, a "local" target of "safe-enough" decisions in the neighborhood of their current operating conditions. This target is a ZFR, within which decisionmakers are willing to select alternatives according to another criterion such as profit- or utility-maximizing of the usual kind. This zone is determined adaptively by behavioral rules that incorporate previous experience. If the zone contains no feasible alternative, decisionmakers then choose a feasible alternative as close as possible to those that are safe-enough. In this case caution dominates their behavior. As the ZFR intersects the feasible region, the effect of caution is to constrain choices that are eventually made according to other criteria. Because target planning—as we have seen—is a special case of utility maximization, the cautious optimizing behavior based on target planning is likewise a special case of utility-maximizing behavior. It is based directly on the concept of distance rather than probability and expected utility.

As target planning and cautious optimizing are merely special cases of conventional utility function maximization, we may as well allow general utility functions to be specified in the theory. In the adaptive approach followed here, however, the adaptive modification of the utility function on the basis of past experience of the decisionmaker must be allowed. If satisficing of such utility functions is to be allowed, then satiation, aspiration, or satisficing levels of utility must be defined. Their effect is exactly the same as before. The derived, satisficing utility function becomes flat at the aspiration level. This is because all choices, whose utilities, according to the original utility function, are above the given satisficing level, are regarded—by the satisficer—as equivalent to the satisfactory or satisficing choice.

Most behavioral scientists regard behavior as determined by many different goals, and these are arranged according to some (perhaps temporary) hierarchy or priority order. A first goal dominates comparison of alternatives until a satisfactory solution is obtained according to this goal; then a less important goal is maximized over the remaining alternatives, and so on, until a single, unique set of choices is reached.

Sequential attention to goals is of considerable use in understanding microeconomic behavior. It explains, for example, why people in one stage of development might behave very differently from individuals in other stages of development even if their respective psychological constitutions are similar. It also explains why many of the considerations that dominate wealthy peoples' choices often seem trivial to poor people or why when survival considerations dominate, a profit goal may be unimportant.

A farmer may have many goals. Presumably, a farmer's past decisions were dominated by some given goal or utility function. If this goal is being satisfied, then the farmer maximizes the next goal subject to the conditions that the decisions do not prevent accomplishing the dominant goal. Environmental conditions may bring about a current situation in which the past decision is no longer feasible or the aspiration level of the goal must be reconsidered. Thus, we see that a number of routes are possible in coming to final choices among alternative goals.

The decision procedure described above is illustrated in Figure 2.1. On the basis of past experience and current data, the technological, resource, and financial constraints are determined. Any adaptive con-

FIGURE 2.1
Recursive priority programs

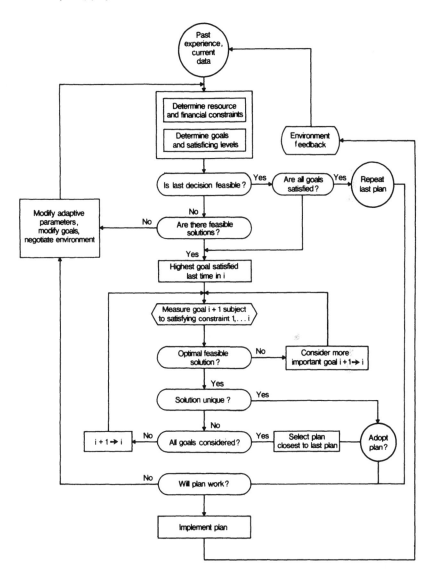

straints, such as those implied by cautious behavior and current satisficing or aspiration levels, are also determined. Then, the past decision is examined for feasibility. From the adaptive point of view current decisions are regarded as departures from past actions. If the past action is not feasible, then a search for feasible alternatives must ensue. If none is found, goals must be modified or the environment negotiated. If the past solution is feasible, it must be checked to see which, if any, of the high-order goals are satisfied. Let us suppose goals 1, ..., i are satisfied. Then the next goal, $i + 1$, is maximized. If there is no further room for choice, the procedure ends and the plan is implemented. If the solution to the last maximizing problem is not unique, then the next goal is maximized and the process continues step by step until there is no further room for choice or until all the goals have been considered. If all goals have been considered and choices are still available, we invoke the principle of conservative selection. This asserts that, among the solutions that satisfice all the goals, the one closest to the past decision is selected.

This process of sequential maximization of goals is called L*, lexicographic, or priority programming. Several important features stand out. First, as long as a given number of high-order goals are satisfied, their priority order has no effect on the solution. Second, satisficed goals act as constraints on choices determined by less important goals.

In spite of the general plausibility of priority programming as a description of behavior, the model still appears to be too complex. Peasant farmers, for example, or even corporate executives probably do not think about their decision procedures in exactly this way. Certainly they would not articulate them using these terms. Nonetheless, we believe this general characterization is much more realistic than standard economic theory, and it does approximate with some accuracy practice on the farm field and in the factory.

Methodologies Related to Adaptive Economics

The ideas outlined above emerged through a synthesis of ideas that were introduced in five distinct lines of development—behavioral economics, system dynamics, recursive programming, adaptive or dual control theories, and evolutionary models.

The first strand of thinking, behavioral economics, had its origin nearly half a century ago during the marginal-cost pricing controversy. It was discovered that business managers used some version of a markup formula to determine prices and that few of them had even heard of what economists have thought to be the rational way business managers behave. Theorists then debated the consistency of these findings with marginalist principles.

The controversy itself need not concern us here. The significant thing is that it launched a new line of inquiry using direct observation as a source of hypotheses about economic behavior. Subsequently, comprehensive efforts to establish a new economics based on observed patterns of behavioral regularity emerged in the behavioral economics school founded by Herbert Simon in his seminal work, *Administrative Behavior*

(1957), and developed in collaboration with him by James March, Richard Cyert, and some of their students, such as Oliver Williamson (March and Simon, Cyert and March, and Williams). The idea behind their approach was to observe how people behave, systematize and formalize their rules of behavior, and study the dynamics of models based on these rules using analysis and computer simulation. Although an impressive number of studies appeared, this approach, until very recently, remained in the shadow cast by the ruling equilibrium paradigm.

While the new line of work focused on behavioral economics was underway, a parallel development was launched when Jay Forrester decided to shift from a career as an inventor and manager of large-scale engineering systems to a career in scientific management. His conceptual approach, in effect, was the same as the behavioral school, but in its execution he developed an ingenious modeling language that allowed these rules to be programmed, documented, and simulated within a single coherent system. The language itself, either in its visual, flow-charting form or in its computer dialect, followed a consistently formulated set of rules for rigorous dynamic modeling (Forrester, 1961).

For purposes of the present discussion, the most significant features of this system dynamics approach are: (1) its presumption that behavior is goal adaptive, using feedback, servomechanism type rules (Forrester, 1966); (2) its belief that response to environmental changes often involves thresholds and discrete switching rules; and (3) its view that nonlinearities pervade economic systems.

The most significant properties of system dynamics models are: (1) their capacity to change behavioral modes during the course of a computer run, with the model result perhaps exhibiting growth, oscillation, or decay in turn; (2) their capacity to generate endogenous shifts in "structure" through systems of feedback loops, and (3) their tendency to illustrate that small differences in coefficients can lead to instability of endogenous variables so as to suggest that quantitative prediction is an inappropriate objective and, instead, qualitative understanding is the proper objective in dynamic analysis.

Explicit economizing or optimizing behavior was demoted to a mere vestige in the behavioral economics and system dynamics schools just summarized. My own work on recursive programming (RP) models, which was begun in 1957, may be viewed as an effort to retain optimizing behavior within the general structure of economic analysis while at the same time placing it in a dynamic, disequilibrium context with a primary objective of incorporating essential features of behavior.

Recursive programming models are, formally, models that represent economizing as a sequence of recursively connected, "local," approximate or behaviorally and constraint-conditioned optimizations or suboptimizations (Day and Cigno). The recursiveness is provided in part by a feedback structure that represents the external environment just as in Bellman's (1957) dynamic programming or in conventional optimal control models. But the optimizing is assumed to be performed by the agent being modeled without the complete knowledge of this external environment so that an optimal strategy in the Bellman sense is not used to represent behavior. Instead, the economizing agent is represented as making more or less cautious departures from current conditions in the best direction according to current perceptions, approximations, and constraints.

The solution of RP models shares the essential features of system dynamics and behavioral simulation models: the evolution of qualitative mode, the shift in structure, and the sensitivity of trajectories to initial conditions and constraints. Of special interest here is that in RP models the shift in structure occurs as constraints that restrain choice switch over time. In this way different variables and equations may govern the evolution of the system at different points in time. This in effect means that different feedback loops are involved. Each such structure is called a phase and a given model contains a set (possibly a very large one) of potential phases and phase sequences or phase evolutions. The result is not only an endogenous theory of changing modes, as in any nonlinear dynamic system, but also an endogenous theory of structural evolution based on explicit economic tradeoffs.

At roughly the same time that the behaviorally oriented schools of Simon and Forrester were being established, Bellman (1957) was formulating a new approach to optimal control theory that emphasized the derivation of optimal rules or strategies. Instead of investigating the dynamic implications for behavior of a given planning procedure in a given "true" environment, as in the RP approach, dynamic programming emphasizes the derivation of a best way of deciding when the structure of the environment is assumed to be known. It is thus a normative or prescriptive approach to decisionmaking in a dynamic context in contrast to the positive or descriptive focus of the behavioral, system dynamics, and recursive programming approaches.

Some years later Fel'dbaum, Marschak, and others extended the idea of dynamic programs to the case where the decisionmaker knew something of the environmental structure but recognized the need to learn more about it. Thus, at each period in a sequence of decisions, the decisionmaker might choose actions in part to achieve current benefits and in part to improve information on which to base future choices.

Some of the mathematical representations of models of this kind lead to quite complex applications of Bayes Rule, the principle of optimality, and econometric estimation methods. Dual control theories are our most advanced theory of how we should approach the building of micromodels if the costs of building and solving models are low. However, they provide a less compelling theory of how real world decisionmakers actually behave.

Some adaptive programming models, especially those involving "model referrent control," acknowledge that control strategies are approximations of the real world and must be updated in response to feedback. They thereby take on the essential features of the recursive programming approach with the special properties that both adaptive and strategic considerations are explicitly incorporated in the optimizing operator.

Although Alfred Marshall is remembered primarily for his exegesis of marginalist equilibrium relationships, his famous text is dominated by biological analogies and explicit reference to competition and selection in an economic context. Firms compete for a market, vary their behavior at the margin, and are selected on the basis of profitability. Losers go bankrupt and winners gradually approach a normal profit equilibrium. It is an evolutionary theory explicitly based on competition, variation, and survival of members of a population that are most efficient. Although this Smith-Malthus-Darwin concept of development has been told as a background story for justifying the equilibrium paradigm and the comparative static method (Alchian), it has seldom been incorporated explicitly in dynamic economic modeling.

In a seminal dissertation published in the *Yale Economic Essays* in 1964, Sidney Winter chose the evolutionary paradigm as the appropriate one for understanding how markets work. In a series of papers coauthored with Richard Nelson, he has followed this line in studying specific evolutionary models, where a population of technological alternatives can be selected or discarded on the basis of simple cost comparisons and according to rules of the type advocated by the behavioral school. Distributions of firms with differing cost structures emerge, but conventional equilibria need not come about. Again, computer simulations have been used extensively in the search for qualitative generalizations, but Winter has shown that analytical results can be obtained as well (Winter [1971] and Nelson and Winter).

All of the approaches outlined have contributed important insights into dynamic economic processes and are important foundations for adaptive economics.

Modeling Agricultural Change

Modeling Farm Decisions and Macroeconomic Feedback

The farm sector is a system of decentralized decisionmakers that includes numerous more or less independent farms. Each farmer makes decisions, usually in a disequilibrium situation, based on limited information about markets and in response to constraints or incentives imposed by markets as well as government policies. The structure of a given farmer's environment—which consists of all other farms, markets, political organizations, and nature—is neither controlled nor well understood by farmers. They adapt by responding to variables such as prices, revenues, quotas, and the past behavior of neighbors. Rules of adaptation link current decisionmaking to past experience and to the past behavior of neighbors both because the farmer may imitate others and because past actions of others do in the aggregate have an impact on the current market situation. The fundamental concepts that must be taken into account in any theory intended to explain actual events are decision and feedback.

A given farmer may imitate his neighbors. In contrast, a farmer may provide services in the form of successful examples that neighbors having less managerial ability may seek to emulate. The farmer interacts directly or indirectly with agents in other sectors through markets for inputs and products. The environment is also strongly influenced by weather and other exogenous forces over which the individual farmer exerts little influence. But, farms also affect their physical environment through water pollution and soil depletion or salination. Thus, the farm is part of a complex system.

It is impossible in quantitative work to apply a microeconomic theory as such to each firm in an economic sector. In agriculture, thousands of individual microcomponents would be required. Individual agents' interactions may be suppressed to overcome this problem. Representative agents must be modeled and used as the basis for describing behavior for the sector as a whole.

I have not used the term "typical" because farm types that must be identified for purposes of understanding change may be atypical and representative of the relatively rare innovator and management leader rather than the majority that may be slow adopters or reluctant decision imitators. To understand these important distinctions, one may need to distinguish representative farms not just on the basis of common geography and similar resource endowments but also on the basis of managerial skills and propensities for innovation or imitation.

Even a representative firm approach may still require too much detail given (1) the purposes for which the information is used and (2) available resources for modeling. Yet it may be desirable to retain an explicit representation of farm decisionmaking to obtain a qualitatively realistic picture of micro-macro interaction. In this case an entire farm region or sector may be represented by a single average farm. Likewise, the farm-sector environment or nonagricultural economy may be treated coarsely so as to minimize detail. This may be done by using a macromodel component of the nonagricultural economy or a set of econometric demand and supply equations for off-farm demand and nonfarm inputs as in the generalized cobweb model approach.

When individual microunits are suppressed—as they must be—and modeled by means of representative farms, aggregation biases will be present. This raises the question of how to select representative farms so as to minimize aggregation bias. Considerable progress has been made in answering this question. Clearly, however, bias cannot be eliminated and, even if minimized, may be substantial. Precise quantitative estimates and accurate forecasts of the farm sector will rarely be achieved with the use of micromodels. Nonetheless, if the representations of microstructure are plausible, the models may be used for gaining an understanding of qualitative dynamics, for determining approximate responses to changes in government policy, and for identifying the strategic details that are likely to impinge significantly on future developments.

From the adaptive economics point of view, models of farm behavior should be based on the central ideas of (1) decision or planning and control with feedback; (2) division and approximation into relatively simple economizing computations such as separate household-firm decision modules; (3) local search and cautious economizing; (4) adaptive expectations; (5) financial feedback and internal as well as external credit rationing; (6) imitation and experience dependence; (7) distinct propensities for innovation, self-management, and imitation; (8) multiple household-firm objectives; and (9) sequential attention to goals. Adaptive micromodels of the farm sector will have the general appearance of Figure 2.1.

An Example

To see how some of these ideas compare with the usual neoclassical, or linear programming approaches, we may briefly consider an adaptive economizing model developed by Day and Singh.

Because agriculture is characterized by multiple outputs and during periods of transition by multiple technologies, activity analysis is used to represent production, financial, consumption, and marketing activities. Linear constraints on resources, finances, and adoption of different techniques are then specified. The choice among available production, consumption, investment, and marketing alternatives (as given by available technology and within the limits that learning allows) is guided by multiple goals arranged in priority order. One set of such goals may be: (1) farm subsistence, (2) cash consumption and savings combination, (3) safety, and (4) profit maximizing. We assume that they are arranged in an absolute priority order and that the first 3 can be satisfied with finite activity levels. This order implies that these goals enter the model as constraints rather than as utility functions to be maximized explicitly.

The subsistence goal represents the objective of meeting immediate survival needs, including a margin for error. The consumption-savings combination goal reflects the tradeoff between a desire for commercial consumer goods and a desire for enhanced future cash earnings (and hence future commercial consumption). The safety goal reflects the unwillingness to modify economic activities drastically during any one time period because of uncertain market feedback. This latter goal is a behavioral tactic that protects against the uncertainty that current choices will really turn out to be the best way to meet lower priority goals. It describes the degree of caution or, from the opposite point of view, the amount of flexibility farmers are willing to exercise in modifying existing patterns of production and investment to take advantage of current opportunities.

The separation of subsistence and profit maximizing is somewhat arbitrary but convenient from a modeling point of view and more or less realistic. Immediate survival is a goal that can be controlled to a considerable extent by the direct exercise of traditional subsistence-farming activities or by holding back a certain amount of cash for commercially obtained necessities. The longer run threat from future exigencies of the market cannot be controlled directly but can only be provided for by means of adaptive rules such as caution in responding to new economic conditions.

Given that (1) subsistence is allowed for, (2) the cash consumption goal can be met, and (3) the amount of working capital allows safe enough changes in production and investment activities, it is assumed that short-run profits are maximized. This way of representing behavior under uncertainty can be thought of as a separation of the overall farm decision problem into a sequence of simpler problems.

A simplified version of our assumptions is shown in Figure 2.2. This diagram is based on the assumption that only two crops are grown. Crop T is a traditional crop with which farmers have had considerable experience. Crop M is a relatively modern crop, recently introduced, that farmers are still learning to adopt. Such modern crops typically include new high-yielding varieties with unfamiliar cultural practices. Technological and financial feasibilities are shown in Figure 2.2(a). Leaving out behavioral considerations, we could choose any combination of the traditional and modern crops within the shaded region of Figure 2.2(a). Both crops can contribute to subsistence needs, crop T directly and crop M indirectly by providing cash to purchase necessities from outside the farm. The marginal rate of substitution between crops M and T in terms of satisfying subsistence needs is determined by the

FIGURE 2.2
Sequential attention to goals or priority programming

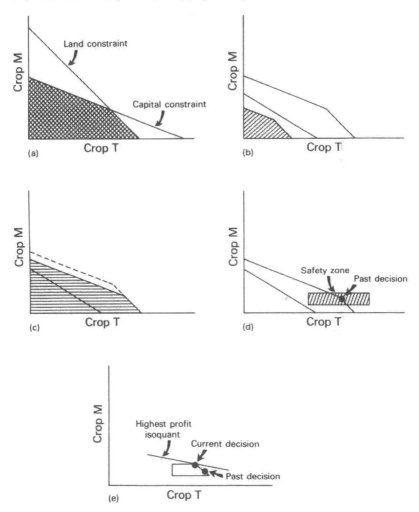

(a) Technically and financially feasible decisions.
(b) Subsistence goal satisficed.
(c) Cash consumption goal satisficed.
(d) Feasible choices that satisfice high-order goals and are safe enough.
(e) The profit maximizing decision.

prices of the two crops and given by the slope of the new constraint in Figure 2.2(b). The shaded area denotes the set of decisions that would satisfice the subsistence goal. If working capital or land or both were available in sufficiently reduced amounts, the subsistence goal could not be satisfied and would instead be maximized. This could lead, for example, to a combination of M and T within the cross-hatched area and a choice that maximizes but does not satisfice consumption. In this case there would be no further scope for behavior.

Returning, however, to the larger feasible region, we can consider the cash consumption goal. Consumption moves the working capital constraint inward and further restricts the production possibilities. The region of decisions that is feasible and satisfices both subsistence and consumption targets is shown in Figure 2.2(c). Choice within this region is further circumscribed by the safety criterion which, when satisficed along with the higher order goals, gives the new region or ZFR shown in Figure 2.2(d). The final choice is one in this region that yields the highest anticipated profit attainable without violating the safety and other high-order goals.

In the Day-Singh model the safety zone is always adapted to and centered on the most recent past decision.

If it should happen that all other aspects of the decisionmaking environment were constant, then ZFR shifts through time to encompass the then past decision. Thus, the choice of crops will move period after period to the left until the farm has completely specialized in the modern crop. We believe that this cautious and constrained suboptimizing reflects decisions at the farm level under conditions of transition—given its uncertain, dynamic nature—in a quite realistic way. The approach also allows for explicitly accounting for noneconomic behavior. For example, the farmer may retain some minimum acreage in crop M simply because he "likes" it.

Thus, this approach seems more realistic than the simpler neoclassical model or the more recently developed expected utility models. When all the goals are satisfied except the last, the decision model boils down to an ordinary maximization problem with behavioral and technical constraints. The effect of adaptation in response to feedback is to modify from period to period the position of the various constraints and objective functions and hence the position and shape of the final choice problem.

The Day and Singh model embodying this microeconomic structure was applied at the regional level under the assumption that the regional model can approximate with useful accuracy the results that would be obtained by adding up individual farm estimates based on a similar model.

This model represents the development process in several ways. Given initial conditions of low or nonexistent capacity in a highly productive technology, farm behavior is dominated by subsistence goals. External demand conditions and internal productivity permitting, commercial sales lead to cash income for which consumption and farm investment compete. Subsistence considerations are gradually pushed into the background. As cash farming grows in importance, caution in response to market forces and profit maximizing come to dominate farm production and investment decisions. Depending on the initial situation, some farmers might adopt new technology rapidly. Others may not at all. Indeed, many alternative histories with different phases or stages of development arranged in alternative sequences are possible in such a model.

Equilibrium of a sector at a stationary state may come about in the absence of technological change, although nothing in the theory guarantees that possibility. Further, current empirical evidence suggests that agriculture, in very diverse situations, is inherently unstable once the state of the arts change and commercial farming activities become important. The cause seems to lie in the highly inelastic demand for agricultural products and its feedback effect through price and working capital supplies. A realistic micromodel should be able to reflect these dynamic and destabilizing properties.

Macrofeedback

Although individual farms may act as if they had no effect on prices, the aggregate effect of farm output and the aggregate demand for inputs does in fact influence prices. This is the aggregate feedback effect. Since such aggregate effects influence each individual farm, the dynamics of an individual farm cannot be understood without accounting for the aggregate feedback effects of the farm sector. Therefore, even if the entire focus of an investigation is on the dynamics of various representative farms, a satisfactory understanding of many aspects of farm-level changes will require modeling some aspects of macrofeedback.

The simplest way in which this can be done is the closed-sector or generalized cobweb model suggested by Day (1963, chapter 3, and 1964). Here the external economy is assumed to establish a temporary equilibrium in which prices emerge that clear agricultural markets in the shortrun. Inverse nonfarm market demand and supply functions are used to represent market feedback by determining prices and revenues.

Another approach is to combine an adaptive economizing model of the farm sector with some other kind of simulation model. This approach is taken in the generalized systems simulation models of Manestsch

FIGURE 2.3a

A dynamic agricultural model: hogs, West Germany (demand growing)

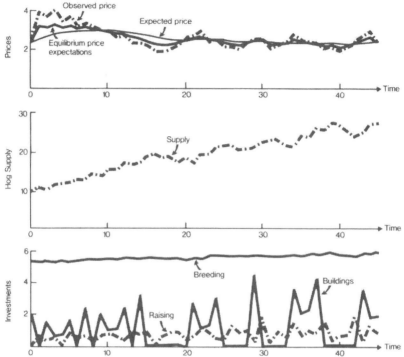

and others. An example using recursive programming is the study of DeHaan. Others are the recursive interactive programming models of Baum, Sposito, and Heady.

A new multisector model under development by Jay Forrester is based on a single coherent modeling philosophy compatible with the adaptive economics approach and offering a model environment for studying macrofeedback effects on an adaptive model of the agricultural sector.

The author is currently developing a new, adaptive multisector model that shares with Forrester's national model the unifying structure of a single modeling philosophy emphasizing adaptation in response to feedback, behavioral rules, and stock-flow adjustment mechanisms to mediate intersectoral disequilibrium. It may be distinguished from the Forrester work by its incorporation of typical features of recursive programming: an activity analysis of production, explicit adaptive economizing, and a distinction between plans and control.

FIGURE 2.3b

A dynamic agricultural model: hogs, West Germany (demand stationary)

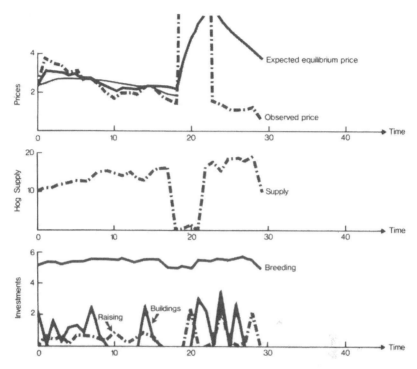

Examples

We will now briefly summarize 2 examples of adaptive economizing models with macro-feedback. The reader must refer to the studies cited to get an adequate understanding of individual model assumptions and performance.

The first is the closed or generalized cobweb sector model developed by Muller and Day (1978) involving cautious, rolling plans with fore-casting and market feedback. It is assumed that farmers' plans are drawn up for a finite, multiperiod horizon. After the first period passes and the first stage plans are executed, new information becomes available through market feedback, and previous plans are revised accordingly. Uncertainty is reflected in more or less pessimistic forecasts and behavioral constraints providing for more or less cautious modification of current activities in response to the evolving market context.

Two simulations with the model, designed to reflect conditions in West German agriculture are shown in Figures 2.3a and 2.3b. In Figure 2.3a, a base case is shown in which demand for agricultural produce is growing. In Figure 2.3b, all parameters are the same except that demand is assumed to be stationary. In this case an extreme cycle erupts with the model becoming unviable in period 29. We see clearly how the internal dynamics of a sector can be modified by events going on outside, but impinging on, the sector.

A second example drawn from models of the RP genre, but involving an explicit multisector structure, is a model developed by Y.-K. Fan in his doctoral dissertation and summarized in Fan and Day (1978). This model incorporates a rural farming sector and an urban-industrial sector. Within each sector economizing production decisions are represented using the cautious optimizing with feedback approach. The production sectors are linked by a financial sector that mediates the flow of credit, by a marketing sector in which prices are reflected, and by a migration sector that provides for moving people in response to socioeconomic conditions and to income and quality of life differentials between the rural and urban areas.

In Figure 2.4 a sample of results is shown that is designed to reflect development in a hypothetical initially underdeveloped, overpopulated country. Of special interest is the overlapping wave character of technology shown in Figure 2.4a and the cusp switch from growth to decay in fibre and export crop production in Figure 2.4b. These drastic changes are occurring while aggregate capital stock accumulates according to the smooth, classic Sigmoid pattern of Figure 2.4d. The z-good represents the allocation of labor in the urban sector to labor intensive, traditional domestic, and personal services, shown in Figure 2.4c.

These examples illustrate how model-generated behavior mimics salient features of historical experience: the unbalanced growth, overlapping wave syndrome, irregular oscillations endogenously generated by deterministic behavior-environmental feedback interactions, the sudden emergence of extreme fluctuations, and breakdown in the model system. Evidently, the type of model under consideration has the flexibility to represent the rich variety of time paths that socioeconomic variables display in reality.

These examples show relationships between (1) the farming sector or a major portion of the sector and (2) the larger economy. Logic suggests similar relationships between microunits such as farms and the farm sector and between microunits and the macroeconomy.

Complex Dynamics and Policy Design

Adaptive Economics and Complex Dynamics

Theoretical and empirical models of the kind that are discussed in the preceding pages point to a coherent, general theory—a new paradigm

FIGURE 2.4
A dynamic model of urban-regional development

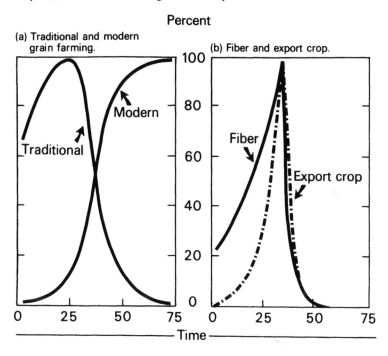

Percent

(a) Traditional and modern grain farming.

(b) Fiber and export crop.

Time

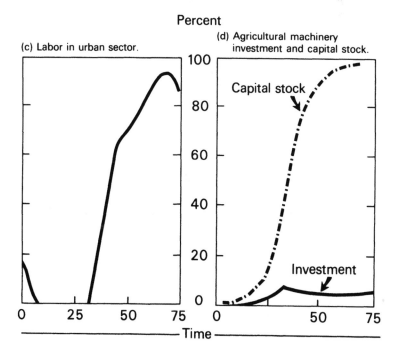

Percent

(c) Labor in urban sector.

(d) Agricultural machinery investment and capital stock.

Time

of economic development and cultural evolution. Adaptive behavior implies disequilibrium; disequilibrium requires the existence of disequilibrium mechanisms; the dynamics of these mechanisms are complex and involve overlapping waves, structural change, irregular oscillation, decline, demise, and breakdown. Let us review and summarize these thoughts in slightly more detail.

Economic adaptation involves an interaction of people (decisionmakers or agents) with their surrounding environment including their organizational mileau. The behavior of agents is, in part, rational and involves explicit choice, economizing and, more generally, optimizing. But people's knowledge of environmental structures is incomplete, their store of data and memory is limited, and their powers of rationalization and computation are severely circumscribed. Rational choice therefore involves problem simplification and problem solution and often must involve elementary adaptive learning. Actual behavior by adaptive decisionmakers involves a blend of rational choice, of traditional or standard operating procedures, and servomechanistic, negative feedback adjustment mechanisms.

Coordination among the members of a group, economic organization, or culture is imperfect. Disequilibria should therefore typify actual situations—unfulfilled expectations, inequations in supply and demand, and worsening situations for some while others improve. Because of disequilibrium, survival must be a key concern in adapting economic systems. The specific homeostatic or adaptive motivated feedback behavior is one type of strategy to bring about general homeostasis, that is, to maintain critical variables within bounds that define survival. Learning and explicit optimizing models must have come into being to serve this same purpose. Special economic institutions—stores, order mechanisms, monetary systems, and resource transfer schemes—also have come into existence to mediate economic activity in disequilibrium. These strategies and organizations do not always succeed. For this reason we should expect to observe variation and selection in rules, individuals, and organizations. In short, an economy will evolve.

Although equilibria, balanced growth, or stationary states may be brought about within a given dynamical system, even simple examples of equilibria are capable of overshoot, oscillation, and more extreme forms of catastrophic change. These unstable modes of behavior are to be expected and may be the rule rather than the exception in adapting processes. Particularly when resources are scarce we must expect to find either deterioration and eventual demise of resource-dependent activities or overlapping waves of technology as new methods, products, and resources replace old ones in a succession of technologically based epochs.

A type of instability inherent in the process is endogenous structural change, or structural evolution reflected in the progression of multiple phase solutions that emerge when longer run development is modeled. Each phase is associated with a unique set of active activities and constraining constraints. These combinations define a way of life characteristic of a given time and place. Since change is occurring somewhere in the overall systems, these characteristic phases eventually switch. New activities are pursued, some old ones may be abandoned. New constraints become binding while some old ones may cease to be important. Certain goals may dominate choices during a given phase while others may guide behavior when a different phase governs change.

The richness of the potential progression through alternative phases depends on the number of choice alternatives, choice criteria (goods), and constraining factors in a model. These, in turn, depend on the number of representative firms, regional situations, and nonfarm sectors included.

A typical characteristic of these models is nonlinear feedback. This phenomenon occurs, even when behavior is represented by linear decision-rules, whenever financial feedback is present. Revenues play a fundamental role in financial flows. Revenues involve the product of prices and quantities, even in the simplest of recursive programming models (Day, 1967; Day, Morley, and Smith).

Recent developments in dynamical systems theory (Li and Yorke, Gürel and Rössler) based on the seminal work of Lorenz (1963) and Smale have led to a deeper understanding of one of the most remarkable properties of deterministic nonlinear feedback systems—namely, their ability to generate irregular fluctuations that resemble the behavior of stochastic processes.

Typical time paths of economic variables in such models seem to evolve through apparently different regimes even when no structural change in the sense of phase switching has occurred. Together with phase switching, which can have the effect of reversing trends and modifying dynamic modes, these complexities suggest that past behavior of a nonlinear system may be a poor guide for predicting even *qualitative,* let alone *quantitative,* patterns of changes in the future.

A still more extreme form of complex dynamics is the phenomenon of breakdown and unviability. This occurs when the regimes defining workability or feasibility of the system are empty. No feasible actions are available. In simulating adaptive economizing using recursive programming models, the message "no feasible solution" signals such a breakdown or unviability. Of course, this signal may occur because of modeling errors of various kinds which have to be corrected. Stupid mistakes aside, those events happen because crucial alternatives are omitted from the model. These omissions, in turn, happen because of a desire to keep it simple, because of an oversight, or because of ignorance of the existence of important activities, constraints, and feedback structures in the real process being studied.

But it can also happen for other reasons. Inviability can also occur because the interactive dynamics under consideration are correctly represented in essential detail. The occurrence of unviability then reflects the underlying longrun inconsistency of the socioeconomic system being modeled. That system could not persist but would have to be reconstituted by changing the rules of behavior governing change. This requires creative morphogenesis, i.e., the purposive invention of new institutions and decisionmaking strategies.

Implications of Complex Dynamics for
Explanation and Policy Design

The theoretical tenets of adaptive economics provide a foundation of plausible assumptions for the construction of operational models. These assumptions are plausible because most of them can be verified by direct observation, interview, and introspection. They are, in a sense, obviously true. Consequently results derived from models that incorporate them should compel attention in explaining economic change and effects of policy.

Complex dynamics associated with adaptive economic models do indeed pass the test of realism because the real world exhibits complex behavior of exactly the same kind. This is especially the case within agriculture.

It should be evident after roughly half a century of development that econometric models are poor predictors. To conclude that models, therefore, have no use is to throw the baby out with the bath water.

Certainly adaptive economics cannot be used any more than any other quantitative approach for accurately predicting the future and its response to current policy changes. But by providing a plausible basis for understanding the qualitative aspects of complex economic dynamics, it should provide realistic insights for the improvement of policy design.

Thus, adaptive economic models can be useful in explaining qualitative features of economic change: how growth and decline occurs and under what conditions; how trends reverse and why; how fluctuations emerge; what aspects of structure lead to irregularity; how phases switch and structures evolve; how unviabilities or system breakdown occurs. These models can be used to explore the features of technology and behavior that favor one qualitative mode or another, increase instability, moderate extremes of fortune, and favor efficacious change.

Economic policy involves such questions as: What should tax rates be? At what level should price supports be set? These are of a quantitative nature. More fundamental questions include: What kind of tax should be used? Should price supports be used? Such questions are qualitative.

Without attempting to provide quantitative predictions of the economic effects of alternative quantitative and qualitative policy changes, adaptive economic models can be used to study qualitative effects of policies. They can help anticipate conditions with and without alternative policies and thereby better understand potential effects of policies. Will a given policy accelerate growth? Will it lead to or prevent instability, decline, or breakdown? Will its effects reverse direction over time so that longer run repercussions will differ from short-run effects? What types of behavior are likely in the absence of policy changes? If under given conditions a model will eventually break down (go infeasible), what new institutions represented by specific new model structures prevent such breakdowns and preserve viability? This latter role involves the use of models to guide institutional development.

Further, the innovative design of micromodels will (1) enhance the design of aggregate adaptive economic models and (2) provide important insights as to how policies will affect different economic agents differently.

Epilogue: Adaptive Economics as a Weltanschauung

Adaptive economics should not be thought of as *the* theory or even *a* theory. It is not *the* model, *a* model, or even a group of models. It should not be thought of as a new paradigm to replace conventional wisdom and standard economics.

Rather, adaptive economics is a way of thinking about and an approach for understanding economic change both in terms of explanations and of policy design.

As such, it is and must remain an evolving field. Many models may be explored. Some may be found useful for one purpose or another. Others may be discarded in the process of competition for intellectual attention and practical relevance. But, if it came to pass that this field should one day flourish, that it may one day vie for attention and influence with the standard apparatus of optimality, equilibrium, comparative static economics, then the prospects for building a more effectively operating society will be greatly improved and the chance for individual opportunity and improved social welfare enhanced.

For, if the world is truly unstable, as the complex dynamics of adaptive economic processes lead one to suspect, then the imperative for human organization would seem to be, evolve or perish.

If efficacious evolution is to take place, then study of it should occupy a central place in economics. Effective study must surely involve consideration of models that incorporate the salient features of real world human processes as they exist and as they work, however complex that might be. Adaptive economics is devoted to the development of models of this kind. With such models adaptive economics will contribute to the design of new institutions and the more effective operation of those that exist.

Acknowledgments

Comments by Kenneth Baum and Lyle P. Schertz were most helpful.

References

Alchian, A. "Uncertainty, Evolution and Economic Theory," *Journal of Political Economy*, Vol. 58, 1950, pp. 211–21.

Baum, K., V. Sposito, and E. Heady. "Recursive, Interactive Programming: Computer Methodology and Procedures." Seventh International Association of International Economists, Banff, Canada, Sept. 1979.

Bellman, R. *Dynamic Programming.* Princeton, N.J.: Princeton Univ. Press, 1957.

————. *Adaptive Control Processes: A Guided Tour.* Princeton, N.J.: Princeton Univ. Press, 1961.

Cyert, R., and R. G. March. *The Behavioral Theory of the Firm.* Englewood Cliffs, N.J.: Prentice-Hall, 1963.

Day, R. H. *Recursive Programming and Production Response.* Amsterdam: North-Holland Publishing Co., 1963.

————. "Dynamic Coupling, Optimizing, and Regional Interdependence," *Journal of Farm Economics*, Vol. 46, May 1964, pp. 442–51.

————. "Profits, Learning and the Convergence of Satisficing to Marginalism," *Quarterly Journal of Economics*, Vol. 81, May 1967, pp. 302–11.

————. "Adaptive Processes and Economic Theory," in *Adaptive Economic Models* (ed. Day and Groves). New York: Academic Press, 1975.

————. "Unstable Economic Systems," in *Systems Dynamics and the Analysis of Change* (ed. A. Paulre). Amsterdam: North-Holland Publishing Co., 1981.

Day, R. H., and A. Cigno (eds.) *Modelling Economic Change: The Recursive Programming Approach.* Amsterdam: North-Holland Publishing Co., 1978.

Day, R. H., and I. Singh. *Economic Development as an Adaptive Process: The Green Revolution in the Indian Punjab.* New York: Cambridge Univ. Press, 1977.

Day, R. H., S. A. Morley, and K. R. Smith. "Myopic Optimizing and Rules of Thumb in a Micro-Model of Industrial Development," *American Economic Review*, Vol. 64, Mar. 1973, pp. 11–23.

DeHaan, H. "A Model of Interregional Competition in West Germany," *Modelling Economic Change: The Recursive Programming Approach* (ed. Day and Cigno). Amsterdam: North-Holland Publishing Co., 1978.

Fan, Y. K., and R. H. Day. "Economic Development and Migration," *Modelling Economic Change: The Recursive Programming Approach* (ed. Day and Cigno). Amsterdam: North-Holland Publishing Co., 1978.

Fel'dbaum, A. A. *Optimal Control Systems.* New York: Academic Press, 1965.

Forrester, J. W. *Industrial Dynamics.* Cambridge, Mass.: MIT Press, 1961.

————. "Modelling the Dynamic Processes of Corporate Growth," in *Proceedings of the IBM Scientific Computing Symposium on Simulation Models and Gaming.* White Plains, N.Y.: IBM Data Processing Division, 1966.

————. "An Alternative Approach to Economic Policy: Macrobehavior from Micro Structure," *Economic Issues of the Eighties* (ed. Kamrany and Day). Baltimore: Johns Hopkins Univ. Press, 1979.

Gurel, O., and O. E. Rossler. "Bifurcation Theory and Application in Scientific Discipline," *Annals of the New York Academy of Sciences,* Vol. 316, 1979.

Li, T. Y., and J. A. Yorke. "Period Three Implies Chaos," *American Mathematical Monthly,* Vol. 82, 1975, pp. 985–92.

Lorenz, E. "Deterministic Nonperiodic Flow," *Journal of the Atmospheric Sciences,* Vol. 20, Mar. 1963, pp. 130–41.

Manestsch, T. S., et al. *A Generalized Simulation Application to Agricultural Sector Analysis with Special Reference to Nigeria,* Michigan State Univ.-East Lansing, Dept. of Agricultural Economics, 1971.

March, J. G., and H. A. Simon. *Organizations.* New York: John Wiley and Sons, 1968.

Marschak, J. "On Adaptive Programming," *Management Science,* Vol. 9, July 1963, pp. 517–26.

Muller, G. P., and R. H. Day. "Cautious Rolling Plans with Forecasting and Market Feedback," *Modelling Economic Change: The Recursive Programming Approach* (ed. Day and Cigno). Amsterdam: North-Holland Publishing Co., 1978.

Nelson, R., and Winter S. "Forces Generating and Limiting Concentration under Schumpeterian Competition," *Bell Journal of Economics,* Vol. 9, 1978, pp. 524–48.

Rausser, G., and E. Hochman. *Dynamic Agricultural Systems: Economic Prediction and Control.* Amsterdam: North-Holland Publishing Co., 1979.

Simon, H. *Administrative Behavior.* 2nd ed. New York: Free Press, 1957.

Smale, S. "Differentiable Dynamical Systems," *Bulletin of the American Mathematical Society,* Vol. 73, 1967, pp. 747–817.

Williams, J. B. "The Path to Equilibrium," *Quarterly Journal of Economics,* Vol. 81, 1967, pp. 241–55.

Winter, S. G. "Economic Natural Selection and the Theory of the Firm," *Yale Economic Essays,* Vol. 4, Spring 1961, pp. 225-72.

————. "Satisficing, Selection and the Innovating Remnant," *Quarterly Journal of Economics,* Vol. 85, 1971, pp. 237–61.

Discussion of Chapter 2

Some Thoughts on Adaptive Economic Policy and the Behavioral Science of Modeling the Farm Firm/Household

Kenneth H. Baum

Although each of the ideas and behavioral precepts advanced by Professor Day's "adaptive economics" has been intuitively recognized by many economists, it is not clear their combined use in a cohesive and inclusive theory of economic (system) behavior has been widely accepted or has been given adequate and thoughtful attention. Although many of these ideas are quite important when modeling observed economic behavior, they are often given cursory attention with footnotes. I believe their significance for the construction of quantitatively accurate representations of economic systems used for agricultural policy analysis cannot be underestimated.

The paradigm of adaptive economics is based on the assumptions of: (1) "satisficing" behavior by individuals due to incomplete information, (2) nonlinear feedback among the defined causal variables and "strategic detail" in the system, and (3) continuous disequilibrium in markets and among markets moderated by inventory or stock adjustments and related institutions. Day suggests that the behavioral assumptions and disequilibrium adjustment mechanisms are extremely important for the interpretation of research relating to economic change. In addition, Day argues that the study of the adaptive adjustment processes and economizing mechanisms for the allocation of resources dictates that qualitatively different analytical approaches be used when modeling economic systems. My understanding of this qualitative change is that questions relating to stability ("when?"), phase switching and oscillatory ("how often?") behavior of inherently unstable, dynamic systems are more important than questions typically asked of systems assumed to be stable and moving toward equilibrium.

The immediate issue at this conference, though, addresses the problem of how agricultural economists should proceed to model the farm firm or household in order to provide additional useful or corroborative

The author is an agricultural economist with the National Economics Division, Economic Research Service, U.S. Department of Agriculture.

information when analyzing national policies. Given Day's thoughts, it is less clear which, if any, of our conventional tools are appropriate. The proper use of analytical tools assumes that we are also using correct economic theory when developing the models, asking the correct questions, and interpreting the results. Day suggests, and I agree, that adaptive economics, as he has defined it, should be given careful consideration when selecting tools for micro- as well as macromodeling.

Day's paper compares adaptive economics to conventional neoclassical economic theory. Neoclassical analyses usually involve use of comparative statics, an analytical technique focusing on how individual and competitive equilibrium optima change when the underlying parameters of preference, technology, prices, and government policy change. Although this technique is widely used and accepted, Day suggests there are many analytical problems relating to process of change characteristics that are not easily analyzed with comparative statics because it does not adequately incorporate (1) nonlinear negative feedback, (2) target planning, (3) satisficing behavior by actors, or (4) the imperfect state of knowledge throughout the system. Consequently, these policy analyses often have difficulty explaining disequilibrium conditions, unanticipated participatory behavior by farmers, oscillatory price, quantity and inventory behavior, extreme fluctuations of endogenous variables, paths of system adjustment to exogenous variable changes (or internal instability), and the emergence of new institutions. However, each of these topics may be important to decision- and policymaking.

The first question for the economist is to decide whether actual behavior of the economic system is stable and converging to a competitive equilibrium or whether real world experience and observations suggest otherwise. My own experience leads me to believe that change is continuous, that my knowledge will always be imperfect, and that I expect to make fewer errors in judgment as I grow older. Consequently, I begin with an intuitive belief in adaptive economics while reevaluating my training and experience in comparative statics. I do not believe that Day asks agricultural economists to dispense with our large human capital investment in traditional analytical paradigms and the neoclassical economic theory. Rather, he suggests the adoption of adaptive economic paradigms and methods of analysis for studies of economic change and the dynamics of economic systems whenever appropriate.

Day's discussion of agriculture as a disequilibrium rather than as an equilibrating system should, in some fashion, be a testable hypothesis. A recent article by Mass offers some additional thoughts on disequilibrium phenomena based on destabilizing impacts of price and quantity adjustments to market conditions. Mass analyzes the comparative advantages of the neoclassical and dynamic (adaptive economics) theories of firm behavior in explaining complex changes in production, consumption, and resource use over time. Although it would be rather simple if these theories were complementary and compatible rather than competitive, each theory is different in terms of assumptions made about the nature of the individual's or system's economic behavior. As Mass and Day suggest, the comparative static analysis of the micro unit almost always assumes that disequilibrium conditions will be eventually, and perhaps quickly, eliminated through price and quantity adjustments. The adaptive economics approach, however, regards economic change as a disequilibrium process. Consequently, economic analysis using an adaptive economics analytical approach places a great deal more emphasis on flows, stocks, the time paths of adjustment of endogenous variables, and the nonlinear feedback effects from information flows when answering dynamic questions.

Day argues further that feedback in adaptive economics does not necessarily imply an attainment of system equilibrium found in optimal control systems. Rather, adaptive economics suggests that fluctuations or oscillations around an equilibrium are not intuitively or quantitatively predictable. Much of Day's earlier research suggests that the behavior of the adaptive system may appear random even when clearly specified logic and decision rules are used (1969, 1975). These results may be a primary reason why predictions made with conventional models are so often inaccurate and constantly revised, despite use of sophisticated econometric or programming techniques.

Many of us, though, have been trained to believe that we can predict and quantify economic behavior. Consequently, when we run into empirical difficulty, we may eventually experience what psychologists refer to as cognitive dissonance—what you perceive is different from what you want to perceive and the two "realities" are irreconcilable. I have found that the only way to solve this problem is either to change the environment (the problem) or change my thinking. Thus, I suggest that we might find it appropriate to adapt our thinking toward our practice of economic analysis rather than keep trying to "change the environment" by refining our mathematical and statistical approaches to predicting economic behavior. As a profession, we do not seem to be increasing prediction accuracy although many more predictions are being made more frequently. This situation may reduce our credibility with policymakers or producers even further. In turn, the number of policymakers and producers who seek us out for advice may begin to decrease, perhaps at an increasing rate.

How would an adaptive economic model of the farm firm/household be constructed differently than an optimal control or recursive programming model? My interpretation of Day's definition of adaptive economics suggests that such models are qualitatively different from the various types of optimal control models that Rausser and others discuss. This apparent problem may be partly a result of Day's philosophical style in this paper and the absence of rigorous mathematical development evident in his previous publications (for example, 1974 and 1978). However, Rausser's references to a dual adaptive optimal control system with active learning appears to be similar to the adaptive economics model. Thus, I would appreciate a more complete discussion to explain these modeling differences, particularly with regard to development and incorporation of nonlinear behavioral relationships, information flows forward from previous periods to the current period, or from future periods back to the present, endogenous system control variables, and goals or the objectives of the system or actor.

Day's essay is presented from a point of view evidently synthesized from systems science, cognitive science, and neoclassical economics. The interdisciplinary synthesis of systems science with neoclassical theory is fascinating and could become increasingly important to many economists working with multivariate models. It should be noted that Day's use of an interdisciplinary paradigm of behavior is not new to agricultural economics, nor should it prove to be unsettling. After all, econometrics has been used for well over half a century, and bioeconomics (pest management) has been attracting increasing attention and research during the last decade. Historically, though, agricultural economists' reliance on and acceptance of interdisciplinary research paradigms has been predicated on the assumption that these paradigms could make it easier to obtain answers to quantitative, methodological, or theoretical problems, and increase reliability of important behavioral insights provided.

My research and discussions with other social and biological scientists have convinced me that economics, as a professional discipline, need not always be clearly defined, since the theory or paradigm of economic decisionmaking is constantly evolving, sometimes, as Kuhn and Ladd suggest, after a "creative struggle." Economics is commonly defined as the study of how limited resources can or may be used to achieve certain objectives. However, I tend to agree with Day that economics is more importantly a behavioral or social science with a primary purpose of •attempting to explain the behavior of individuals or groups and the process of their producing, exchanging, and consuming goods and services.

The problem, then, as I suggested earlier, is how to best model the farm firm/household to provide both quantitative and qualitative economic intelligence. The implications of Day's remarks are that equilibrium-oriented approaches to modeling the micro unit may be quite simply incorrect and consequently provide misleading results. Now, this is not to say that these modeling approaches are analytically incorrect through misuse of the latest sophisticated statistical techniques or computer software. Rather, their use results in incorrect information because they are unable either to portray actual (disequilibrium?) household economic and decisionmaking behavior or to test hypotheses concerning that behavior. Dozens of references in the literature can be found in studies attempting to find the optimal, profit-maximizing solutions, allocation of resources, or decision criteria for a producer or sector given certain internal and external conditions. However, there are relatively very few research studies relating, describing, or testing actual farm behavior, decision criteria, or how changes in the farm's resource base actually occur. Nevertheless, the assumptions made by the researcher about the behavior of the actor being studied are critical to the realism of the problem's solution and the interpretation of that solution. Consequently, I would have to strongly agree with Day that a more qualitative analytical approach, such as adaptive economics, could prove to be informative and also greatly expand the disciplinary knowledge base for policy analysis at both the micro and macro level.

My final comments center around the type of information product we will want to present to those in a position to formulate and implement policy. What kind of information can models of the farm firm/household provide with some degree of credibility? Second, what information do we want them to provide? Day suggests that qualitative information has a comparative advantage over quantitative information. Perhaps agricultural economists could make a more strenuous effort to become more heavily involved in the policy development process rather than the policy evaluation process. But, would our profession's ability to maintain or increase research resources be enhanced if we primarily provide only qualitative information? To some extent, our priority in the disbursement of social science research funds is dependent on our service ability to provide economic intelligence to justify various qualitative policy positions with quantitative facts—treasury outlays, number of insolvent farms, level of loan rates, and so forth. One suggestion is that our conventional models might be used at a comparative advantage for these shortrun quantitative estimates and predictions, while adaptive economics models might then be used for complementary, longer run planning and qualitative analyses. It is clear, though, that the proper combination of models and modeling approaches will ultimately be decided through exposure, publication success, and the market for economic intelligence.

In closing, I believe there is an increasing tendency for agricultural economists to view their work objectives as research in quantitative methods rather than explaining or hypothesis testing of economic behavior particularly as a process. In addition, the belief, perhaps implicitly assumed, that knowing more quantitative facts will somehow provide either an understanding of dynamics of change in agriculture or the rationale for different types of government policies is just not sustainable. Policy analysis questions are hardly ever directed toward what the farm should do to maximize profits or more efficiently organize their resources. Rather, policy concerns are far more likely to be directed toward information about what farmers are likely to do, when, and why. Adopting the adaptive economics paradigm in our micromodels may help resolve many of these issues in the future.

References

Day, Richard. "Adaptive Processes and Economic Theory," in *Adaptive Economic Models* (ed. R. Day and T. Groves). New York: Academic Press, 1975.

————. "The Structure of Recursive Programming Models" in *Modelling Economic Change: The Recursive Programming Approach* (ed. R. Day and A. Cigno). Amsterdam: North-Holland Publishing Co., 1978.

————, S. Morely, and K. Smith. "Myopic Optimizing and Rules of Thumb in a Micro-Model of Industrial Growth," *American Economic Review*, Vol. 75, No. 1, Mar. 1974.

————, and E. H. Tinney. "Generalized Cobweb Theory of Economic Growth, Decay, and Cycles," *Economic Journal*, Vol. 79, No. 1, Mar. 1969.

Kuhn, Thomas S. *The Structure of Scientific Revolutions.* 2nd ed. Vol. 2, No. 2. Chicago: Univ. of Chicago Press, 1970.

Ladd, George W. "Artistic Research Tools for Scientific Minds," *American Journal of Agricultural Economics*, Vol. 61, No. 1, Feb. 1979.

Mass, Nathaniel. "Destablishing Impacts of Price and Quantity Adjustments to Relative Supply and Demand," *Journal of Economic Behavior and Organization*, Vol. 1, No. 4, Dec. 1980.

Rausser, Gordon. "Active Learning, Control Theory and Agricultural Policy," *American Journal of Agricultural Economics*, Vol. 60, No. 3, Aug. 1978.

Discussion of Chapter 2

A Proper Perspective for Considering Adaptive Economics

David Zilberman and Gordon C. Rausser

Richard Day's presentation outlines an interesting set of issues directed at micromodeling in agriculture. How could we not concur with his recommendations that actual behavioral patterns of farmers should be studied and documented more intensely and carefully, that more efforts are needed on conceptual models of agricultural firm behavior, that the assumptions underlying these models should correspond more closely with reality, and that more emphasis should be given to understanding the behavior of the agricultural sector rather than its estimation and prediction? To be sure, these recommendations have a familiar ring. The real question is balancing the tradeoff between simplicity and accuracy in micromodel specifications for various purposes. The purpose for a particular model effort will largely dictate the tradeoff between complexity and inaccuracy in the modeling of the reality. Armed with this perspective, we are more optimistic than is Day on the profession's ability to predict and estimate the impact of policies on the evolution of the agricultural sector and on the potential of economics to determine the shape and form of agricultural policies.

Before launching into an evaluation of Day's arguments, let us first outline his views. He criticizes the static equilibrium approach taken by the majority of microeconomic studies. He emphasizes the need to incorporate dynamic feedbacks, disequilibrium, imperfect markets, and bounded human abilities in microeconomic models of the farm sector. He outlines a framework that he has advocated for some time—namely, "adaptive economics"—as capable of incorporating these features in microeconomic model representations.

The realization that assumptions of conventional neoclassical models of perfect competition do not hold for U.S. agriculture is hardly novel. Actually, many of the models of the U.S. farm sector that provide the foundation for conventional wisdom are heuristic; these models have

The authors are, respectively, assistant professor of agricultural and resource economics, and chairman and professor of agricultural and resource economics, Department of Agricultural and Resource Economics, University of California, Berkeley.

dominated other models in influencing and shaping U.S. agricultural policy. Moreover, they include—admittedly in a loose fashion—some of the important elements that Day argues must be incorporated into useful micromodels. Unfortunately, much of this work is overlooked in Day's survey. The particular models we have in mind are:

T. W. Schultz's model of U.S. agriculture. This model emphasizes the inherent disequilibrium in the farm sector induced by the dynamics of technological change, the low elasticities of demand for agricultural products, and the rigidity of agricultural supply response. In this formulation, the frequent technological innovations in modern agriculture initiate a process of natural selection, and farmers with superior abilities to deal with disequilibrium survive and prosper.

Cochrane's model of the "technological treadmill." This model assumes that the dynamics of agriculture is driven by exogenous technological change. The ability to adopt new technologies discriminates among or distinguishes various farmers. Farmers operate with limited foresight of future prices; thus, the rates of return from adoption of new technologies by late adopters is below static equilibrium levels. Furthermore, some of the late adopters—and certainly the nonadopters—may face shutdown conditions.

G. L. Johnson's model of asset fixity. Johnson perceives agricultural input markets to be imperfect with spreads between purchase and sale prices of inputs causing asymmetric supply responses for farm products. These asymmetric responses are more elastic for price increases than for price decreases.

The heuristic frameworks provide the foundation for developing and applying microeconomic models in agriculture. They can be, and have been, used as points of departure for rigorous modeling of the agricultural sector. In addition, they provide the threads to sew together a framework that is consistent with Day's notion of adaptive economics.

One of the key elements of Day's adaptive economic approach is its focus on behavioral frameworks. However, there are 3 major types of model frameworks that are used in microeconomic analysis: (1) traditional neoclassical micromodels assuming full information, profit maximization, perfect competition in both input and output markets, and neoclassical production functions; (2) behavioral models assuming bounded rationality (limited computational and informational capacities of economic agents), satisficing rather than optimizing behavior, and ad hoc decision rules; and (3) new industrial organization models assuming optimizing behavior subject to informational and computational constraints or cost and imperfect markets. Many of the latter models focus on institutions and firm behavior resulting from imperfect information and uncertainty. For example, Spence demonstrates the role education plays as a signal in labor markets; Stiglitz alerts us to the role that risk ratings and deductibles play as signals in insurance markets; and Akerlof explores various mechanisms for discriminating among workers with different abilities.[1] To be sure, we have very little experience with the more advanced new industrial organization models that have appeared in the literature. The traditional neoclassical models are the most frequently employed in agricultural empirical applications. This preference is understandable; they are easier to implement and are consistent with the simple intuition of most researchers. In contrast, behavioristic models are more specific and, as a result, permit less general inferences to be drawn. Moreover, their empirical counterparts generally require specific software and, thus, their implementation is more costly than neoclassical models. For many specific applications, the gain in accuracy resulting from behavioristic frameworks does not warrant the additional computational cost associated with such models. For these applications, the simplicity offered by the traditional neoclassical models outweighs their greater inaccuracy.

Compared with the new industrial organization models, behavioristic models are less general and do not offer the same richness of insights. Actually, the new industrial organization models might be viewed as an outgrowth of the conflict between the neoclassical and behavioristic paradigms even though these two paradigms have different focuses (the neoclassical paradigm focuses on the market, whereas the behavioristic paradigm focuses on the internal organization of individual firms). This focus of the new industrial organization models is to attempt to explain the decision rules and choices that behavioristic models take as given.

What the foregoing discussion suggests to us is that the exclusive use of behavioristic models limits the potential value of the adaptive economics approach. It is our view that the effectiveness of adaptive economics will be enhanced if more efforts are devoted to the formulation of new industrial organization models (assuming maximizing behavior and recognizing imperfect markets, uncertainty, and costly information). These formulations offer much promise in explaining behavioral patterns and emerging (declining) institutions in the U.S. agricultural sector. In many empirical situations, the behavioristic approach can be employed in juxtaposition with a new industrial organization formulation. This process can occur when researchers use the behavioristic approach to establish behavioral rules to be investigated and explained by the new industrial organization-type models. As a result of this combined approach, quantitative models can be constructed that predict and evaluate the effects of alternative policies, particularly in the short run where the rules of behavior are presumed to be constant.

The development of useful quantitative policy is based partly on behavioristic rules for individual firm behavior and requires aggregation over farms to yield sectoral behavioral functions. Unlike Day, we are not overly concerned about microdata availability in the future. It seems to us that the proliferation of microcomputers will result in very extensive data networks and simulations that use samples for actual distributions of the key parameters. The information explosion that will result over the next few decades should allow more accurate quantitative modeling of farmers' responses to alternative policies. A general conceptual framework that takes us in the direction of such quantitative evaluations has been developed by Rausser, Zilberman, and Just (1981a).

The Rausser, Zilberman, and Just framework emphasizes the distributional effects of commercial agricultural policies. This model simplifies the individual farms' multiperiod dynamic optimization problem by imposing myopic optimization.[2] This simplification allows the introduction of a number of realistic elements, for example, imperfect capital markets, land-price speculation, varying asset qualities, and asset fixity in farmers' decision problems. Analytically derived aggregate relationships and industrywide outcomes are captured. This model has been generalized to include risk aversion and dynamic adjustments as well as tax and monetary policy effects on U.S. agricultural production (Rausser, Zilberman, and Just (1981b)). When this basic model formulation is empiricized, it will represent an application of the adaptive economic approach for American agriculture which incorporates some of the key ingredients that have been advanced by Schultz, Cochrane, and G. L. Johnson. A number of interesting theoretical propositions have been derived from this formulation, and it awaits empirical confrontation and actual policy-impact analysis.

In summary, Day has provided a provocative set of issues that are indeed important in future micromodeling efforts for U.S. agriculture. He outlines some interesting directions which, if placed in proper perspective, offer much promise for future microeconomic research in agriculture. We are more optimistic regarding the potential outcomes of such research efforts, apparently, than Day is himself.

Notes

1. The new industrial organization approach is similar to the "post-Bayesian paradigm" in statistics (Faden and Rausser). In this paradigm, simplicity—or the cost of complexity—plays a crucial role. In contrast to the conventional Bayesian approach, the cost of information sampling, data collection and summarization, and so forth, are treated explicitly in the formulation. Moreover, in contrast to classical statistics, the "level of significance" is part of the choice set rather than predetermined.

2. Testafsion has developed conditions for estimating the accuracy of this approximation.

References

Akerlof, G. A. "The Market for 'Lemons': Qualitative Uncertainty and the Market Mechanism," *Quarterly Journal of Economics*, Vol. 84, 1970, pp. 488–500.

Cochrane, W. W. *Farm Prices, Myth and Reality.* Minneapolis: Univ. of Minnesota Press, 1958.

Faden, Arnold, and Gordon C. Rausser. "Econometric Policy Model Construction: The Post-Bayesian Approach," *Annals of Economic and Social Measurement*, Vol. 5, 1976, pp. 349–62.

Johnson, G. L. "Supply Function—Some Factions and Notions." *Agricultural Adjustment Problems in a Growing Economy* (ed. F. O. Heady, H. G. Diesslin, H. R. Jensen, and G. L. Johnson). Ames: Iowa College Press, 1958.

Rausser, Gordon C., David Zilberman, and Richard E. Just. "An Equilibrium of Distributional Effects of Land Controls in Agriculture." Working Paper No. 80. Univ. of California, Berkeley, Dept. of Agricultural and Resource Economics, 1981a.

————. "The Distributional Impacts of Agricultural Programs," in Proceedings from Perspectives on Food and Agricultural Policy Research Workshop, Univ. of Maryland, College Park, Md., 1981b.

Schultz, T. W. "The Value of the Ability to Deal With Disequilibria," *Journal of Economic Literature*, Vol. 13, 1975, pp. 827–46.

Spence, M. *Market Signaling.* Cambridge, Mass.: Harvard Univ. Press, 1974.

Stiglitz, J. "Monopoly, Nonlinear Pricing, and Imperfect Information: The Insurance Market," *Review of Economic Studies*, Vol. 44, 1977, pp. 407–31.

Testafsion, Leigh. "Global and Approximate Global Optimality of Myopic Economic Decisions," *Journal of Economic Dynamic and Control*, Vol. 2, 1980, pp. 1–26.

Discussion of Chapter 2

The Challenge of Chaos: A Plea for Parsimony in Theoretical Constructs

Barry Richmond

Professor Day's paper is broad in scope and very stimulating. I have 3 points to make in response.

The first point is of a technical nature. Day creates 2 misimpressions in his discussion. Neither is very serious. He seems to equate "equilibrium" with "achievement of goal," or some notion of a "desirable state." The following passage illustrates the blurring of the distinction between the 2: "The flows among various agents, based as they are on individual adaptive processes, are in general imperfectly coordinated— they are thus out of equilibrium." An economic system can be "at equilibrium" far from the goals of individual agents. The equilibrium level of unemployment among inner-city ghetto teenagers is certainly not desirable. A high birth-death rate equilibrium is reached in many subsistence agricultural societies, where no one could claim to desire high rates of infant mortality. Equilibrium is a condition in which net rates of change are zero. An equilibrium state may, or may not, coincide with the achievement of economic or societal goals.

A second set of misimpressions created by Day relates to his characterization of system dynamics. He states that ". . . the most significant features of system dynamics approach are (1) its presumption that behavior is goal-adaptive, using feedback, servomechanism type rules . . . (2) responses to environmental changes often involving . . . discrete switching rules. . . ." Both types of feedback loop, positive as well as negative, are goal-seeking, not just the negative feedback loop as Day implies. Positive feedback loops are distinct from negative feedback loops only in that the goals which they seek are receding in the direction of search. Second, responses to environmental changes in system dynamics models do not, as Day states, often involve discrete switching rules. Instead, responses are almost always represented as continuous processes.

The author is an assistant professor of engineering, Dartmouth College.

The second point in response to Day's paper is an amplification of a point he makes. The theoretical impossibility of point-in-time, numerically accurate prediction in "chaotic" systems is well made. Day does not, however, leave the reader with much hope. But, indeed, there is hope. "Chaos" arises from the structure of the equations used to represent a particular economic system. If the equations approximate the structure of the actual system well, the structural arrangements in the real system must give rise to the desired instability. Hence, it is possible to make changes in the structure of the equations—and in the structure of the actual system through policy intervention—that can mitigate or eliminate "chaotic" tendencies. In short, "chaos" may not need to be taken as a necessary condition of economic systems, but rather one which can arise and can be addressed through policy intervention.

The third, and final, issue that I want to raise with respect to this paper is a plea for parsimony. Richard Day has set out in search of a comprehensive synthesis—one which draws together classical and neoclassical economics with dynamics and adaptive control. His efforts in this challenging pursuit should be applauded. However, as a model builder, I fear that his conceptual landscape is becoming too littered with theoretical constructs. The more such a priori constructs are identified, the more restricted is a modeler's vision when that person seeks to conceptualize a problem. I urge Professor Day to be as conceptually stingy as possible in forging his challenging synthesis.

These minor qualifications are not intended to distract the reader from the major issues addressed. His work should be mandatory reading for economists in general and for economic modelers in particular.

3. Complementarities Between Micro- and Macro-Systems Simulation and Analysis

C. Robert Taylor

Abstract This paper discusses the complex art of modeling complex systems. Special emphasis is given to modeling complementarities between micro- and macromodels of the agricultural sector. These complementarities can be viewed either in terms of interpretation of model results or the structure of model specification. Historically, a few complementarities between micro- and macromodel results have been described. Rather, most of the results are inconsistent. A complementary micro-macromodel specification is the key to eliminating inconsistent results.

Keywords Complementary macromodels and micromodels, macroeconomic models, microeconomic models

Decisionmakers continually demand information on equity and distributional effects in addition to economic efficiency of agricultural policies. This demand for information and the complexity of agricultural sector relationships has increased interest in developing policy-oriented quantitative models of the agricultural sector. Previous modeling efforts directed toward policy evaluation have tended to focus on either the micro-systems (the firm) or on the macro-system (the aggregate). Separating these approaches has not proven entirely satisfactory for policy analyses, in part because of: (a) invalid estimates from micromodels attributable to assumptions that some aggregate variables, such as price,

_____

The author is a professor of agricultural economics and economics, Montana State University.

remain unaffected when in fact they are affected by the resulting behavior of the sector as a whole; (b) macromodels' lack of structural specification or flexibility to estimate detailed (distributional) impacts; and (c) macromodel results often being invalid because of possible aggregation or composition bias.

Consequently, my motivation for writing this paper is the perceived need for a quantitative and qualitative framework to use in estimating relative effects on representative micro-units as well as the aggregate sector effects of various agricultural policies. Thus, the paper discusses the complex art of modeling complex systems, with special emphasis given to complementarities (or their lack) between micro- and macromodels of the agricultural system. A few facts are provided, a few straw men are set up, and several opinions are given.[1] However, given that modeling and interpretation of modeling results are and will continue to be largely art, I do not presume to make definitive statements about which models are complementary and which are not, and what is the best way of synthesizing micro- and macro-models. Nevertheless, this paper should stimulate thought and discussion on these topics.

Micro- and Macromodel Complementarities

Micro- and macromodel complementarities can be discussed from several definitional perspectives including model specification, model results, and quantitative modeling techniques. Micro- and macromodeling efforts are obviously complementary in terms of developing quantitative techniques. In terms of interpretation of or actual model results, complementarities between micro- and macromodel results have existed historically. In addition, these model results are not substitutable without glaring inconsistencies simply because micromodels are inappropriate for estimating aggregate effects and macromodels are obviously inappropriate for determining micro-effects because of their lack of structural detail.

Thus, when micro- and macromodels are used separately to analyze policies, we would expect inconsistent results for a variety of reasons. For example, a particular agricultural policy may be designed to influence commodity prices. Obviously, micro-effects estimated under the assumption of prices unaffected by the policy can be expected to differ in magnitude and direction compared with results obtained with prices that are affected by the policy. Even using prices obtained from a macromodel in a micromodel can lead to inconsistent results because of differing supply responses in the models.

Another reason for inconsistent results is that micro farm growth models are often specified without accounting for the fact that firm competition for the same resources (land, credit, and so forth) will directly influence the distributional and supply response outcome of policies. Macromodels can be similarly misspecified, but they are not always expected to provide the same information as micromodels. Thus, inconsistencies between micro- and macromodel results are to be expected because of invalid assumptions that may not be important at a specified level of aggregation, but may be critical at the other level.

Even in situations where representative farm models have been linked to macromodels, inconsistencies in model results were either not reported or masked with fudge factors. Furthermore, many micro- and macro-models are incompatible without a certain amount of ad hoc massaging of the models or results before interpretation. Thus, micro- and macro-models will not be complementary unless both are well thought out and specified as a system. The linking of micro- and macromodels afterwards and their respective economic and behavioral assumptions will most likely give nonsensical results, while carefully thought out, compatible models should at least tend to give consistent results. Of course, the latter approach may give results that are consistently inaccurate, but inaccuracies and inconsistencies are inherent in the former approach.

Model Specification

The key to developing complementary micro- and macromodels is model specification. Specifically interrelated issues related to specification that should be addressed include the conceptual approach to macro-modeling and sources of aggregation biases, positive versus normative versus hybrid models, types of micro- and macromodel linkages, compromises in model specification and construction necessitated by data and computational considerations, and stochastic versus deterministic models.

Macromodeling Approaches

Let us consider 2 possible macromodeling approaches. One conceptual approach is from the top down, whereby two equations are used to estimate the value of an aggregate price index that will equate indices of aggregate supply and demand. While this approach may be acceptable for making highly aggregative projections of policy effects, it has at least 3 serious deficiencies for making detailed distributional and efficiency estimates. First, the aggregation bias inherent in the 2-equation approach is likely to be quite large. Second, there are almost insurmountable problems in determining behavioral, production function, and interregional competition relationships that are necessary for a meaningful disaggregation of the aggregate estimates. Third, with the top-down approach, we lose sight of microeconomic theory that provides considerable guidance for macromodel specification.

A second conceptual approach to macromodel development is from the bottom up: The macromodel comprises submodels for each firm and consumer in the sector. While this is an ideal approach, it is obviously not empirically or computationally feasible. Practical compromises must be made, either by constructing the macromodel with specialized representative firm and consumer submodels or by aggregating these and other firms and consumer submodels. This approach introduces aggregation bias, but this bias can be minimized in many cases.

With both the top-down and the bottom-up approaches, aggregation bias is pervasive due to empirical and computational compromises. However, these biases are expected to be much less severe with the bottom-up approach because aggregation bias is apparent at each step of the aggregation process and can thus be minimized, while with the top-down approach, aggregation bias cannot be determined. To use a metaphor to compare the 2 approaches, you do not build a barn from the roof down; rather, you build it from the foundation up even if the foundation is not as strong as desired. For this reason, the remainder of this paper will consider only a bottom-up approach. Since aggregation bias is a major criticism aimed at macromodels estimated with aggregate data, this issue will be examined more fully for positive models and then for normative models.

Aggregation Bias in Positive Models

Although the conditions for exact aggregation in an econometric model are well known (Theil), it is useful to examine these theoretical conditions in the context of the production side of a macromodel. As noted previously, the key to having complementary micro- and macro-models appears to be in model specification. For simplicity, consider a single-product, 2-variable factor quadratic production function, which is a second-order approximation to any function:

$$y_i = a_{0i} + a_{1i}x_{1i} + a_{2i}x_{2i} + \frac{1}{2}b_{1i}x_{1i}^2 + \frac{1}{2}b_{2i}x_{2i}^2 + b_{3i}x_{1i}x_{2i} \tag{1}$$

where y_i is output for the ith firm, and x_{1i} and x_{2i} are factors of production. Under the assumption of profit maximization, the associated product supply equation and factor demand equations, which are of primary interest in macroeconometric models, are:

$$y_i^* = \gamma_{0i} + \frac{1}{2}\beta_{1i}\left(\frac{r_1}{p}\right)^2 + \frac{1}{2}\beta_{2i}\left(\frac{r_2}{p}\right)^2 + \beta_{3i}\left(\frac{r_1}{p}\right)\left(\frac{r_2}{p}\right)$$

$$x_{1i}^* = \eta_{1i} + \beta_{1i}\left(\frac{r_1}{p}\right) + \beta_{3i}\left(\frac{r_2}{p}\right) \tag{2}$$

$$x_{2i}^* = \eta_{2i} + \beta_{3i}\left(\frac{r_1}{p}\right) + \beta_{2i}\left(\frac{r_2}{p}\right)$$

where r_1 and r_2 are factor prices, p is product price, and γ_{0i}, β_{ji}, and η_{ki} are related to the parameters of equation (1). If firms face the same prices and if inputs and outputs are homogeneous, supply and factor demand equations in equation (2) can be aggregated without bias:

$$y_{1\cdot}^* = \gamma_{0\cdot} + \frac{1}{2}\beta_{1\cdot}\left(\frac{r_1}{p}\right)^2 + \frac{1}{2}\beta_{2\cdot}\left(\frac{r_2}{p}\right)^2 + \beta_{3\cdot}\left(\frac{r_1}{p}\right)\left(\frac{r_2}{p}\right)$$

$$x_{1\cdot}^* = \eta_{1\cdot} + \beta_{1\cdot}\left(\frac{r_1}{p}\right) + \beta_{2\cdot}\left(\frac{r_2}{p}\right) \tag{3}$$

$$x_{2\cdot}^* = \eta_{2\cdot} + \beta_{3\cdot}\left(\frac{r_1}{p}\right) + \beta_{2\cdot}\left(\frac{r_2}{p}\right)$$

where dots represent summation over all firms. Thus, if these assumptions are valid, aggregate product supply and factor demand equations can be estimated from aggregate time series data without bias, even though we cannot estimate the parameters for individual firm relationships. If the production functions are not quadratic, then the specification in equation set (3) is still valid to the second order of approximation, which should be acceptable for most micro- or macro-studies.

Of course, there are valid reasons for questioning profit maximization, price, and variable homogeneity assumptions, but equation sets (2) and (3) are useful for highlighting potential micro-macromodel specification complementarities. If we had a macroeconometric-simulation model based on (3) along with product demand and factor supply equations, the macromodel could be used to simulate certain types of policies. Then, price impacts obtained from the macromodel could be used in representative firm models, equation set (2), to estimate distributional impacts. Thus, the macro- and micromodel results would be consistent because of the compatible model specification.

Two features of (2) and (3) deserve special comment after having been completely ignored in an alarming number of econometric studies, including some I have done. First, the product supply and factor demand functions are homogeneous of degree zero in real prices. Second, the symmetry or reciprocity relationships hold under profit maximization; that is:

$$\frac{\partial y_i^*}{\partial r_j} = \frac{-\partial x_{ji}^*}{\partial p} \quad \text{for } j = 1, 2 \text{ and}$$

(4)

$$\frac{\partial x_{ji}^*}{\partial r_k} = \frac{\partial x_{ki}}{\partial r_j} \quad \text{for } k \neq j$$

Without symmetry, product supply and factor demand functions imply a nonsensical production function. Also, β_1, β_2, and β_3 appear in all 3 equations in (2) or (3) and should be used as a priori information in econometric estimation of these relationships. In addition, the production function is strictly concave only if $\beta_1 < 0$, $\beta_2 < 0$ and $\beta_1\beta_2 > \beta_3^2$. Even if firms do not maximize profits, symmetry may hold (for example, see Pope). Statistical tests should be conducted to determine if symmetry can be imposed on the estimated relationships. Imposition of symmetry also clarifies measurement of economic surpluses, because surplus is then independent of the path of integration. Thus, surpluses are not well defined where symmetry does not exist.

Indirect estimation of supply functions for crops by estimation of acreage response functions is often preferred in econometric studies. For such model specification, symmetry between cross-prices or cross-net returns can be shown to hold for profit-maximizing firms (Collins). In estimating regionalized acreage response functions for major U.S. field crops, there is a high probability that symmetry holds, even for fairly large regions such as the corn belt and delta states. Furthermore, there appears to be a high probability that symmetry holds for national agricultural commodity demand relationships.[2] In addition, imposition of symmetry for demand and supply functions eliminates inconsistent signs of cross-price coefficients and sometimes eliminates nonsensical signs. For these reasons, symmetry should be considered in estimation of micro- and macroeconometric models.

The preceding theoretical set of equations (model) illustrates that, under certain conditions, aggregation bias is not a serious concern with a properly specified macroeconometric model. However, in the case of the agricultural sector some of these conditions are questionable. First, the results of detailed normative models of agricultural production indicate that producers at least come close to maximizing profit, as would be expected from the competitive nature of the industry. However, additional positive microeconomic modeling can shed more light on this issue and thus aid in formulation and specification of macroeconometric models.

Existence of heterogeneous outputs or inputs can lead to significant aggregation bias. For most agricultural commodities the problem of heterogeneous outputs does not appear to be a serious problem as long as the econometric model is crop-specific. However, heterogeneous inputs, especially capital items, may create substantial aggregation biases. This type of bias could be minimized by estimating product supply and factor demand equations for small, relatively homogeneous regions rather than by estimating only national relationships. Of course, this will increase the cost of estimating a macroeconometric model, but will not have much impact on the primary cost of solving an econometric-simulation model for market clearing prices. Again, complementary micromodel studies would be of great value in determining the severity of aggregation bias attributable to heterogeneous input variables.

Perhaps the most serious aggregation bias in a macroeconometric model occurs if firms do not have the same expectations of prices (Pope). From the set of equations in (2), if expected prices are not the same for all firms, a bias exists and if average expectations are used in the aggregate relationships, then a bias exists in (3). A similar problem exists in some micromodel specifications if a firm does not have single-point price expectations. So additional examination of how expectations are formed and how they differ among firms would be useful to both micro- and macromodeling efforts.

The above discussion has dealt with static, deterministic concepts, except for price expectations. Yet both stochastic and dynamic factors, including irreversibilities, are extremely important in the agricultural sector. Although many econometric studies have considered stochastic and dynamic factors, they have usually been introduced into the model in an ad hoc manner. A more fruitful approach is to introduce these factors first into a theoretical model, and then derive a set of equations to be estimated.

Aggregation Bias in Normative Models

Since nonlinear formulations of detailed normative macromodels are not computationally practical, we shall discuss only linear programming (LP) models. Although the necessary and sufficient conditions for exact aggregation in LP models have been widely discussed (Day; Paris and Rousser), we do have sufficient data to determine the extent of aggregation bias in national LP models. However, regional analyses (e.g., Sheehy and McAlexander; Frick and Andrews; and Barker and Stanton) indicate that the bias can be serious.

National LP models tend to give normative supply functions that are relatively more elastic and to the right of positively estimated supply functions. This tendency indicates either that "what is" is different from "what should be" or that aggregation bias is serious. I believe the differences are attributable in large part to the latter, aggregation bias. The bias problem is also manifested in model results for producing regions that are far from regional observed data and from less aggregative LP model results. The deviations for small regions often average out when summed. It is doubtful that inconsistencies between LP micro- and macromodel results will ever be eliminated. Also, the empirical and computational compromises required to specify a solvable and operational national LP model strongly suggest that complementarities between micro- and macromodeling efforts will continue to be limited.

What Type of Model to Use?

An important question pertaining to micro-macromodeling is: "What is the most appropriate type of model for policy analyses—positive, normative, or hybrid?" The answer to this question may be conditional on the type of policy to be analyzed. Thus, a single set of micro- and macromodels will not be sufficient for all policy analysis questions. Unfortunately, definitive answers are not available because the agricultural economics profession has not compared ex ante model results with occurrences afterwards. Until such comparisons are provided, the type of model to use can only be discussed in terms of intuition and modeling experience.

For particular types of policies, i.e. policies that result in changes in specified variables within the range of historical observations, it appears that positive macromodels would give the most accurate results for policy analyses. However, because we have inadequate time-series data for representative micro-units, we are limited in positivistic analysis of similar policy effects on micro-units. The second best alternative is likely to be a positive macromodel linked to normative or hybrid micromodels, particularly if the macromodel has not been specified to analyze the effects of nonhistorical or financial policies.

Despite many objections to static, normative micro-macromodels, they may be the only realistic quantitative framework for investigation policies outside the range of historical observations. However, the appropriate macromodel for evaluating policies which have not been implemented in the past, but which result in relatively small changes, is a moot issue. In some cases analysts may prefer synthetically or artificially modifying a positive macromodel rather than resorting to a normative macromodel.

In conclusion, development of complementary and accurate micro- and macromodels requires feedback based on ex post facto evaluation of ex ante estimates, irrespective of the policies analyzed. Until such comparisons become a part of the model development process, assessing the accuracy of various types of models or fully exploiting potential complementarities between micro- and macromodels will not occur.

Concluding Remarks

As noted previously, micro-macromodel complementarities can be viewed in terms of model results and specification. Until theoretical and empirical complementarities in model specification are exploited, micro-macromodeling efforts and model results will be inconsistent and incompatible.

In terms of model specification, complementarities between micro- and macromodels are likely to be inversely related to the size of the models; with very large models it is difficult if not impossible to fully understand the reasons for inconsistent model results. Another reason for keeping models as simple as possible is that if a model or set of models gets too large, there will be few if any impartial validity checks of the models; noninvolved economists are rarely willing to invest the time required to fully understand each component of the models and their linkage. Thus, with large models there will be little interaction with nonmodelers, and the models will be viewed as "black boxes."

There is also a tendency for modelers to pursue grandiose plans for large, highly complex models. Quite often, the modeler is asking too much of the available data. Rather than attempting to collect more data to pursue this type of effort, we should first make better use of existing data by estimating simpler models whenever possible. In addition, fully exploiting possible complementarities between micro- and macromodel specification requires that a single researcher or a small group of researchers working closely together develop the models. Conversely, if micro- and macromodels are developed separately, there will be little of the kind of interaction that fosters complementary models and model results.

Finally, quantitative micro- and macromodels should not be viewed as a framework for obtaining "the" answers. Rather, models should be viewed as learning tools that sharpen the qualitative professional judgment of economists.

Notes

1. Current opinion holds that the free market is more efficient and arguably more equitable than agricultural markets constrained by farm programs. If this

argument is accepted, agricultural economists should let the market work and quit wasting time and other resources on policy evaluation models!

2. In terms of a utility-miximizing consumer, symmetry holds only for compensated demand functions. If the income effect in the Slutsky equation is small, then we have near symmetry for ordinary demand functions.

References

Barker, R., and B. F. Stanton. "Estimation and Aggregation of Firm Supply Functions," *Journal of Farm Economics*, Vol. 47, No. 3, Aug. 1975, pp. 701–12.

Collins, G. S. "An Econometric Simulation Model for Evaluating Aggregate Economic Impacts of Technological Change on Major U.S. Field Crops." Unpublished Ph.D. dissertation, Texas A & M University, Dec. 1980.

Day, R. H. "On Aggregating Linear Programming Models of Production," *Journal of Farm Economics*, Vol. 45, Nov. 1963, pp. 797–813.

Frick, G. E., and R. A. Andrews. "Aggregation Bias and Four Methods of Summing Farm Supply Functions," *Journal of Farm Economics*, Vol. 47, No. 3, Aug. 1965, pp. 696–700.

Paris, Q., and G. C. Rausser. "Sufficient Conditions for Aggregation of Linear Programming Models," *American Journal of Agricultural Economics*, Vol. 55, No. 4, Nov. 1973, pp. 659–66.

Pope, R. D. "Supply Response and the Dispersion of Price Expectations," *American Journal of Agricultural Economics*, Vol. 63, No. 1, Feb. 1981, pp. 161–63.

Sheehy, S. J., and R. H. McAlexander. "Selection of Representative Benchmark Farms for Supply Estimation," *Journal of Farm Economics*, Vol. 47, No. 3, Aug. 1965, pp. 681–95.

Theil, H. *Principles of Econometrics*. New York: John Wiley and Sons, Inc., 1971.

4. Incorporating Aggregate Forecasts into Farm Firm Decision Models

Donald O. Mitchell and J. Roy Black

Abstract This paper discusses the information requirements of farmer decisionmaking, the types of forecasts which are available, and forecasting methodology for incorporating aggregate forecasts into farm firm decision models. The results of using USDA outlook statements in making the decision to sell corn at harvest versus to store corn past harvest is examined over the 1952–77 period. The results show a net increase in returns due to basing decisions on the outlook statements rather than on following an always store rule.

Keywords Agricultural forecasts, agricultural sector models, farm decision analysis, information delivery systems

Farm firm decisionmaking involves many decisions that require forecasts of future input and output prices and other variables. Agricultural economists in the public sector have provided such forecasts of agricultural input and output prices and other variables since the late 1930s. University extension economists have used newsletters, outlook columns in farm magazines, and radio and television programs to inform farmers of their forecasts. The U.S. Department of Agriculture (USDA) also provides forecasts and reviews of the agricultural markets in their Situation and Outlook reports.

Events of the 1970s, particularly the tightening of world feed and food grain and oilseed markets, resulted in the entry of private firms into the forecasting business. Forecasts of agricultural variables such as prices, production, trade, and net farm income are provided for subscribers. Well-known firms using econometric models include Data

The authors are associate professor and professor of agricultural economics, respectively, at Michigan State University, East Lansing.

Resources, Inc., Chase Econometrics, and Wharton Econometric Forecasting Associates. Sparks Commodities, while using less formal models, provides shorter run production, trade, and price forecasts. Farmers have not typically been subscribers to these services because of the high fees involved. However, universities are now beginning to provide similar types of forecasts for subscriber clients, and these forecasts are likely to be more accessible to farmers. For example, the authors are involved in operating an econometric forecasting service through the Department of Agricultural Economics at Michigan State University. This activity is called the *Michigan State University Agriculture Model Project*, and it provides longer run (10-year) forecasts of average annual variables to agribusinesses on a semiannual basis. This project uses a large econometric model that combines a detailed U.S. model with a 9-region world grain trade model. This model is also used for policy studies and extension programs in farm-level marketing, management, and agricultural policy. The University of Missouri is developing a similar type of project.

Many private firms also provide current market information. Doane's Agricultural Service has a long tradition of providing a market newsletter to farmers. More recently, new firms, including the Iowa Grain Livestock Hedging Corporation, Top Farmer, Pro Farmer, and Illinois Farm Bureau, have begun providing market information. Commodity brokerage services have begun to expand their array of services. Typically, these firms offer a weekly newsletter outlining commodity analysts' price expectations within a short time horizon (next 7–10 days) as well as for a longer period. Suggestions are offered for pricing and sales management. For a higher fee, many firms provide a "hotline" that enables farmers to speak directly with commodity analysts. A few forecasting firms such as CATTLE FAX are involved in both gathering data and forecasting activities.

Developments in communications and data banking in the 1970s complemented the developments in forecasting. For example, inventory reports, crop production reports, and assessments by USDA Situation and Outlook economists are available on a time-share computer to state extension economists within hours of their official release. Outlook statements by selected State Cooperative Extension Service economists are also made available through the time-share computer. A number of university agricultural economics departments also subscribe to wire services, such as the Commodity News Service. These developments have substantially improved the timeliness of state outlook activities.

New and emerging technology in computer information systems may dwarf the adjustments already seen. Dr. Robert Cramer of Kellogg Foundation estimates that 80 percent of the commercial farms in the

United States will have substantive onfarm computing capacity by the end of the decade. Computers will be used to support accounting packages, optimal control models, and automatic generation of management information. Configurations will include onfarm general-purpose computers, specialized onfarm systems using micro-processors in irrigation scheduling and other control activities, and communication systems. Some of these computer systems will facilitate use of national-level data banks. Currently, for example, Michigan farmers can access commodity price information for purposes of technical analysis. The purpose of these communication systems is to link onfarm systems to other information systems. A number of other states have similar systems.

With this growing availability of forecasts and the electronic capabilities for delivering these forecasts quickly, the possibilities for using forecasts in farm decision models are greatly enhanced. In fact, the electronic capabilities for delivering information have grown more rapidly than our knowledge about how to use the information. The balance of this paper reviews a normative approach to farmer decisionmaking under uncertainty, developing some of the implications for incorporating aggregate forecasts into farm firm decision models. Our method is based on our experiences at Michigan State University in developing and applying a large-scale econometric model for forecasting and supporting Cooperative Extension Service marketing and farm management programs. A case study will be reviewed to illustrate how aggregate forecasts have been used in decisionmaking at Michigan State University.

Farm Firm Decision Framework

We consider first how farmers make decisions and present a taxonomy of the nature of farmers' decisions. Many individuals, particularly extension economists, have developed ad hoc perspectives on farmer decisionmaking under uncertainty, but relatively little is known about how farm families actually make decisions. Our focus will be on the normative literature, particularly the decision analysis paradigm that has been used to describe decisionmaking under uncertainty and the economic value of forecasts (for example, Anderson et al., 1977; Ladd and Williams, 1977; Raiffa, 1968; Weinstein and Fineberg, 1980; and Winkler, 1972).

Management economists typically sort decisions into those that are strategic (what should be done?) and those that are tactical (how should we do it?) (Harsh et al., 1981). As an example of the types of decisions which must be made by a farmer, consider the case of a Michigan cattle feeder. Beginning with the strategic decision of how many head of cattle to feed, a series of tactical decisions follow (see Table 4.1). By

TABLE 4.1
Selected strategic and tactical decisions made by Michigan cattle feeders
over the course of a year

Date	Decisions
August 15	Decide how many cattle will be fed during the coming year and the average corn and corn silage disappearance per animal.
	1. Determine the acreage of corn to be harvested as silage versus grain.
	2. Decide when to use protein/mineral additives:
	a. Add urea and anydrous ammonia to silage using the cold flow method or aqua-ammonia mineral mixture.
	b. Add later at feeding time.
August 20- September 15	Decide when to start cutting corn silage and how much to silage, based on date, weather, and crop development. If it is getting late and the corn plant is still relatively immature, then frost dangers may cause the cattle feeder to begin to harvest silage. Harvest soybeans.
September 1-30	Reevaluate silage acreage decision based on developments in the corn, feeder cattle, and fed cattle markets.
September 25- November 25	Decide whether to leave corn stalks, harvest within 3 days after high-moisture grain harvest and ensile, or harvest (bale or stack) dry stalks for bedding.
December- November	Decide which protein/mineral supplement to use.
November- December	Decide how many animals to feed on high silage diet and how much corn to feed or sell.
December	Review timing of cash purchases or sales for tax management purposes.
December 15	Decide to purchase supplies for spring planting now, or delay until later in winter or early spring.
January 15	Decide on preliminary acreage mix between corn and beans for purchase of chemicals.
November 15- March 15	Select and verify for yield, standability, corn-to-stalk ratio, and moisture content at harvest.
April 20- May 10	Monitor soil conditions for planting.
May 10-June 1	Conditionally decide on mix of corn and soybean acreage based on market developments and spring weather. Determine conditional switch date.

August 15, crop acreage decisions have been made, and the farmer must ask: "How much of my corn acreage should be harvested as silage for feeding cattle versus as grain?" At this time, the farmer must also decide whether to add nonprotein nitrogen products, such as urea, to the corn silage at harvest rather than add urea and soybean meal at feeding during the year. Thus, forecasts are required for corn and soybean meal prices and for the expected profits from feeding cattle.

The farmer continues to evaluate the corn-grain versus corn-silage decision as the silage harvesting process continues, based on the corn, feeder cattle, and fed cattle markets. In the post-harvest November-December period, the farmer reevaluates the decision of whether to feed fewer animals on a high corn silage diet and sell more corn or to feed more animals and sell less corn. Again, the farmer must obtain information to make a rolling assessment of the price prospects for corn and the profitability of feeding cattle. In the November-December period, the farmer begins to evaluate tax implications of alternative purchase and sale strategies, and to discount structures for purchasing in the fall versus in winter or early spring. Within-year price patterns and anomalies now become the focus of interest.

In January, the farmer begins preliminary planning on the acreage mix between corn and soybeans, partly to establish seed and chemical purchases. The operator is initiating the when-to-price decision for new crop commodities. In the May 10–June 1 period, the farmer must determine the corn versus soybean acreage, depending on both market developments and spring weather. The farmer is determining a conditional switch date. The forecasted relative prices of corn and soybeans are critical for the acreage decision. Absolute price levels are important for pricing and financing decisions.

Cattle replacement decisions are described in Table 4.2. The farmer routinely assesses when to sell existing cattle, and whether or not to replace them. If they are to be replaced, then the farmer must consider the type of cattle to replace them with.

Table 4.3 shows examples of strategic planning decisions made by Michigan farmers who feed cattle. These include decisions concerning whether or not to increase facilities, whether to add capacity for drying corn, and whether to acquire or sell land. Forecasts are needed to estimate the profitability of the cattle feeding enterprise over the prospective life of the capital investments. The corn-drying facility decision requires forecasts of the economic value of being able to sell dry cash corn. These forecasts require a general assessment of the economic viability of the cattle feeding sector. Is the sector in good economic health, in a growth phase, or overexpanded?

TABLE 4.2
Cattle replacement decisions

Decisions	Criteria
Decide whether to replace existing cattle	Current status of cattle, and projection of next period (weight, carcass grade, and yield grade)
	Feed and feeder market price projections for next 6 weeks and fed market price projection for the potential replacement animal (4-6 months ahead)
	Weather conditions
	Availability of transportation
If so, what type of animal to feed?	Frame size
	Sex
	Weight
	Previous nutritional treatment (compensatory performance, preconditioning)
	Breed
	Age

TABLE 4.3
Selected strategic decisions

Decisions	Criteria
Maintain present feeding operation, expand, or discontinue	Housing, including feed storage and waste handling
	Cropping system impacts, including mechanization
	Specialization of the marketing function
Reevaluate corn handling system	Dry part of corn for flexibility in cash crop sales
	Ensile corn for energy savings
	Snap ear corn
Buy or sell land	

This rather exhaustive example illustrates the number of decisions which must be made by a farm decisionmaker. Assessments of both the short-term and long-term input and output markets are required. Many potentials exist for incorporating aggregate forecasts into farm-level decisionmaking. We would now like to turn to some of the specific details of how a firm could adopt aggregate forecasts.

Incorporating Aggregate Forecasts Into Farm Firm Decision Models

Incorporating an aggregate forecast of a variable such as a commodity price into a farm firm decision model involves several considerations. First, the aggregate forecast must be localized to correspond with the market available to the firm. This typically involves adjusting a national average price to account for transportation costs, quality factors, and local demand conditions. Secondly, this localized aggregate forecast must be used to obtain the firm's expected value of the variable. If no other forecasts or information are available, then the localized forecast could be used directly. If other forecasts are also available, then weights must be given to each forecast and a single expected value must be obtained. The third and final step in incorporating an aggregate forecast into the decisionmaking of an individual firm is to use the forecast variable and adjust the decisionmaking process accordingly. This may be viewed as making a decision under the assumption that the forecast is the actual value of the variable, or the decision rules may be modified to reflect a decision based on a forecast variable.

An aggregate forecast is taken to mean a forecast of a single variable such as corn price for an aggregate region. An example of such a forecast would be the national average corn price received by U.S. farmers. This type of forecast is available from a number of private, government, and university sources. In order to be useful to an individual producer, this aggregate forecast must be expressed in terms relevant to the producer. Adjustments must be made to reflect the local variable given the forecast of the national average variable. Since factors such as transportation costs are readily available between a specific producing area and a central terminal point, it is possible to adjust for these factors. Price differences due to quality can also be accounted for since aggregate forecasts for a specified commodity quality are available in many cases. A final adjustment which might be considered is to adjust local prices to reflect local demand conditions. Whether the individual factors are considered in obtaining a localized forecast from an aggregate forecast or whether a simple ratio is used, it is possible to obtain a price forecast which reasonably reflects the situation faced by an individual producer.

Given the localized forecast, the individual producer must make two additional decisions: how much weight to give to the forecast and how to use the forecast in decisionmaking. Many decisions in agriculture require action to be taken based on values of variables which cannot be observed. The planting decision is an example. The producer must decide on his resource allocation 6–9 months prior to observing the price available at harvest time. This sequence of events suggests that farmers who attempt to maximize profits must have an expected price for each commodity when the planting decision is made. This expected price may be as simple as the harvest time price in the previous year or the price available at planting time. When an additional estimate of price becomes available, then the producer must assign a weight to this new price forecast and review his expected price accordingly.

Denote the price expected by the individual producer in period t by $E(P_t)$. The problem of obtaining $E(P_t)$ becomes a problem of determining the weights, w_i, used to weight each price forecast:

$$E(P_t) = w_1 P_{t-1}^{f_1} + w_2 P_t^{f_2} + \ldots + w_i P_t^{f_i}$$

where $0 \leq w_i \leq 1$ for all i. In this case

$$P_{t-1}^{f_i}$$

is the actual price observed in period $t-1$ and other forecasts

$$P_t^{f_1}$$

are available in year t. A producer who determines the expected price, $E(P_t)$, based solely on last year's price, would have $w_1 = 1$ and $w_i = 0$ for $i \neq 1$.

A number of alternative procedures can be used to obtain the weights w_i. One procedure is to assign equal weights to all forecasts. An alternative procedure would be to select the weights based on the past performance of each forecast. (As an example of several weighting schemes to obtain a composite forecast, see Brandt and Bessler, 1981.) However, the relevance of past forecast accuracy may be questionable. As Just and Rausser have shown in a recent paper, the track record of the major commercial firms and the USDA shows that these groups have not forecast price as well as the futures market (Just and Rausser, 1981).

Evaluating the accuracy of a forecast is a difficult task. We do not propose to address this topic in any depth; however, we do want to identify a problem we see with the common procedures. The historical track record of past forecasts is often used as an estimate of future forecast accuracy. This is, unfortunately, a less reliable method than it appears since the past track record is only useful for predicting future performance if the basic forecasting structure remains unchanged. This requirement means that both the forecasting model and the operating (interpretative) personnel remain unchanged. This conditon is rarely met for three months, let alone the years of forecasts necessary to develop a track record which can be used to predict the future accuracy.

From our own experience, we believe that forecasting is still very much an art—not a science. Ratajczak of the Georgia State University forecasting group shares this view and recently stated, "Before anything goes out, I look at the results and make changes." Ratajczak holds the distinction of having the most accurate forecast of inflation over the last 4 years (*Wall Street Journal*, October 1981). Whenever the forecasting personnel change, the old track record is invalidated. Even when the personnel and forecasting model remain stable, the track record may still be a poor measure of future performance because it does not take into consideration learning on the part of the forecaster. The composite track record of the major commercial firms, however, should be a reasonably good measure of the forecast accuracy which can be expected from large econometric models. For example, the average forecast errors identified by Just and Rausser for Data Resources, Inc., Chase Econometrics, and Wharton Econometric Forecasting would seem a reasonable measure of the accuracy of that type of forecast. Finally, the Just and Rausser results cannot be used to argue that future forecasts of one group will consistently outperform another group unless the stability of the forecasting systems can be defended.

The final step in incorporating aggregate forecasts into the individual firm's decisionmaking involves using the forecasts. Few individual decisionmakers will wish to treat a forecast as if it were true and act accordingly. Rather, some modification will usually be made in the firm's decisionmaking rules to reflect the uncertainty of the information. This topic of profit maximization under uncertainty constitutes an extensive literature and will not be reviewed in this paper.

A Case Example

A question even more fundamental than how should a firm incorporate aggregate forecasts into decisionmaking is the question of whether the firm should use forecasts at all. Since it is difficult to predict the accuracy of forecasts, as previously noted, we have chosen to examine the results of basing a grain storage decision on USDA forecasts over a 26-year period. While such an example cannot address the question of the success of future forecast accuracy, it does provide a valuable perspective on the use of aggregate forecasts in farm decisionmaking.

Outlook statements from the USDA for the 1952/53–1977/78 period were assembled, and a grain storage decision was made based on the outlook statement. One problem of using these outlook statements is that they are often ambiguous citing both positive and negative factors which will influence the grain market but not drawing a final conclusion. To simulate the interpretation which a farm decisionmaker would have made, two students were presented with the forecasts for the crop years 1952/53 through 1977/78. The information provided was camouflaged to ensure that the students could not recognize particular years. The students were then asked to decide whether to sell corn at harvest or store until February based on the forecast information. There were only two exclusive options, sell all corn at harvest or sell all corn in February. Sales could not be made between these dates nor subsequent to February. A net return function was developed based on use of commercial off-farm storage to evaluate the results of following the USDA outlook statement. These results were compared with the decision to always store past harvest.

The results of the students' decisions are depicted in Table 4.4. Situation 1 shows that to store from harvest through February the average net returns from storage would have been $0.02 per bushel, with median returns of $0.086. The biggest net loss would have been $0.88 per bushel, while the biggest net gain would have been $0.66 per bushel. Situations 2 and 3 depict the results in the store versus sell-at-harvest decision based on their decision to sell at harvest in 11 or the 26 years and to store from harvest to February in 15 of the 26 years. The 11 sell decisions would have been correct in 6 years but wrong in 5; the average net loss that would have resulted from storage in the years in which the decision was to sell at harvest was $0.086 per bushel, with a median loss of $0.04 per bushel. Thus, the decision to sell at harvest was wrong nearly 50 percent of the time, 5 of the 11 years. The students did not store during the year in which prices declined the most from harvest to February, but the decision was to store for the years in which the second and third largest price declines occurred. The fourth, fifth, and sixth biggest price declines occurred during sell-at-harvest years.

The pattern was similar in the years in which the decision was to store past harvest to February. This was the wrong decision in 4 of the 15 years. The average net return was $0.097 per bushel and the median net return was $0.14 per bushel. The second and third largest price declines from harvest to February occurred during the store decision years. However, this decision did prevent the largest loss, as well as a

TABLE 4.4
Results of grain storage decisions, ranked by profitability

Case	Earnings from storage (1979 dollars)					Summary statistics		
	– – – – – Dollars per bushel – – – – –					Mean	Median	Standard deviation
1. Always store past harvest								
26 observations	-0.88,	-0.28,	-0.25,	-0.24,	-0.15,	-0.15	0.020	0.086	0.264
	-.12,	-.04,	-.04,	-.02,	.01,	.03			
	.09,	.10,	.11,	.11,	.12,	.14			
	.14,	.15,	.16,	.19,	.19,	.23			
	.27,	.66							
2. Years for which students decided to sell at harvest								
11 observations	-.88,	-.24,	-.15,	-.15,	-.12,	-.02,	-.086	-.040	.291
	.01,	.11,	.11,	.14,	.27				
3. Years for which students decided to store until February								
15 observations	-.28,	-.25,	-.04,	-.02,	.09,	.11,	.097	.140	.211
	.14,	.14,	.15,	.16,	.19,	.19,			
	.23,	.66							

number of the smaller losses. The largest price increase and the third through twelfth largest price increases occurred in years for which the students decided to store.

The average per bushel net return for storage for the 26 years, based upon use of the USDA market information to make the store or not to store decision, was $0.056 per bushel as contrasted to $0.02 per bushel for an always store rule. An increase in net earnings of $0.035 per bushel resulted from the use of the USDA forecast and application of storage concepts. Additionally, the coefficient of variation of earnings was cut in half. Thus, use of forecasts increased net earnings and reduced price risk.[1]

Implications of New Concepts
on Decisionmaking Under Uncertainty

Studies by Robison and King and other researchers on decisionmaking under uncertainty have the potential of improving farmers' decision-making capacity. When combined with developments in communications and computer technology, the extension of the concept of stochastic dominance by Meyers, and the elicitation of risk preferences by King and Robison permit the commercial application of these new concepts in an applied setting. It is possible, within the context of the expected utility hypothesis, to develop farm management actions and/or strategies that are optimal with respect to economic or policy scenarios and their respective probabilities. Because interval procedures are used in eliciting preferences, a list of candidate actions/strategies is developed from which the farm family can choose.

These concepts have been tested in an extension context via the pricing and sales management workshops. The upper 40th percentile of farmers are comfortable with the use of decision trees and other analytic tools underlying the decision analysis methodology. The ana-lytical methodology is sufficiently similar to many sports, particularly football, so that the method of analogy can be used to teach the subject matter quickly. Extension economists typically argue that farmers ought to begin with knowledge of the historical tendencies of the market and then ask if there are keys in the same sense that a quarterback in football calls audibles at the line of scrimmage that would lead them to deviate from their long-term game plan. Here, extension economists have developed rules of thumb such as increasing the percentage of the portfolio that a farmer should price during the late July and early August season in short crop years where the key is the existence of the short crop year. Materials, such as those outlined in the case study, are useful to get across the concept that the keys are imperfect and the objective is to achieve certain long-term goals while minimizing the likelihood of costly outcomes. When farmers are comfortable with the underlying concepts developed in the discrete decision tree framework, the expected utility hypothesis can be used in a straightforward manner. Implications of concepts outlined in Rausser and Hochman appear feasible by mid-decade, particularly if consortiums of farms with similar interests join together to exploit new technology and some of the required specialized management skills.

Summary

The analysis of the incorporation of aggregate forecast information into farm firm decision models must recognize: (1) the sequence of decisions made by farm families, (2) management information control and decision aid systems, (3) managerial skill, and (4) risk preferences. The decision analysis framework provides a taxonomy for studying these issues. The distinction between tactical and strategic decisions, a framework used by management specialists, is also useful.

Developments in communication and computer technology will significantly affect the activities of both forecasters and users during the 1980s. Specialists have estimated 80 percent of commercial farms will be making substantive use of these new technological alternatives by 1990. This will permit farm-level decisionmakers' access to more complex decisionmaking models and heighten the need for forecasts.

Farmers are expected to begin with the application of simpler computerized decision aids and electric spread sheets. Some form of stochastic dominance with respect to a function in the selection of candidate strategies will be applied by mid-decade.

Notes

1. The rules used in developing the "keys" to look for in market information were developed using the same period for which the forecasts were made; thus, the results are an upper bound on the economic value of forecasting. However, since the forecasts are based on general ERS forecasts, the definitiveness of the forecasts would have been expected to be higher from sources more focused on local market conditions and less sensitive to a political "cost of being wrong" function. The exercise because of its scope and choice of subjects is illustrative, not definitive, but it suggests a fruitful area of research.

References

Anderson, J. R., J. L. Dillon, and J. B. Hardaker. *Agricultural Decision Analysis.* Ames: Iowa State Univ. Press, 1977.

Armstrong, D. "Probability Forecasts Based on World Crop Yields." M.S. thesis, Michigan State Univ., Dept. of Agricultural Economics, forthcoming.

————, D. Mitchell, and R. Black. "Probability Forecasts for 1981/82 and Low Yield Scenarios," *MSU Agriculture Model Quarterly Report*, Vol. 1, No. 4, Michigan State Univ., 1981.

Bessler, D. A. *Some Theoretical Considerations on the Elicitation of Subjective Probability*. Indiana Agr. Expt. Sta. Bulletin 332. 1981.

————, and J. A. Brandt. *Composite Forecasting of Livestock Prices: An Analysis of Combining Alternative Forecasting Methods*. Indiana Agr. Expt. Sta. Bulletin 265. 1979.

Brandt, Jon A., and David A. Bessler. "Composite Forecasting: An Application with U.S. Hog Prices," *American Journal of Agricultural Economics*, Vol. 63, 1981, pp. 135–140.

Connor, L. J., and W. H. Vincent. "A Framework for Developing Computerized Farm Management Information," *Canadian Journal of Agricultural Economics*, Vol. 18, 1970, pp. 70–75.

Crowder, R. T. "Statistical vs. Judgement and Audience Considerations in the Formulation and Use of Econometric Models," *American Journal of Agricultural Economics*, Vol. 54, 1972, pp. 779–83.

Eisgruber, L. M. "Management Information on Decision Systems in the U.S.A.: Historical Developments, Current Status on Major Issues," *American Journal of Agricultural Economics*, Vol. 55, 1973, pp. 930–37.

Granger, C.W.J., and P. Newbold. *Forecasting Economic Time Series*. New York: Academic Press, 1977.

Harsh, S. B., L. J. Connor, and G. D. Schwab. *Managing the Farm Business*. Englewood Cliffs, N.J.: Prentice-Hall, Inc., 1981.

Just, Richard E., and Gordon C. Rausser. "Commodity Price Forecasting with Large-Scale Econometric Models and the Futures Market," *American Journal of Agricultural Economics*, Vol. 63, No. 2, May 1981, pp. 197–208.

King, R. P. "Operational Techniques for Applied Decision Analysis Under Uncertainty." Unpublished Ph.D. dissertation, Michigan State Univ., Dept. of Agricultural Economics, 1980.

————, and L. J. Robison. "An Interval Approach to Measuring Decision Maker Preferences," *American Journal of Agricultural Economics*, Vol. 63, 1981, pp. 510–20.

————, and L. J. Robison. "Implementation of the Interval Approach to the Measurement of Decision Maker Preferences." Michigan State Univ., Dept. of Agricultural Economics, forthcoming.

Ladd, G. W., and J. E. Williams. "Application of Net Energy System and Base and Decision Theory for Determination of Cattle Rations and Rates of Gain," *Journal of Animal Science*, Vol. 45, No. 6, 1977, pp. 1222–30.

Nelson, A. G., G. L. Castler, and O. L. Walker. *Making Farm Decisions in a Risky World: A Guidebook*. Oregon State Univ. Cooperative Extension Service, Corvallis, 1978.

Purcell, W. *Agricultural Marketing: Systems, Coordination, Cash and Futures Price*. Reston, Va.: Reston Publishing Co., 1979.

Raiffa, H. *Decision Analysis: Introductory Lectures on Choices Under Uncertainty.* Reading, Mass.: Addison-Wesley Publishing Co., 1968.

Rausser, G. C., and E. Hochman. *Dynamic Agricultural Systems: Economic Production and Control.* New York: North-Holland Publishing Co., 1979.

Robison, L. J. "Lecture Notes on Decision Making Under Uncertainty." Agricultural Economics Staff Paper 81-86. Michigan State University—East Lansing, 1981.

Weinstein, J. C., and H. V. Fineberg. *Clinical Decision Analysis.* Philadelphia: W. B. Saunders Co., 1980.

Winkler, R. L. *Introduction to Bayesian Inference and Decision.* New York: Holt, Rinehart and Winston, 1972.

Zentner, R. B., D. D. Greene, P. L. Hickenbotham, and B. R. Eidman. "Ordinary and Generalized Stochastic Dominance: A Primer." Staff Paper P81-27. Univ. of Minnesota—St. Paul, Dept. of Agricultural and Applied Economics, 1981.

Discussion of Part 2

Complementarities Between Micro- and Macro-Simulation Models

James W. Richardson

Taylor's thesis is that "the key to eliminating inconsistencies and developing complementary models is in model specification." Basically I agree that appropriate model specification would make micro- and macromodels more consistent. Even though the theory of model specification and aggregation has been in the literature for quite some time, an abundance of consistent micro-macromodels—or even a single one in agricultural economics—is not available. These 2 facts lead me to conclude that inconsistent models are due to reasons other than model specification. I submit that 3 things explain the inconsistent results of micro- and macromodels: how models are developed, who uses them, and how model development is funded. These reasons can be addressed directly to reduce micro-macromodeling inconsistency.

Quantitative economic models are generally developed by graduate students working on thesis research. These models are either micromodels or macromodels. In addition, a graduate student's model is not general; it is unable to quantify the effects of a large number of economic situations since it usually has only one purpose. Typical purposes might include modeling estate planning, farm growth, price supports, or international trade. A disaggregated, micro-macro policy model is beyond the scope of a normal graduate student's thesis. Models developed by graduate students also lack continuity; that is, they are frequently shelved after the thesis requirements are met. These models are seldom reused by the faculty or another graduate student, or reestimated. Consequently, increased attention to model specification at the development stage may correct some problems with these types of inconsistencies, but unless the same research team is involved in all stages of development, the final micro-macromodel will again produce inconsistent results.

Second, models are generally used by policy analysts in government not involved in their development. These analysts frequently fail to understand the generality or lack of it and the quantitative limitations

The author is an associate professor of agricultural economics at Texas A & M University.

of the model or to completely understand how to interpret how the model actually arrives at the printed results. Thus, policy analysts may unintentionally create inconsistent results with well-specified models. Since policy analysts' work is by nature a short-run effort, they often lack or are prohibited from developing the appropriate model for the current problem. The next best thing to having the right model is to create an "add-on" model, that is, to make slight or radical changes in an existing model or link a "good macromodel" to a "good micromodel," or do both. Most of the time these changes and linkages are made without contacting the original model developer(s). Given the way these second generation ad hoc models are often constructed and used, it should not be surprising that the results are inconsistent. It is surprising, however, that these "add-on" models actually work.

Third, the system of funding model development may explain the lack of consistent micro-macromodels needed by policy analysts. Federal funding for model development, generally provided by contracts, has been sporadic over the past 15 years. A contract that has sufficient money at the outset may be terminated prematurely due to insufficient funding before completion. Funds are frequently spread throughout various universities rather than concentrated in one place where a small team of researchers can completely develop a consistent model. State experiment stations, which also finance some models, are also guilty of supplying modeling funds sporadically. We cannot expect, however, that the experiment stations would fund the development of large-scale micro-macro policy models.

I believe the chances that we can correct these problems and develop consistent micro-macromodels are good if we make a number of changes in the way models are developed, used, and funded. I strongly recommend the following changes:

1. Graduate student advisors must avoid the one thesis–one model syndrome; that is, each thesis should be an integral piece of an overall model to be pieced together by a single modeler or small group of modelers.
2. Model builders must develop models that can handle a wider range of policy issues. The era of one-problem models is over.
3. Public funding agencies must insist on more complete documentation for models they finance.
4. Policy analysts should minimize the number of "add-on" models or be required to contract with the developers to make the necessary modifications.

5. Federal agencies should fund a single modeler or a small research group to develop a model rather than divide the money and responsibilities among a number of researchers at different universities.
6. Capital outlays for model development should be long-term and continuous commitments.

Without these basic changes in the development, use, and funding of micro-macro policy models, micro- and macromodels that produce inconsistent results will continue to be widespread.

Discussion of Part 2

Understanding and Human Capital as Outputs of the Modeling Process

Thomas A. Miller

It is difficult to evaluate the contribution of these papers to the specific objectives of this session, mainly because these objectives are necessarily imprecise. The conference organizers had suggested to me that the session was intended to address the question of how to hookup or link firm-level models and macromodels with the goal of improving the policy analysis capability of ERS and other institutions.

Accordingly, my discussion will take a close look at the question of hookup or linkage, building on the comments of John Lee, Richard Day, Neill Schaller, and Barry Richmond as well as on the papers presented in this session. In reference to this session, most of my discussion will relate to Taylor's paper.

Need for Synthesizing Micro- and Macromodels

Taylor describes a general "need for a quantitative framework that can be used to estimate relative impacts on microunits as well as aggregate impacts of various agricultural policies." This need should be emphasized. Increasingly, agricultural policy analysis requires consideration of distributional impacts of alternative policies.

Recent experience with research on the structure of agriculture is an important example. Policies such as price supports affect aggregate price and income levels—but they also have a differential impact on economic well-being and survivability of different farms—and therefore a longer term impact on the structure of farming. Never has the need for complementary linkages of micro- and macromodels been so great. But with a few exceptions, it appears that our research has failed to develop these linkages in a technical or qualitative sense. A more important criticism is our failure to explicitly consider the question of how to hookup the knowledge gained from micro- and macromodels and the policy analysis process. Models become useful for policy analysis only

The author is an agricultural economist, Economic Research Service, stationed at Colorado State University.

to the extent that we make explicit arrangements for using the knowledge gained from models in the policy analysis process.

Understanding as a Result of the Modeling Process

If the hookup or linkage activity is going to be successful, we must have a rather extensive view of the outputs from the model-building process and explicitly consider how to use these outputs in policy decisionmaking. Earlier today Richard Day stated that the most essential step in policy analysis is to understand the world—what will happen and why. Barry Richmond also observed that the purpose or desired product of modeling is not just estimates; it is understanding. The last paragraph of Taylor's paper expands on this idea and suggests that quantitative micro- and macromodels should not be viewed as a framework for obtaining "the" answers. Rather, in Taylor's opinion, models should be viewed as learning tools that sharpen the professional's economic judgment.

This understanding, sometimes referred to as the qualitative result of modeling, is critically important if farm and aggregate models are to be hooked up to the policy analysis process. It is, in fact, the key ingredient. In policy analysis, important questions often have a short time frame—too short to determine another model solution—and are often not in the exact form represented by existing solutions. This situation forces analysts to draw upon their understanding of the real world situation being modeled rather than completely upon the empirical results of quantitative models. To be useful in policy analysis, models must help analysts understand the system and comprehend why the results turn out as they do, and allow them to use this information when preparing the policy analysis along with the quantitative results.

Understanding Often Related to Size

Taylor discusses aggregation error and specification error and describes a number of factors affecting the optimum level of aggregation. He observes that to fully develop the understanding that models can provide and to avoid the number crunching syndrome, modelers should adhere to the KISS philosophy, which is "keep it simple, stupid." I think what he is suggesting here is a possible tradeoff. As we increase detail and complexity to reduce aggregation and specification errors, we often reduce our ability to understand the model and the system being modeled. Stated another way, increasing the size and complexity of both farm models and macromodels may reduce aggregation and specification errors, but may also reduce the understanding of the real world that can be gained from the modeling activity.

The possibility of a tradeoff between accuracy and understandability is important as we look at the level of aggregation and prospects for a hookup between micromodels, macromodels, and policy analysis. As I look at the models developed by USDA and the land-grant universities, I conclude there has been a general tendency to pursue detail in model construction beyond the point where the value of an additional detail equals the cost of providing it. The pursuit of detail to minimize aggregation and specification error obviously has costs in terms of the difficulty of quality control, lack of subjective tests for validity, increased data and labor requirements, and so forth. Such models can easily become so complex that empirical results must be taken as given, and it becomes impossible to understand the underlying characteristics and relationships of the system being modeled. Even if they provide accurate empirical results, such models provide little understanding useful to policy analysis.

To the extent that there is a tradeoff between accuracy and understanding, an understandable level of aggregation is of primary importance. Minimizing aggregation error should be carefully pursued if it decreases our ability to understand why and how the system works. As Taylor suggests in the last paragraph, the training and expert judgment derived from the process of building models can often provide more important output from model building than the actual quantitative results.

Development of Human Capital

These observations lead directly to a second aspect of modeling that is often overlooked—the human capital output of the model-building activity. Research dollars allocated to model-building activities represent 2 types of capital investment: (a) an investment in the physical model, and (b) an investment in developing human capital. This human capital development can be viewed as a joint product of the model-building activity.

One part of this human capital amounts to a general educational component. Models constructed at universities are generally undertaken with recognition of the general educational component of human capital. An intended product of most of these models is the training of new Ph.D's and a general education in the principles of model construction and use.

A second form of human capital is less often recognized; it is the unique human capital related to a specific modeling project. Unique human capital is inseparable from the understanding and qualitative results defined earlier by Richard Day and others. We must consider this unique human capital as we attempt to hookup or use micro- and macromodels in policy evaluation. After working in the ERS agricultural policy analysis group for nearly 10 years, I am convinced that the unique human capital created in model development is more important to policy analysis than the physical model capital and that the best way to use models in policy analysis is for the specific model builder to work directly in a policy analysis capacity. Very simply, model builders must be involved in policy evaluation to make use of this unique human capital. Their unique understanding of how the system works, what will happen, and why it happens is a necessary ingredient for policy analysis.

Institutional and Administrative Linkages

This comment is directed primarily toward ERS research managers and administrators although it contains ample food for thought for model builders in universities. Better firm and aggregate models can improve ERS policy analysis only to the extent that ERS can acquire the unique human capital that is developed as a joint product of the model-building activity. This unique human capital at universities and other institutions as well as in ERS is a critical resource for policy analysis. The final aspect of hookup is the administrative or organizational question of how to employ this unique human capital directly in policy analysis.

From the standpoint of policy analysis in ERS, available models have obvious deficiencies. However, as I hear the discussion today and look at the models described in this conference, I am impressed about the detail, realism, and capability of the models being built today. I do not think that the deficiency in our ability to analyze policy lies in the models themselves or in the technical links between models. Rather, the analytic deficiency lies in our failure to be able to link up the policy evaluation function in ERS and the understanding gained in the modeling process by individuals in ERS, at universities, and at other institutions. The linkage of micro- and macro-modeling and the policy analysis process can be accomplished only through institutional changes that explicitly consider the continued creation of unique human capital and the placement of this unique human capital directly in a policy analysis role.

The word "continued" warrants additional comment. Unique human capital, particularly the capital represented by the observation and data collection process behind model building, depreciates rapidly. As a result, the unique human capital developed in constructing a model is most useful to policy evaluation for a relatively short time, say 1 or 2 years. Thus, the process of creating this unique human capital and moving it into the policy analysis process must be continuous. Analysts must be constantly building models to develop unique human capital and then be given or assume a policy analyst's role to use this understanding. Since most model building is done at land-grant universities, significant institutional changes must be made if we are to really hookup the human capital developed in modeling and the policy evaluation process.

Discussion of Part 2

Modeling and Evaluating Policy and Institutional Impacts on Farm Firms

Marvin Duncan

The papers by Taylor and by Michell and Black have taken quite different approaches to the subject of farm firm behavior and sectoral performance. Both papers provide competent analyses and serve as springboards for discussion.

Taylor's paper documents the limited complementarity between micro- and macromodels in common use. It then argues that proper model specification provides the key to achieving their complementarity. Taylor puts forth a strong argument for using a systematic approach to macro-modeling that is based on a "bottom up" type of construction. He argues that appropriate attention should be given to model specification relying on microeconomic theory. A model at the micro level, specified in accordance with the theoretical understanding of firm behavior, can then be aggregated without undue bias if certain assumptions about homogeneity of inputs and outputs are met and if the same prices are relevant at both the micro and macro levels. Taylor then sets forth some procedures and a research agenda designed to demonstrate that, in those cases where general assumptions are not met, the biases can be minimized.

The question of whether positive or normative modeling provides more useful information is also raised. Taylor seems to choose positive modeling. However, I think it would be a mistake to dismiss normative models as less useful. Many of the more interesting current economic policy questions may be more amenable to normative, than to positive, analysis. Finally, very correctly in my judgment, Taylor stresses the need for feedback on model performance based on ex post facto evaluation of ex ante estimates.

Mitchell and Black's paper takes a different direction. They address operationality of models of farm firm decisionmaking. They correctly stress the need for farm firm decision models that incorporate aggregate forecast information on production, price, and policy consistent with

---

The author is a vice president and an economist with the Federal Reserve Bank of Kansas City.

the sequence of farm firm decisions. The authors, I expect, did not unwittingly refer to farm family decisions, to their management-information control system, and to their managerial profiles. Indeed, it is important to remember that economic decisions by the farm firm typically encompass the resources and the preference function of the farm family. To lose sight of this relationship is to misunderstand the dynamics of decisionmaking in American agriculture.

Mitchell and Black forecast a remarkable increase in farmers' use of computer technology as an aid to decisionmaking. Their forecast may be overly optimistic unless useful analytic procedures and appropriate training are available for those who want to use the technology. Mitchell and Black distinguish between tactical and strategic decisionmaking. They note that much simpler analytic procedures will suffice for tactical decisionmaking than for strategic decisionmaking. However, it is not clear to me that farmers as yet understand these differences in decision processes, nor is it clear that economists agree as to where tactical leaves off and strategic begins.

The authors suggest appropriate aggregation levels to achieve workable models and useful output for decisionmaking. Price and quantity output information from the macromodel becomes input into the micromodels. However, we do not yet understand how the response functions of producers may be changing as they begin to incorporate forecasts of demand and price into their management process. This ought to be a productive subject for additional research.

I do not share the authors' apparent satisfaction with the capacity of aggregate models to generate correct and/or reasonable forecasts of variables needed by firm-level decisionmakers without the use of "add factors" or substantial tinkering with the model. Even with this tinkering, I have grave doubts about the value of the 10-year forecasts that Mitchell and Black propose to develop for management decisionmaking.

What can we deduce as agricultural economists and modelers from these two papers? I can suggest two useful propositions. First, real world decisionmakers have become increasingly suspicious of big models— even of models generally—and of the integrity of model results. Concurrently, although economists have become more enamored of and more skilled in model construction, model forecasts seldom prove correct. Moreover, decisionmakers who use model results are often not appropriately aware of the impacts that changes in assumptions can have on model output. Rapid changes in structural or behavioral parameters at both the micro and macro levels limit the ability of modelers to forecast the future based on past data. There is little reason to think that this limitation will change; indeed, greater interaction among economic agents may exacerbate the dilemma. Moreover, the decision process by farmers is likely more complex and less uniform across farm firms than economists have assumed. Even at the micro level, there is both more diversity in input use and less homogeneity of inputs than economists usually recognize.

Hence, on balance, although micromodels will become more important to farm firm decisionmaking and although macromodel output will continue to be a vital part of the model input, models are increasingly less useful (by themselves) to foretell what "will be."

Nonetheless, it seems to me that firm-level decisionmakers increasingly need to test their own hypotheses about the impact of events on business outcomes. Models generally—and in our context, microeconomic models—will be far more useful to the firm in a "what if" application to examine alternative outcomes. Generalized models into which individual firm-level data can be incorporated will be particularly useful to farm firm decisionmakers. In examining a variety of scenarios, farmers will be particularly interested in the downside risk as a result of a particular management decision.

Second, micromodels to support firm decisionmaking—if effective and practical—share some general characteristics with aggregate policy models. These include the need for readily available and timely data, quick response time for the user, the need for cost control, readily understood assumptions, and effective filtering and translation of results by the analyst (Farrell). The most useful output from models may be relatively simple multipliers, price elasticities, and so forth. Abner Womak has used this approach at the University of Missouri in interpreting the results of a large crops model for an extension audience.

Third, the proliferation of mini-computer, remote-terminal-based, analytic capacity by farm firms represents an exciting opportunity to modelers and to extension personnel. Enough interaction between economists and model users has not yet occurred to make such analytic capability usable for a general farm audience. The availability of a data system that is current, has integrity, and has wide application is critical to successful use of firm-level models. Consequently, Rausser and Just have urged development of an all-purpose data set, a data management system, and reliable software for estimation. These tools could then be used in a portfolio of specific-purpose models for prompt turnaround on a wide range of policy questions.

Finally, Rausser and Just have suggested that credibility will be enhanced in policy modeling if modelers give the most serious attention to: (1) economic theory, (2) nonsample information such as expert judgment, (3) recent sample data, and (4) the entire sample of data sets. Johnson appears to agree, especially on the first two points. I suggest that these prescriptions are applicable to micromodeling as well.

References

Farrell, Kenneth R. Discussant remarks for "Principles of Policy Modeling in Agriculture," in *Modeling Agriculture for Policy Analysis in the 1980s*. Kansas City, Mo.: Federal Reserve Bank of Kansas City, 1981.

Johnson, Stanley R. "Alternative Designs for Policy Models of the Agricultural Sector," in *Modeling Agriculture for Policy Analysis in the 1980s*. Kansas City, Mo.: Federal Reserve Bank of Kansas City, 1981.

Rausser, Gordon C., and Richard E. Just, "Principles of Policy Modeling in Agriculture," in *Modeling Agriculture for Policy Analysis in the 1980s*. Kansas City, Mo.: Federal Reserve Bank of Kansas City, 1981.

PART 3

National Policy Perspectives on Modeling Farm Behavior

5. Implications of National Agricultural Policies in Modeling Behavior of Farm Firms

Kenneth R. Farrell

Abstract Improved microeconomic data and analysis are needed to assist in the formulation of public policy and in expert evaluation of public programs. Particular areas that might benefit are production, price, and income policies; macroeconomic policy linkages to agriculture; structural policies; and regulatory policies. The land grant universities and the Economic Research Service should consider the development of a micromodel and data consortium to make more efficient use of available resources, to serve as a repository for selected farm simulation models and data, and to enhance communication among researchers.

Keywords Agricultural policy, institutional relations, macroeconomic policy, micromodel, regulatory policy, structure of agriculture

To my knowledge this is the first national conference in more than a decade addressed specifically and solely to microeconomic modeling of agriculture. Why hold such a conference at this time? Was it conceived to demonstrate once again our very considerable abilities to formulate elegant mathematical-statistical models with the farm firm as a convenient medium for the coronation? Was it that we felt pangs of guilt from having departed from Leontief's accolades of 10 years ago in which he described agricultural economics as "a branch of economics in which

Senior fellow and director, Food and Agricultural Policy Program, Resources for the Future, Washington, D.C.

scientific methods are applied to observable facts"? Or has the topic, through our mysterious professional ways, become the latest intellectual fad in the cycle of fads we are wont to create and pursue?

I have been assured that the planning of the conference derived from none of those sources, but from a judgment that to better understand and analyze economic change in agriculture we need additional and better quality microeconomic data and research as a basis for predicting farm firm behavior in response to economic and policy stimuli. I subscribe to that judgment. And I subscribe to the judgment implicit in the title assigned to this paper—that better understanding and improved data and economic models of farm firm behavior are needed in the formulation and administration of national agricultural policies.

Having expunged my conscience of any doubts about the motives and objectives of the conference, I will separate the remainder of my remarks into 2 parts: (1) policy and program issues for which there is a particular need for microeconomic data and for better models of farm firm behavior and (2) discussion of the roles of various research institutions in developing and maintaining data and models.

Some Major National Public Policy and Program Issues

I submit that national public policy officials require microeconomic data and analysis for 2 major reasons: to assist in formulating policy and program options and in choosing among such options and to assist in ex post facto evaluating the effectiveness of these policies and programs in achieving specified objectives. I might add, only partly facetiously, that policy officials need such data and analysis to stay in closer touch with the realities of agriculture—realities often obscured by highly aggregated data and macroeconomic models.

I have noted on numerous occasions the tendency at the national level to implicitly view agriculture as if it were a single, homogeneous sector. For example, estimates and forecasts of such variables as farm income and production costs are typically published and discussed as national statistics. But even when such statistics are developed at the regional or state level, they obscure wide variation at the farm level. The classic example of the shortcomings of such highly aggregated data and analysis is illustrated by circumstances at the time of the American Agricultural Movement (AAM) tractorcade to Washington in the late 1970s. Official USDA farm income data depicted a generally healthy farm sector. On average, it may have been healthy, but that was not the case in the lower and central Great Plains according to AAM representatives. Unfortunately, USDA data and economic information provided little insight into the real economic circumstances of Great Plains farms.

Similarly, we regularly design periodic surveys to yield statistically valid estimates at only the national or regional level. As a result, there are large voids in knowledge at the state and substate levels. I suspect that many of the rules of thumb by which program decisions are made at the national level have extremely shaky microeconomic data and analytical foundations behind them.

Nevertheless, we should recognize that it is impracticable to call up and aggregate microeconomic data and analysis each time a policy or program decision is to be taken at the national level. Rules of thumb based on aggregative analysis and data are and will continue to be indispensable to national-level policymakers. But, policymakers should know the magnitude of the variance terms attending those national aggregate figures, whether geographic, by size or economic class of farm, or other variable significant in the policy decision. These arguments strongly suggest the need for a sound microeconomic data and analytic base to undergird aggregative and macroeconomic analysis. One of the purposes of this conference should be that of examining the question of whether we currently have analytical methods and data capable of capturing the complexity of economic and behavioral relations at the farm or micro-level to provide this foundation. We should also explore the tradeoffs in allocation of research and data collection resources for micro-, aggregative, and macroeconomic analytic purposes in the context of limited resources. Four areas can briefly illustrate the importance of microeconomic data and analysis for national public policy and program purposes.

*Agricultural Production, Price and
Income Policies and Programs*

The 1981 Agriculture and Food Act reaffirms commodity price and income programs as the centerpiece of agricultural policy in 1982–85. In addition, it assigns to the Secretary of Agriculture wide discretionary authority to develop and administer programs related to loan rates, target prices, set asides and land diversion, storage and grain reserves, payment limitations, and a melange of other variables. Microeconomic data and analysis of several types and for several purposes will be needed to carry out provisions of the act.

From the Secretary's standpoint, USDA needs reliable assessments of how farmers might respond to various program stimuli, such as alternative levels of loan and target prices, voluntary set asides, alternative levels of paid diversion of land, incentives for participation in the grain reserve program, and others. This information is essential to rational formulation of programs as well as for ex post facto evaluation of program effectiveness. To illustrate further, consider the case of a secretarial decision concerning acreage set aside or diversion. If a voluntary program is being considered, policy officials need sound estimates of how farmers might respond given the availability of program benefits. Participation and production from the standpoint of farmers turn not only on the available program benefits, but on their expectations of market prices, production cost structure, and other management-related variables.

I suspect that we know much less than we should about farmers' behavior in response to economic conditions and program stimuli. It makes a great deal of difference as to whether 30, 50, or 70 percent of farmers participate in the program to a Secretary who wants to achieve a given reduction in program acreage. Farm-level decisions could also significantly affect program costs. Examples of the need for sound microeconomic data and analysis for developing and evaluating commodity production, price, and income programs are extensive. I need not elaborate the obvious, except to reiterate that such data and analysis should undergird aggregative analysis of farm programs as well as be a source of observable facts to the policymakers in their decisions.

There has been considerable discussion about the need for more precise targeting of commodity and conservation programs, about the differential effects of farm programs geographically, by type and size of farm, among commodities and between crop and livestock producers. In the context of a longer time period, basic policy issues turn on the longrun nature of farmers' production, cost and supply functions in the context of risk and uncertainty, new or modified technology, natural resource endowments, and changing price-cost relationships. How do farmers make investment decisions with respect to conservation, equipment, and other capital assets? If such decisions are conditioned by public policies and programs, how and to what extent do such variables affect the decisions? I suggest that better understanding and analysis of farm firm behavior are essential for public policy purposes, for improved farm management and extension purposes, and as a basis for more reliable, comprehensive aggregative models.

Some observers have concluded that the 1981 farm bill will be the last such omnibus agricultural legislation. Whether or not it is, we need anticipatory economic analysis of how agriculture might fare in the absence of commodity price programs or with major modification of commodity programs to reduce federal government budget exposure. The 1981 farm bill directs the Secretary to study the implications of an income insurance program, presumably to replace or modify traditional commodity price programs. The program would be focused on whole farms. Thus, micro-level analysis will be needed to establish benefits and costs in a probabilistic context. Are we prepared to address this issue? I suspect that when we look at the issue in the context of the great heterogeneity of agriculture and our understanding of the economics of the farm firm, we will find our knowledge larder lacking.

Macroeconomic Policy-Related Issues

Emerging issues related to the national macroeconomic environment during the 1980s also raise relevant issues for agricultural economic research. Inflation directly affects the farm sector by changing asset values, primarily land, and indirectly by changing the ability of certain individuals to gain control of those assets. Because agriculture is a relatively capital-intensive industry, high inflation and interest rates act to increase the costs of producing food and fiber. High interest rates can cause severe cash flow problems for many beginning farmers because they usually own a relatively large amount of debt. As their financial leverage increases, so does the likelihood of their bankruptcy. Should special financial assistance be given to beginning or entry-level farms during periods of high inflation or wide price and income fluctuations? Farm-level modeling research describing optimal, prescriptive, and observed growth patterns of different types of farms in different areas of the country would help provide some perspectives for these types of policy decisions.

The state of the national economy also indirectly provides another set of questions that might be addressed more appropriately with micro-level research than with available aggregate sector models. The importance of income from off-farm work to small farms is very important in many areas. What effect would a general recession that eliminated many of these jobs in the rural sector have on these farms? Would there be a wave of bankruptcies? Would the supply flexibility of the farm sector provided by these marginal producers change and become less elastic?

Finally, producers and nonfarmers are becoming increasingly aware of and taking advantage of general tax legislation. The Economic Recovery Act of 1981 changes numerous tax provisions that directly affect the treatment of capital gains, purchase of farmland, investment activities in larger and more efficient machinery, depreciation accounting, and income tax rates. Many of these provisions will likely have an increasing qualitative and quantitative impact on the structure and behavior of the farm sector. Research questions include: What effect will the new tax provisions have on producer financing, debt acquisition, and cash flow behavior? Will more producers be able to increase the rate of investment in new technology or make investments that will increase production capacity? Will land-purchasing behavior of many producers be accelerated or merely bid up the price of land?

Many of these questions raise the related issues of whether anticipated added investment based on increased cash flows will eventually lead to higher average costs of production and whether the farm sector will consist largely of tenant operators who share-crop or cash-rent land. The acceleration of the separation of operation and ownership could be one result of recent estate and inheritance tax law changes, in addition to magnifying farm income problems and increasing the average age of farmers. Finally, rental arrangements between tenants and landlords could undergo large changes as tenancy increases. It is likely many of these changes will affect the economic well-being of producers, landowners, and farm households.

Structure Issues

Despite recent proclamations that the economic structure of agriculture is a matter of critical and immediate national public policy significance for the 1980s, the issue has been with us for several decades. Its antecedents derive from land settlement of the West, the populist movements of the late 19th and early 20th centuries, the great depression and its aftermath in the 1930s, and the so-called "industrialization of agriculture" in the 1950s and 1960s. Rhetorically, the issue has been clothed in various slogans—"save the family farm" being the most persistent. One dimension of the issue is dependent on whether the structure of the farm sector and the organization and ownership of resources are significantly affected by national agricultural policies.

The question remains whether policy and structure relationships can be established empirically because of the large number of other confounding causal factors which affect structure. Bruce Gardner concluded from his review of the literature that "there is no definitive research that isolates the relative importance of various exogenous influences on the economic organization of agriculture." Nevertheless, that the commonly held belief of the exogenous influences on structural change have not yet been established empirically does not deny their existence. Because aggregate econometric and programming models do not typically endogenize the structure of the farm sector, the issue might perhaps be better addressed with micro-level models of individual farmers or with sets of individual types of farms in competition for resources.

It seems clear that several policy variables could directly affect the financial well-being of many individual crop producers and indirectly affect livestock producers and, thus, the structure of the farm sector: the effect of program payment limitations by size and type of farm, imposition of normal cropland acreages instead of acreage allotments, price supports in areas of high soil loss, methods of production controls, the role and use of commodity reserves held by farmers or the federal government, the linkage between conservation programs to reduce soil loss due to wind and water erosion and commodity programs, and research and education. I suggest further research is needed, some of it microeconomic, if for no other reason than to add rigor to discussion of the issue. The time may come when policy officials are prepared to debate seriously whether an explicit structure policy is needed. Dynamic micromodels might contribute to such debates.

Regulatory Policy Issues

Emphasis on regulatory reform by the current administration has generated much demand for estimates of costs, benefits, and their distributions for a wide range of economic, environmental, health, and safety regulations. Without elaborating on the obvious need for such estimates, let it suffice for me to say that microeconomic data and analysis should be essential parts of that activity.

Institutional Relations in Microeconomic Research

Having illustrated the need for enrichment of microeconomic data and analysis, I must hasten to add that demand for economic data and research is expanding in numerous directions. However, resources to support public data development and economic research are likely to be static or to decline in real terms in the years ahead. Thus, it behooves us as economists to practice the principles of our discipline by considering how we can best utilize scarce resources to meet competing demands. In considering that principle, we need to explore relations and possible divisions of labor among research institutions to enhance our collective productivity and output. In particular, I suggest we examine relations between the ERS and universities in conducting microeconomic research of the types previously outlined.

ERS and land-grant universities have a significant interest in research activities concerning micromodels of farm-household behavior for many reasons. At the state level, universities must be responsive in many matters to the clientele who provide resources and pay their wages. Furthermore, universities, through their colleges of agriculture and State Experiment Stations, are responsible for both disciplinary and applied analysis. Many of the university studies have also been able to provide scattered economic intelligence at the regional and national level. ERS, on the other hand, has long had a primary responsibility, either assumed or defined, for providing aggregate policy analysis and quantitative information for national legislative and administrative bodies.

There are opportunities, indeed imperatives, for developing complementary relationships between those 2 publicly supported research
institutions to provide information and research relating to agriculture
and policy. ERS has access to an enormous data base immediately useful
to microanalyses including primary census data, cost-of-production surveys, the farm enterprise data system, aggregate time series data, and
a wide-ranging series of economic indicators. Many of these data are
indispensable to the research needs I have identified. Many states also
have access to a data base for their own particular set of farms. Although
both institutions have competent personnel in quantitative methods of
economic analysis, the university setting may provide a comparative
advantage in basic methodological and disciplinary research. I am not
suggesting that absolute specialization of labor between the two is either
desirable or feasible. But, I do strongly believe that a more logical and
effective coordination of effort is both desirable and feasible.

One such possibility might be the development of an ERS-university
consortium on micromodels and data systems. The consortium would
neither dispense research funds nor approve or disapprove research
activities. Rather, it would act as a collective repository or bank for a
few generalized, farm-level models with modular components that could
include simulation and optimization elements, the relevant aspects of
the income tax code, cash flow and asset accounting relationships, and
a full set of policy and program levers that could be modified for
different types of research problems. The consortium would share or
lend the fully documented and maintained models to any cooperating
institution, with a commitment that the models would be returned to
the consortium with full documentation of any modifications made,
together with documentation of the analyses. The consortium could also
enhance communication among researchers by an occasional seminar
or workshop during which members could present research findings,
discuss ongoing research efforts, and publish proceedings for the public
record.

There may be other more efficient and preferred organizational
suggestions to enhance working relations for these purposes. I encourage
you to discuss such alternatives during this conference. I believe we
can no longer justify expenditure of public research resources that results
in proliferation of specialized models with very short "half lives."
Certainly, we cannot justify continued, similar resource use until we
have tried seriously to develop mechanisms to enhance the quality and
level of relevant microeconomic research and data. The argument is
strong for you to consider arrangements that will result in fewer, more
adaptable, and transportable modular micromodels that will enable us
to conduct reproducible policy-related research on a quick-turnaround

basis. Of course, theoretical and quantitative research to test the realism of behavioral assumptions is also of high priority.

Conclusions

I draw 4 major conclusions:

1. Information concerning the impact of farm programs and policies on different types of representative farms in different regions of the United States is important to policy formulation and development. Aggregate models tend to homogenize results and information and neglect input and financial markets, when strategic detail may be vital to informed policymaking criteria.

2. Micromodels can provide a means of testing many theoretical and empirical assumptions about producer and household consumption, risk, participation, and resource use behavior.

3. The use of farm-level micromodels in policy analysis, at the national level is that of: providing complementary and corroboratory evidence; suggesting mistakes in the information supplied by macro- or aggregative models; and providing structural, financial, regional, and input use data that are either inadequate or absent in macro-sector models. This linkage needs far more research.

4. It is not necessary to rediscover or reinvent micromodels every few years. This process not only wastes resources, but places emphasis on building the models, rather than on creatively using them. Only a few general models with modular elements may be needed. If so, we should explore the possibility of an ERS-university consortium and a bank for maintenance and loan of these models.

Acknowledgments

The extensive assistance of Kenneth H. Baum, Economic Research Service, U.S. Department of Agriculture, in preparing this paper is gratefully acknowledged.

PART 4

Institutions and Implications for Micromodels

6. Financing Growth and Adjustment of Farm Firms Under Risk and Inflation: Implications for Micromodeling

Peter J. Barry

Abstract   A portfolio model illustrates a conceptual framework for modeling the effects of business and financial risks and inflation on the performance of farm businesses. The portfolio model expresses an optimal organization of assets and liabilities for a risk-averse investor. The inflation model accounts for the effects of earnings growth on farmers' total returns, asset values, and liquidity. Implications for financial and policy analysis are illustrated through the strong effects of business risks, financial risks, and inflation-induced liquidity problems on the optimal structure of farm portfolios.

Keywords   Inflation, portfolio model, risk

My paper focuses on the financial components of farm firms and their implications for micromodeling. I briefly review the current setting for agricultural finance and past modeling developments. Then, I establish a conceptual framework for micromodeling of farm firms that accounts for the joint effects of leverage, liquidity, and risk under conditions of static equilibrium. The framework is based on portfolio theory, although its basic components apply to other conceptual and methodological

---

The author is a professor of agricultural finance at the University of Illinois.

approaches. Next, I focus on the implications of inflation for micro-modeling and, last, on policy implications.

Current Setting

Farmers have long used credit and related services provided by financial institutions that differ greatly in their sources of funds, specialization in farm lending, legal and regulatory environment, and government affiliation. Included are the Cooperative Farm Credit System, government lending agencies and credit programs, farm input suppliers, life insurance companies, and the commercial banking system. Considerable financing by individuals, especially sellers of farmland, occurs as well. Some parts of the farm sector attract outside equity capital, although most equity comes from farmers' retained earnings and unrealized capital gains.

These institutions and related financial markets have been vital in the development of U.S. agriculture and should become increasingly important in the future. Their organizational structure and design have enabled the smaller scale, noncorporate structure of most farm units to survive, yet adjust over time to a changing economic and social environment. Rapid growth in farm debt has contributed greatly to efficient production and marketing of commodities; to replacing, modernizing, and expanding the capital base of farming operations; and to meeting household needs. Moreover, credit reserves are one of farmers' most important sources of liquidity in responding to risks. Such reserves provide the capacity to carry over loans, defer payments, and refinance high debt loads during times of financial distress. Hence, financial performance and growth of farming operations depend heavily on loan funds that are made available to farmers in a timely fashion, for various purposes, and in amounts, costs, and maturities that compare favorably with terms available to other economic sectors. Moreover, the credit programs of government agencies can be tailored to meet specific liquidity or income maintenance needs of farmers, often on concessionary terms.

Farmers and their lenders have adjusted to a long series of events in recent decades that brought new opportunities and problems. A transition occurred from a heavy earlier emphasis by farmers on production and productivity to emphasis in the 1950s and 1960s on mechanization and growth in farm size (with related demands for capital and credit), then to dealing in the early 1970s with more volatile commodity markets, and finally to responding to higher, volatile rates of interest and inflation. Financial risks for many farmers were increased substantially by the increased volatilities of commodity and financial markets and by individual decisions to acquire farm debt. The higher average prices, although more volatile, of commodities, the persistence of general inflation, and increases in land values stimulated even greater demands by some for credit to buy land, machinery, livestock, and other capital goods.

The strong growth in farm debt, especially during the 1970s, implies that borrowing by farmers has become more aggressive, more sophisticated, more permanent, and more complex in credit evaluations. Loan losses in farming have been low, and overall financial performance has been strong. However, much of farmers' total returns has occurred as capital gains on farmland and other assets. Operating income also increased, but by smaller amounts. When subject to financial analysis, this pattern of returns in farming indicates a highly solvent industry due to unrealized capital gains on farm assets that are collateral for the growth in debt, yet also show liquidity problems and cash flow pressures resulting from the continued low, but volatile current rates of return to production assets.

These contrasting positions are illustrated by measures for stock and flow concepts of leverage for the farm sector over the 1950–80 period (Figure 6.1). The stock concept, measured by the debt-to-asset ratio, shows relatively low, stable leverage in the farm sector. The flow concept, measured by the ratio of interest-paid to returns to farm assets, shows higher, more volatile leverage due to the combined effects of growing debt use, higher interest rates, and strong capital gains. Both measures exclude nonfarm income and assets. Additional funds from these sources may be available to households involved in these nonfarm activities.

In farm credit markets, the instabilities created by stringent regulations for some lenders, growing competition among financial institutions, and change in monetary policy, all occurring in an inflationary environment, have brought new sources of financial risk to farmers. Costs and availability of farm credit are now much less insulated from forces in national markets. Inflationary conditions are transmitted to farmers in a quicker, stronger fashion through their effects on interest rates. The stresses from these conditions fall on all lenders, but heavily on agricultural banks and their highly leveraged farm borrowers. Since 1975, these banks experienced steady decreases in their market shares of nonreal estate debt and sharp increases in their costs of funds and loan rates. In contrast, other farm lenders with access to loan funds in national markets have maintained or improved their market shares. The increase in farm lending by the U.S. government was especially strong in recent years, indicating a shift to public credit sources.

The current pattern of deregulation in financial markets and its effects on competitive behavior among financial institutions should continue and will likely result in relatively high volatility in interest rates and

FIGURE 6.1
Annual coverage ratios and debt-to-asset ratios, 1950-80

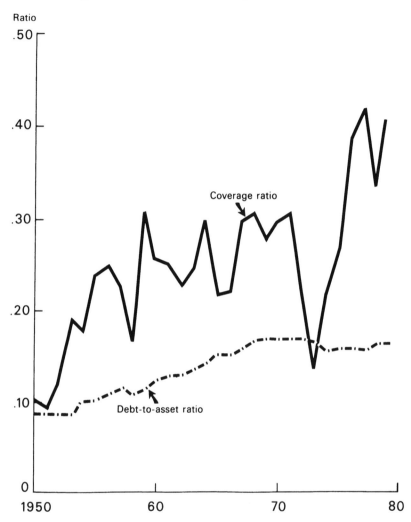

greater precision in pricing farm loans by lenders. The increased swings in interest rates are an additional source of financial risk for farmers; however, greater dependability of availability of loan funds, especially from agricultural banks, should offset in part the risks added by greater volatility in interest rates. Moreover, because interest rates are easier to measure than is fund availability, farmers should have clearer signals about changing conditions in financial markets. Finally, the balance is again shifting away from credit provided by government programs. The Reagan administration is seeking to reduce the volume and degree of subsidy in Farmers Home Administration programs and largely return it to its role as a lender of last resort for farmers. The result is a pullback of the public safety net for farm lending and a shift in the availability of farmers' credit away from public sources toward greater reliance on commercial sources.

These factors—high and volatile interest rates, high variability in farm income, swings in funds available from some lenders, and the strong role of capital gains in farmers' total returns—have magnified the financial problems in those farming operations that choose to be highly leveraged, and they have added to the complexity of financial planning and analysis in agriculture. They warrant new directions in managing liquidity, risk, and finances. Micromodels used in counseling with farmers will be challenged to indicate effective ways to respond to these new conditions. Similarly, micromodels used to support policy analyses will be challenged to provide insight about farmers' responses to alternative policies and how these policies might be implemented.

Past Developments in Concepts, Methods, and Modeling

Developments in concepts, methods, and modeling for evaluating the financial components of farm firms have closely paralleled the evolution of managerial issues cited in the preceding section. The growth environment of the 1950s and 1960s stimulated substantial research on the internal growth processes of family-size firms in a competitive market environment (Washington State University). Investment analysis, capital accumulation, credit evaluation and financing strategies, leverage and liquidity tradeoffs, financial control, and their effects on the rate and direction of firm growth over a farm's life cycle were emphasized. Developments in capital theory yielded important analytical insight about the separation properties of investment and financing decisions under certainty conditions that yielded time-efficient sets of portfolio choices for utility maximizing decisionmakers (Hirshleifer; Weingartner; Cocks and Carter). Static, deterministic models of firms made much use of multiperiod linear programming and simulation as the tools of analysis (Irwin; Barry). Optimal control theory provided a rich dynamic framework for conceptualizing a firm's path through time, including its response to the length of horizon, learning and feedback processes, and to adaptive control (Dorfman; Boussard).

A major empirical finding of these studies was the strong influence of financial choices on firm growth (Baker). Moreover, the form of risk response by growth-oriented farmers with different levels of risk aversion was strongly exhibited through differences in holdings of credit and other liquid assets, thus implying the importance of liquidity premiums as indicators of a farmer's risk aversion. Risk responses in the production and marketing organization of farm firms were much less prominent. In turn, the size, structure, and other attributes of credit reserves were highly sensitive to various strategies followed by farmers in borrowing and debt management (Barry and Baker).

The increased market risks of the 1970s shifted the emphasis more toward the risk components of micromodels. Sources of risk, measures of risk, attitudes toward risk, methods of managing risk, and their implications for financial structure came to the forefront in empirical analysis (University of Illinois). Greater attention was given to the government's role in risk bearing as large swings in farmers' financial conditions shifted the balance away from public programs for risk bearing early in the 1970s, and then back again as emergency loan programs gained prominence later in the decade (Gardner).

Finance theory again yielded analytical insight about the interrelationships between investment and financing decisions under risk conditions and the derivation of risk efficient sets for expected utility maximizing decisionmakers (Markowitz; Tobin). Farm firms were often modeled in terms of risk programming and stochastic simulation in order to evaluate tradeoffs between financial gains, risks, liquidity, and survival (Mapp and others). Linkages between sources of risk and risk responses in the production, marketing, and financial organization of the firm were identified and evaluated in empirical research (Barry and Baker, 1976). Interrelationships were observed between the adoption of numerous strategies in marketing (contracts, hedges, storage, and spread sales) that improve both expectations on loan repayability and the firm's financial capacity to support growth and liquidity (Barry and Willmann).

Continued emphasis was observed on lenders' use of nonprice responses in evaluating a farmer's credit in which differences in the borrower's loan limit, security requirements, loan maturities, loan supervision, and documentation outweigh adjustments in interest rates (Robison and Barry). Innovations were also evaluated in financial programs and instruments that enable lenders to participate more substantially in risks associated with random variations in farmers' debt servicing capacity (Baker, 1976; Lee; Hanson and Thompson). An example is the variable amortization plan, which formalizes the credit reserve concept by allowing a farmer's debt servicing commitment to vary in response to changes in the farmer's economic environment.

A Conceptual Framework

The mean-variance portfolio model provides an appealing framework for evaluating the combined effects of leverage, financial risk, and business risk on a farm's total risk, and for evaluating optimal portfolio choices for risk-averse decisionmakers. By optimal, I mean that a farm's activities in production, marketing, investment, consumption, and financing are organized to yield the highest possible level of well-being for its decisionmakers as the unit moves through time in an uncertain environment. Well-being is modeled in utility terms with expectations and preferences cast as expected utility maximization for a risk-averse decisionmaker. Expected utility is thus a function of the desired tradeoff (λ) in utility terms between the expected level ($\bar{r}_e$) and variance (σ_e^2) of returns to the equity holder.

$$E[U(r_e)] = f(\bar{r}_e, \sigma_e^2, \lambda). \tag{1}$$

The mean variance (EV) approach is well known and much debated, especially about the limited generality of its assumptions. It assumes either that decisionmakers' expectations can be modeled by normal distributions which are fully specified by their mean and variance, or that their utility function is expressed by a quadratic reflecting preferences only toward the means and variances of the returns distribution. If one of these conditions is met, then an expected utility-maximizing choice occurs among the mean-variance efficient set. Hence, risk is treated in probabilistic terms with variance used to measure the likelihood of events occurring that yield results less than expected.

Despite these limitations, the mean-variance approach is widely used in economic and financial analysis (Robison and Brake). Its strong foundations in microfinance theory (Markowitz; Tobin), explicit measures of risk, and rigorous demonstration (Tsiang; Levy and Markowitz) of its usefulness as an approximate method for portfolio selection help make it an acceptable model for financial analysis. Moreover, the approach can also be adapted for analysis by other criteria as in the case of stochastic dominance or the safety-first rule.

Risk-Efficient Portfolios

The traditional portfolio model defines a risk-efficient set as combinations of risky assets (or activities) that provide minimum variance for alternative levels of expected returns. In farming, these risky assets are expressed as choices in production (crops, livestock), marketing (contract and open market sales, vertical integration, storage, and sequential selling), and investment (land, machines, breeding livestock). In empirical analyses, the risk-efficient set is also subject to other resource constraints and business requirements that characterize the problem setting; however, I focus now on the fundamental components of the portfolio model.

Suppose that the choice is between 2 activities, X and Y. Both generate risky assets having rates of return expressed as normally distributed random variables with known means ($\bar{r}_x$, $\bar{r}_y$), variances (σ_x^2, σ_y^2), and covariance (σ_{xy}). If P_x and P_y represent their proportions in the farm portfolio and $P_x + P_y = 1$, then the expected portfolio return $\bar{r}_{x/y}$ is

$$\bar{r}_{x/y} = \bar{r}_x P_x + \bar{r}_y P_y \tag{2}$$

and the portfolio variance is

$$\sigma_{x/y}^2 = P_x^2 \sigma_x^2 + P_y^2 \sigma_y^2 + 2 P_x P_y \sigma_{xy} \tag{3}$$

An EV efficient set is found by minimizing variance (3) subject to alternative levels of expected returns (2) and subject to other resource constraints and business requirements.

Risk-Free Borrowing

The portfolio model was extended to include financing by introducing the concept of a risk-free asset with zero variance. Positive holdings of the risk-free asset represent lending, and negative holdings represent borrowing or leveraging, each occurring at the same risk-free rate. Combining the risk-free asset with the portfolios of risky assets increases the size of the efficient set and makes it more risk efficient.[1] It also yields one portfolio (A) of risky assets that dominates all others and that can be held in varying combinations with the risk-free asset in order to determine the optimal capital structure. This analytical approach yields the separation theorem that implies that a firm's investment decision in risky assets is independent of its financing decision involving holdings of the risk-free asset (Tobin; Van Horne).

To illustrate numerically, suppose that an "optimal" portfolio (A) of risky assets is identified with an expected return $\bar{r}_a = 0.16$ or 16 percent, subject to a standard deviation $\sigma_a = 0.06$ or 6 percent. Suppose that the interest rate on the risk-free asset, for example, a fixed-interest rate loan to finance the purchase of land, is $i_d = 0.12$ or 12 percent. Since this rate does not vary over the term of the investment, its standard deviation is $\sigma_i = 0.00$.

The expected rate of return ($\bar{r}_e$) to equity capital then is

$$\bar{r}_e = \bar{r}_a P_a + i_d P_d \tag{4a}$$

where $P_a + P_d = 1.0$. P_a and P_d are the respective proportions of risky assets and the risk-free asset in the total portfolio. Leverage implies negative holdings of the risk-free asset. Thus the expression for returns to equity is modified to

TABLE 6.1
Risk and returns to equity for alternative debt-to-equity ratios

Debt-to equity (1)	P_a (2)	P_d (3)	Expected return (r_e) (4)	No credit risk σ_e (5)	C.V. (6)	Credit risk σ_e (7)	C.V. (8)
			Pct.	Pct.		Pct.	
0	1.00	0	16.00	6.00	0.38	6.00	0.38
.25	1.25	-.25	17.00	7.25	.43	7.76	.46
.50	1.50	-.50	18.00	9.00	.50	9.22	.51
.75	1.75	-.75	19.00	10.50	.55	10.92	.57
1.00	2.00	-1.00	20.00	12.00	.60	12.65	.63
1.50	2.50	-1.50	22.00	15.00	.68	16.16	.73
2.00	3.00	-2.00	24.00	18.00	.75	19.70	.82
3.00	3.50	-2.50	26.00	21.00	.81	23.26	.89

$$\bar{r}_e = \bar{r}_a P_a - i_d P_d \qquad (4b)$$

where $P_a - P_d = 1.0$.

The standard deviation of the returns to equity is the weighted standard deviation of the risky assets

$$\sigma_e = \sigma_a P_a. \qquad (5)$$

Expected returns and standard deviations for equity are shown in columns 4 and 5 of Table 6.1 for several levels of leverage. Higher leverage increases the expected returns to equity under these assumptions, but the standard deviation increases even faster. Total risk relative to expected returns increases with greater leverage, as indicated by the increasing coefficients of variation.

The effects of business and financial risk on the firm's total risk can be distinguished for analytical purposes (Gabriel and Baker). Business risk is attributed to variability (σ_a) of the returns to the firm's risky assets. It is considered independent of the financial organization. Financial risk arises from the composition and terms of the financial claims on the firm held by its lenders and lessors. Only the financial risks associated with debt financing are considered here.

The following analysis shows how a measure of business risk is magnified by a measure of the firm's financial leverage to determine the total risk to equity holders. In this approach, the measure of financial leverage serves as a direct indicator of the firm's financial risk. If leverage is zero ($P_a = 1.0$), then business risk and total risk are the same. As leverage increases, so does total risks (*TR*) relative to business risk (*BR*);

the degree of increase directly indicates the firm's financial risk (*FR*). This approach to delineating business and financial risk is adapted from Gabriel and Baker. They derive an additive relationship between *BR* and *FR* in determining total risk that shows the absolute increase in total risk that is attributed to debt financing. The emphasis here, however, is on the percentage increase in business risk that is attributed to debt financing. Thus, the two approaches are essentially the same, differing only in their measurement concepts.

Business risk (*BR*) and financial risk (*FR*) combine to determine the total risk (*TR*) to equity holders in the following way (Barry and Baker):

$$TR = (BR) \cdot (FR) \qquad (6)$$

Total risk can be expressed by the coefficient of variation for equity holders, as indicated by the entries in Table 6.1 for the different levels of leverage:

$$TR = \frac{\sigma_e}{\bar{r}_e} = \frac{\sigma_a P_a}{\bar{r}_a P_a - i_d P_d} \qquad (7)$$

Business risk is expressed by the coefficient of variation for risky assets:

$$BR = \frac{\sigma_a P_a}{\bar{r}_a P_a} = \frac{\sigma_a}{\bar{r}_a} \qquad (8)$$

Then, financial risk is found by dividing total risk in (7) by business risk in (8):

$$FR = TR \div BR$$

$$= \left[\frac{\sigma_a P_a}{\bar{r}_a P_a - i_d P_d} \right] \div \frac{\sigma_a}{\bar{r}_a} \qquad (9)$$

$$= \frac{\bar{r}_a P_a}{\bar{r}_a P_a - i_d P_d}$$

Substituting the expressions for business and financial risks into (6) yields:

$$TR = \left(\frac{\sigma_a}{\bar{r}_a} \right) \cdot \left(\frac{\bar{r}_a P_a}{\bar{r}_a P_a - i_d P_d} \right) \qquad (10)$$

The first term to the right of the equality expresses business risk and the second term expresses a measure of financial leverage that serves as an index for financial risk. Higher leverage increases the value of the second term and thus increases total risk relative to business risk.

To illustrate, consider the previous numerical example using a debt to equity ratio of 1.0 ($P_a = 2.0$; $P_d = 1.0$). The coefficient of variation that measures business risk (BR) has a value of $0.375 = (6/16)$; the level of leverage that measures financial risk has a value of $1.60 = (0.16)(2)/[(0.16)(2) - (0.12)(1)]$. The coefficient of variation that measures total risk is then $0.60 = (0.375)(1.60)$. Thus, the financial risk introduced by leverage magnifies business risk 1.6 times in determining total risk.

Effects of changes in leverage and thus financial risk, on total risk can be easily evaluated using equation (10). The distinction between business and financial risk provides a useful framework for evaluating farmers' responses to changes in risk. Business and financial risks may well be tradeoffs in the risk behavior of farmers. A decline (increase) in business risk could lead to farmers' acceptance of an increase (reduction) in financial risk as they seek to restore an optimal structure of assets and liabilities in their portfolio. Similar tradeoffs may occur in the behavior of lenders in deciding on size of loans to extend to farmers.

Risky Borrowing

The analysis to this point has assumed that borrowing is a risk-free activity. This assumption was likely realistic when farmers' debt use was less and interest rates were relatively low, stable, and subject to fixed rates on loans. Even then, however, lenders' nonprice responses to factors affecting farm performance tended to destabilize farmers' credit reserves, thus adding to their financial risks. Moreover, the current environment of high, volatile interest rates along with the growing use of floating rate loans by most types of farm lenders clearly makes borrowing a risky activity. Its cost can be expressed as an expected value that has nonzero variance and covariances (or correlation) with returns on other risky activities. Called "credit risks," these variations are additional sources of financial risk that must be considered along with leverage and business risk in evaluating farmers' optimal portfolio. Thus, the portfolio model must be modified to account for these "credit risks." In doing so, the separation property between investment and financing conditions that occurred for risk-free borrowing no longer holds.

The effects on portfolio risk are shown by allowing short sales, or negative holdings, of a risky financial asset with the proceeds of this borrowing activity used to expand the investment in other risky assets (Fama, p. 226).[2] This short-sale activity causes a reversal in the effects on portfolio risk of the degree of correlation between assets relative to long positions in all assets. The measure of correlation between returns on short and long positions in two different assets is the negative of the correlation value when both are held long. Thus, leveraging causes greater risk when lower levels of correlation occur between assets that are held short and long. The net impact on portfolio risk depends on the variances of the returns to the respective assets and on their correlation, although the usual effect is an increase in portfolio risk.

The expected return to equity for leveraged farm assets involving credit risks is expressed as the expected returns to risky farm assets less the expected cost of borrowing:

$$\bar{r}_e = \bar{r}_a P_a - \bar{\imath}_d P_d \tag{11}$$

Variance of returns to equity is the sum of variances on the returns to assets and the costs of borrowing less their covariance, with each weighted by the proportions of assets and debt in the portfolio:

$$\sigma_e^2 = \sigma_a^2 P_a^2 + \sigma_i^2 P_d^2 - 2P_a P_d \sigma_{ai} \tag{12}$$

Covariances can also be expressed by $\sigma_{ai} = c\sigma_a\sigma_i$ where c is the correlation coefficient ($-1 \leq c \leq 1$).

Thus, variance of the returns to equity in the portfolio is inversely related to the correlation between asset returns and borrowing costs. The higher the asset returns relative to borrowing costs, the lower is the portfolio variance. Returns to a farmer's equity are stabilized if there is positive correlation between returns to assets and borrowing costs. In contrast, opposite movements in asset returns and borrowing costs will destabilize returns to equity.

The numerical example used earlier is extended to include the risk added by variability in borrowing costs. Let the standard deviation of borrowing costs be $\sigma_i = 0.04$ or 4 percent and assume a zero covariance ($\sigma_{ai} = 0.00$) between asset returns and borrowing costs. The impacts of risky borrowing on total risks and coefficients of variation are shown in columns 6 and 7 of Table 6.1. Credit risks add more to total risk as leverage increases, although the actual increase depends on the variances and covariances.

Most portfolio analyses of farm businesses have either ignored credit risks or have assumed that borrowing is riskless. In either case, the values for σ_i^2 and σ_{ai} would be zero. This assumption leads generally to an understatement of a farmer's total risk. In extreme risk situations strongly positive correlations between asset returns and borrowing costs could offset the variance added by borrowing costs.

Developing appropriate measures of risk for costs of borrowing is a complex task. The sources of variation arise from forces in financial markets, lending risks in agricultural markets, and lenders' responses to changes in individual farmer's creditworthiness. Moreover, lenders' risk responses to differences in farmers' creditworthiness primarily occur through nonprice practices rather than through risk premiums in interest rates. While this practice is changing over time, many lenders' nonprice risk responses still occur through differences in loan limits and other credit terms for borrowers. A recent study (Barry, Baker, and Sanint) demonstrated one approach to measuring lenders' nonprice responses to changes in farm income. The results indicated a negative correlation between farmers' costs of borrowing and farm income. Most of the response was associated with the costs of borrowing for capital expenditures rather than operating loans. The negative correlation indicates that, for the situation being studied, increased variation in farmers' costs of borrowing adds to farmers' total risk through both the variance and covariance effects.

These outcomes are less likely when interest rate risks result from conditions in financial markets. These conditions are primarily reflected by changes in interest rates, which clearly signal farmers about changes in money market conditions. Moreover, the growing emphasis on floating rates by farm lenders is accelerating the transmission of interest rate changes to borrowers. Changes in costs and availability of credit that originate in financial markets are far removed from farmers' operating environment. They add variance to farmers' total risk, but are largely independent of changes in financial performance in the farm sector, unless the aggregate effects of this performance contribute significantly to inflation or other national conditions.

Implications of Conceptual Framework for Financial Analysis

This portfolio model provides an integrated conceptual framework for expressing an organization of assets and liabilities in farm businesses that is optimal in utility terms for risk-averse decisionmakers. It has richer, more general content than past financial models that focused on firm growth under conditions of certainty. The portfolio model delineates an optimum organization based on information about the following items: (a) a farmer's preferences (λ) about the trade-off between risk and returns; (b) expected levels and variabilities of returns on farm and nonfarm assets; (c) expected levels and variabilities of costs of financial capital with emphasis on borrowing; (d) correlations among and between returns and costs for assets and liabilities; and (e) a set of decision choices for structuring assets and liabilities.[3] These components are essential to any decision situation and can be refined and extended to fit the particular situation.

The framework also can be used in a comparative static mode to evaluate how an optimal or equilibrium portfolio responds to forces that create nonoptimalities or disequilibrium. Examples of these forces include changes in expectations and in risk aversion, creation of new choices in managing assets and liabilities, and windfall gains or losses that alter the composition of assets and liabilities. Moreover, the framework also serves to trace through the effects of various actions by private decisionmakers or public policymakers in attempting to restore optimality, once a disequilibrium has occurred.

The modeling of microprocesses of portfolio adjustments under risk is still in an early state of development. It has not received much attention in either theoretical or empirical analysis. Some conceptual progress occurred with the development of a general method for evaluating the response of an expected utility-maximizing choice to factors that shift the mean-variance efficient set or that cause changes in the decisionmaker's level of risk aversion (Robison and Barry). Suppose, for example, that farmers experience an increase in the variability of expected returns as a result of more volatile commodity prices while the level of expected returns stays constant. This increased business risk signifies a shift in the risk-efficient set of activities to a lower level of expected returns for the same level of risk carried earlier. The original farm portfolio now is nonoptimal. Thus, the farmer must adjust the portfolio so as to restore an expected utility-maximizing composition of assets and liabilities. For a farmer with constant absolute risk aversion, the revised portfolio will have lower expected returns and risk than did the unrevised portfolio. The reduction in risk and returns would be greater for a farmer with decreasing absolute risk aversion. Similar types of response occur for other changes in optimal portfolios.

Decision criteria, possible portfolio adjustments, and how conditions are evaluated by lenders and borrowers in the decision process are largely empirical questions. They depend heavily on the characteristics of farmers and farming operations, on their markets and legal environment, and on the responses of lenders and other financial claimants on the firm. One kind of portfolio adjustment is to change financial risk by modifying the farm's financial leverage. Similarly, changes in the level of and rental terms on leasing of farm assets may alter the farm's total risk in response to changes in business risk.

Gabriel and Baker demonstrate an approach to the portfolio adjustment process that emphasizes a farmer's trade-offs between business risk and financial risk, subject to a maximum risk tolerance. A decline in business risk may lead to acceptance of greater financial risk, thus offsetting the lower business risk. Their analysis of the aggregate portfolio of the farm sector yields empirical evidence consistent with this phenomenon, including the increasing importance of liquid reserves as risk increases.

A policy-oriented analysis is illustrated by Boehlje and Griffin who evaluated how government price-support programs can influence the investment and financing behavior of risk-averse farmers. They reasoned that cost-of-production indexed support prices will provide greater certainty in farmers' cash flows and thus reduce business risks, with more rapid firm growth resulting from increased bid prices for durable assets, especially land, greater borrowing capacity, and higher financial leverage. Their simulation analysis for crop farms differing in size, capital structure, and farmer characteristics indicates that smaller, more highly leveraged operations are likely to receive fewer benefits from indexed price-support programs than larger operations with lower leverage. Hence, public policies intended to reduce farm business risks may be offset by farmers' subsequent willingness to assume greater financial risk.

Other contemporary illustrations of portfolio adjustments can be provided, too. The reduction in availability and increase in cost of credit provided to farmers through public credit programs should increase and further destabilize farmers' costs of borrowing. Similar effects may occur for costs of borrowing from agricultural banks as the process of deregulating interest rate controls continues to be phased in. These changes in borrowing costs should prompt risk-averse farmers to undertake actions that lower the degree of leverage in their operations. Alternatively, farmers might consider actions that increase or stabilize expected returns to farm assets, thus offsetting at least in part the higher cost and reduced availability of public credit reserves. These latter actions could be tied to availability of other types of public programs such as the new Federal Crop Insurance Program for disaster protection. This program should affect several components of the farm portfolio model. One set of effects is to reduce the risk of loss for returns to farm assets for many farmers, and thus increase the expected level of these returns. It should also indirectly affect farmers' financing by stabilizing and reducing the costs of borrowing from commercial lenders. Moreover, the commercial lenders' response to farmers' use of new insurance programs could differ from their past responses to emergency credit programs. Emergency credit is generally keyed to specific crises, thus creating uncertainties about the continued availability of these funds.

Farmers' levels of risk aversion may also change with changes in their wealth, risk, or other sources of well-being, thus adding further complexities to the analysis. Finally, the recent increase in volatility of interest rates brings a new source of financial risk to farmers that has no historic basis for measuring or for assessing feasible portfolio responses. Again, the most promising approach appears to involve the lender and farmer working together to build wider safety margins into the farm's performance, reduce the rate uncertainty, and still provide the financial capacity for carrying on the operation and responding to the more conventional business risks. In all these cases the portfolio model provides a reasonable framework for identifying linkages between assets and liabilities under conditions of risk, and for designing programming, simulation, and other methods of empirical analysis.

Inflation and Micromodeling

Higher inflation rates in recent years have focused much interest on how inflation affects farmers' income, wealth, and liquidity, particularly for values of farmland and financial feasibility of farmland investments. The asset structure of farming is uniquely characterized by the dominance of farmland and the tendency for farm income to be capitalized into land values that appear high relative to current earnings. Profitability in farming has long been characterized by relatively low rates of current returns to farm assets and a tendency for land values and net worth to increase substantially over time. This situation has made farmland appear over-priced relative to its current earnings. It also makes highly leveraged land investments appear financially unfeasible when expected cash returns in early years fall short of the debt servicing requirements. The results are cash flow pressures, severe liquidity problems, and resort to numerous defensive practices by farmer and lender alike.

Valuation Concepts and Liquidity

New understanding of valuation concepts under inflation and new empirical evidence suggest that farmland likely is more reasonably priced relative to its recent patterns of earnings growth than is commonly believed (Melichar; Harris; Tweeten; Robison and Brake). These developments indicate that unrealized capital gains in land and other assets should logically be treated as part of the total returns to farm assets. In addition, Martin Feldstein has shown how current practices in income taxation—cost based depreciation and ordinary income versus capital gains—may cause values of nondepreciable assets like land and gold to rise faster than values of other assets when inflation rates are rising. These relationships between earnings growth, inflation, and capital gains mean that modifications of financing programs and debt repayment plans may be very appropriate responses to inflation-induced liquidity problems.

The concepts are similar to the growth stock phenomenon in equity markets. They are rather complex, but can be illustrated in a simple asset pricing model under inflation following Fisher's capital theory approach. Fisher suggested that a risk-free market interest rate is composed of a risk-free rate for time preference (r_t) and a premium (i_f) for anticipated inflation. This relationship between interest rates and inflation has had much study and, while the results are mixed, the "Fisher Effect" appears widely accepted (Wood; Friedman). Furthermore the relationship is important to asset valuation in which an asset's current market value is estimated by capitalizing its flow of expected future earnings at an appropriate interest (or capitalization rate).

First, define the following variables and notation: P_n = net earnings attributed to the farmland in period n; A_n = asset value in period n; D_n = value of debt; E_n = value of equity; r_t = average real cost of capital where $r_t = (r_e)(E/A) + (r_d)(D/A)$; r_e = real cost of equity; r_d = real cost of debt; i_f = a premium for anticipated inflation; and g = a real rate of change of earnings flow P. Variable g accounts for earnings that respond to inflation at different rates than the inflation premium (i_f) in the interest rate. Hence, g is an incremental rate that is anticipated for the income flow in addition to its growth at rate i_f. Positive $(+g)$ and negative $(-g)$ values then indicate the margin and direction of difference between the asset's (land) earnings growth and the general inflation rate. The valuation model for a perpetual income flow then is expressed as

$$A_0 = \frac{P_0(1+i_f)(1+g)}{(1+r_t)(1+i_f)} + \frac{P_0(1+i_f)^2(1+g)^2}{(1+r_t)^2(1+i_f)^2} + \frac{P_0(1+i_f)^3(1+g)^3}{(1+r_t)^3(1+i_f)^3} + \cdots \quad (13)$$

Cancelling the expressions for i_f and applying procedures of the Gordon growth model (Copeland and Weston) reduces (13) to

$$A_0 = \frac{P_0(1+g)}{r_t - g} \quad (14a)$$

as long as $g < r_t$ and $g > 0$. When $g < 0$, the expression is:

$$A_0 = \frac{P_0(1-g)}{r_t + g} \quad (14b)$$

The asset value is then found by dividing the asset's real earnings in period 1 by the real capitalization rate which is further adjusted for the asset's own change in real earnings. Suppose for example that an acre of farmland is expected to yield a real rent of $120 in perpetuity

($g = 0.00$) and that the real capitalization rate, net of general inflation, is 0.04 or 4 percent. Then the land value is $3,000 per acre.

In expression (14a) the presence of real change, g, has the effect of partitioning real return, r_t, into a current real return plus a real capital gain.

$$r_t = \frac{P_0(1+g)}{A_0} + g \qquad (15)$$

Higher rates of growth of real payments mean that a higher proportion of real return, r_t, is composed of capital gain (g) and a lower proportion as current return. Finally, if the valuation process is repeated a year later, the asset value A_1 is found to increase at the compounded rates of inflation and real growth.

$$A_1 = A_0(1+i_f)(1+g) \qquad (16)$$

Under these conditions, an asset's total nominal rate of return is $r_t + i_f + r_t i_f$, which consists of both a current return plus a nominal capital gain. The asset's real return is r_t, which may also consist of both a current return plus a capital gain (or loss). The capital gain's portion increases as real earnings increase.

These concepts about asset values and earnings growth lead to a perplexing paradox about efforts to increase earnings in agriculture. Suppose, for example, that a technological breakthrough in productivity causes real farm earnings to grow at a faster rate. Or, earnings growth could accelerate as a result of changes in government policies or more substantial international trade agreements. The following effects might occur: (1) the total rate of return on farm assets will increase; (2) the faster growth of farm earnings is capitalized into higher asset values; (3) current rates of return to farm assets decline; (4) capital gains increase; and (5) liquidity problems are worsened, especially on new investments by highly leveraged farmers.

What does the empirical evidence show about past rates of earnings growth for farmers? E. O. Melichar, who applied the asset pricing model to aggregate data for the farm sector, observed that farmers' returns to assets experienced real growth of about 4 percent a year over the past two decades and prices of farmland increased to reflect this real growth plus the general inflation rate. Similar evidence in Illinois comes from comparing growth rates for various types of farm earnings using data from the Illinois Farm Business Farm Management Association (FBFM). These data are for FBFM grain farms that fell in the 340–500-acre size class in the 1960–80 period. Table 6.2 reports geometric means of annual growth rates for four measures: (a) current returns to farm assets, (b) current returns to farmland, (c) values of Illinois farmland, and (d)

TABLE 6.2
Annual growth rates, Illinois data, 1960-80

Period	Land values	Current returns to		Consumer price index
		Farm assets	Farm-land	
1960-80	9.98	8.13	6.13	5.25
1960-70	4.19	4.46	2.49	2.74
1971-80	16.09	11.93	10.00	7.82

general inflation as measured by the Consumer Price Index.[4] The growth rates for land values, asset returns, and land returns match up closely on a before-tax basis. The average annual growth rate for land values was 9.98 percent over the 20-year period, compared with 8.15 percent for asset returns and 6.13 percent for land returns. All these figures exceed the average annual inflation rate, showing real growth in their respective series. Similar relationships occur for the 1970s, but with a greater margin between the growth rates for land values and current returns. As expected, the annual changes in the series of current returns showed much more variability than annual changes in land values.

Now consider the liquidity problems that arise under inflation. Two types of liquidity problems may arise for leveraged investors who purchase farmland subject to expected earnings growth. One liquidity problem occurs because current earnings are a smaller proportion of the return to equity compared with capital gains. The other liquidity problem arises from differences in the kind of compensation required by lenders and experienced by equity holders of farm assets. Lenders require all of their scheduled compensation (interest and principal) in a cash payment part of which is a real return and part of which is an inflation premium needed to compensate for anticipated loss of purchasing power in their debt claim. The borrower, however, experiences only part of the return as a current payment; the rest is capital gain. Thus, the borrower experiences a financing gap by having insufficient cash from the assets' early earnings to meet the debt payments. The borrower's equity position is growing favorably, but not the cash position.

For the case where the earnings flow and interest rates on debt respond equally to inflation ($g = 0.00$), negative current returns in the first period will occur when:

$$P_0(1+i_f) - (r_d + i_f + r_d i_f)D < 0$$

This condition can be expressed through the equity-to-asset ratio:

$$\frac{E}{A} < \frac{i_f}{r_e + i_f + r_e i_f}$$

Negative current returns are avoided only if the equity-to-asset ratio exceeds the ratio of the inflation rate to the nominal cost of equity.

Modeling and Policy Implications

These valuation concepts, the empirical evidence, and the resulting liquidity problems for farmers indicate the importance in micromodeling of accounting for the combined influences of current returns and capital gains on farmers' total returns. The past pattern of earnings growth along with the dominant role of fixed assets (especially) land in agriculture make it logical to expect financing gaps and liquidity problems on new farm investments financed by borrowed capital. The magnitude of these problems depends on the relationship between farm earnings, inflation, and interest rates; the efficiency with which markets capitalize asset returns into asset values; and the leverage positions of the investors involved. Including these inflation-related phenomena in micromodels should provide a richer setting for understanding farmers' responses to inflation and for enhancing the quality of decisionmaking in inflationary times.

Moreover, these inflation phenomena have important implications for financial analysis in agriculture, for policy considerations, and for innovations in financing programs. For financial analysis, asset markets have rewarded farmers for their growth in productivity and earnings by significant levels of capital gains; these gains are important sources of wealth and long-term borrowing capacity, but are also a cause of illiquidity in cash flows, especially in growing farming operations. Thus capital gains should be considered an integral part of farmers' total returns, rather than treated as separate phenomena largely unrelated to financial performance. Similarly, the analyses of changes in public policies affecting earnings growth for farmers should consider how the possible effects on capital gains are distributed among farms differing in their tenure positions, level of leverage, enterprise mix, and other structural characteristics.

Micromodels may also be useful in evaluating alternative financial policies under inflationary conditions. An example is the possible modification of repayment policies on debt so that farmers can maintain a desired capital structure while still meeting their financial obligations. At one extreme this could be accomplished by indexing the principal of the loan to inflation so that the increase in loan balance offsets the inflation premium in the loan rate. This practice would maintain the real positions of the lender and borrower, and remove the need to divert other earnings to meet loan payments, to arrange short-term loans to meet long-term payments, to bail out the farm operations through periodic refinancing, emergency loans, or other crises responses. An alternative to inflation indexing is to specify full amortization of term debt according to an inflation-adjusted repayment scheme: graduated payments, purchasing power mortgages, shared appreciation mortgages, and so on. These plans may be more administratively feasible to lenders, since they make full amortization of the loan instead of permanent indebtedness. Micromodels could be used to analyze the effects of such plans on both lender and borrower in order to achieve more effective responses to the problems of inflation.

Summary and Conclusions

As this paper has shown, the new sources of risk from instabilities in financial markets along with a new inflationary environment have combined with the traditional sources of business and financial risk in agriculture to bring new directions in managing risks and liquidity. These financial responses are important because of the prominent role of credit for many farmers in financing their business and in providing a source of liquidity to cope with swings in farm income. This new environment in agriculture, in turn, requires a further enrichment of the financial components of micromodels to more fully reflect the effects of financial risks and inflation.

The portfolio model established here illustrates a conceptual framework for expressing an organization of assets and liabilities that is optimal in utility terms for risk-averse investors. Its components (decision criterion, choices in assets and liabilities, sources of risk and return) characterize any type of decision situation; they can be refined, extended, or respecified to fit the particular situation. Moreover, the framework can be used in behavioral or prescriptive analysis to evaluate how changes in a farm's economic, financial, or policy environment influence its organization of assets and liabilities, as well as the effectiveness of various methods for managing risk and liquidity. Especially important are the possible trade-offs that may occur as farmers respond to changes in business or financial risks caused by changes in policy and other factors.

Similarly, the insight provided by micromodeling on how inflation and growth in farm earnings influence the balance between farmers' current income and capital gains should aid in evaluating farmers' liquidity, the design of financing programs, the feasibility of agricultural investments, and the effects of public policies on earnings growth.

Microanalysis of both the risk and inflation phenomena should prove useful in the consideration of farm policies and aggregative analysis. Moreover, the need to combine in the same model the effects of inflation and risk with farmers' management of assets and liabilities is a challenging task for further research.

Acknowledgments

The author gratefully acknowledges the helpful comments and suggestions provided by Lyle Schertz and Stephen Gabriel of the Economic Research Service, U.S. Department of Agriculture.

Notes

1. The concept of risk-efficient sets is often demonstrated using a graphic approach (Van Horne).
2. Fama shows a graphic approach to modeling the effects of short sales of a risky asset.
3. Included in the set of choices are decisions about rates of consumption, withdrawals for tax obligations, and leasing as well as ownership of capital assets.
4. "Asset returns" is a residual figure found by subtracting charges for the farmer's management returns from the FBFM published figure for capital and management earnings per acre. The operator's labor is charged as an opportunity cost, thus a fixed charge on the price that operators place on their own labor in determining the net earnings attributable to their assets. "Land returns" go farther by deducting from asset returns a charge for returns to non–real-estate assets, figured at the U.S. Treasury Bill rate.

References

Baker, C. B. "Credit in the Production Organization of the Firm," *American Journal of Agricultural Economics*, Vol. 49, 1968, pp. 507–21.
———. "A Variable Amortization Plan to Manage Farm Mortgage Risks," *Agricultural Finance Review*, Vol. 36, 1976, pp. 1–6.
Barry, P. J. "Theory and Method in Firm Growth Research," *Economic Growth of the Agricultural Firm*. TB-86. Washington State Univ., College of Agriculture Research Center, 1977.
———, and C. B. Baker. "Financial Responses to Risk," *Risk Management in Agriculture*, forthcoming.
———. "Management of Firm Level Financial Structure," *Agricultural Finance Review*, Vol. 37, 1977, pp. 50–63.
———, and D. R. Willman. "A Risk Programming Analysis of Forward Contracting with Credit Constraints," *American Journal of Agricultural Economics*, Vol. 58, 1976, pp. 62–70.

————, C. B. Baker, and L. R. Sanint. "Farmers Credit Risk and Liquidity Management," *American Journal of Agricultural Economics,* Vol. 63, 1981, pp. 216–27.

Boehlje, M., and S. Griffin. "Financial Impacts of Government Support Price Programs," *American Journal of Agricultural Economics,* Vol. 61, 1979, pp. 285–96.

Boussard, J. M. "Time Horizon, Objective Function and Uncertainty in a Multi-Period Model of Firm Growth," *American Journal of Agricultural Economics,* Vol. 53, 1971, pp. 467–77.

Cocks, K. D., and H. O. Carter. "Micro-Goal Functions and Economic Planning," *American Journal of Agricultural Economics,* Vol. 50, 1968, pp. 400–11.

Dorfmann, R. "An Economic Interpretation of Optimal Control Theory," *American Economic Review,* Vol. 59, 1969, pp. 817–31.

Fama, E. F. *Foundations of Finance.* New York: Basic Books Inc., 1976.

Friedman, B. "Price Inflation, Portfolio Choice, and Nominal Interest Rates," *American Economic Review,* Vol. 70, 1980, pp. 32–48.

Gabriel, S., and C. B. Baker. "Concepts of Business and Financial Risk," *American Journal of Agricultural Economics,* Vol. 62, 1980, pp. 560–64.

Gardner, B. L. "The Farmer's Risk and Financial Environment Under the Food and Agricultural Act of 1977," *Agricultural Finance Review,* Vol. 39, 1979, pp. 123–41.

Hanson, G. D., and J. L. Thompson. "A Simulation Study of Maximum Feasible Farm-Debt Burdens by Farm Type," *American Journal of Agricultural Economics,* Vol. 62, 1980, pp. 727–33.

Harris, D. "Land Prices, Inflation, and Farm Income: Discussion," *American Journal of Agricultural Economics,* Vol. 61, 1979, pp. 1105–07.

Hirshleifer, J. "On the Theory of Optimal Investment Decision," *Journal of Political Economy,* Vol. 66, 1958, pp. 329–52.

Irwin, G. D. "A Comparative Review of Some Firm Growth Models," *Agricultural Economics Research,* Vol. 20, 1968, pp. 82–100.

Lee, W. F. "Some Alternatives to Conventional Farm Mortgage Loan Payment Plans," *Canadian Farm Economics,* Vol. 14, Feb.–Apr. 1979, pp. 12–20.

Levy, H., and H. M. Markowitz. "Approximating Expected Utility by a Function of Mean and Variance," *American Economic Review,* Vol. 69, 1979, pp. 308–17.

Mapp, H., and others. "Analysis of Risk Management Strategies for Agricultural Producers," *American Journal of Agricultural Economics,* Vol. 61, 1979, pp. 1071–77.

Markowitz, H. M. *Portfolio Selection.* New York: John Wiley & Sons, Inc., 1959.

Melichar, E. O. "Capital Gains Versus Current Income in the Farming Sector," *American Journal of Agricultural Economics,* Vol. 61, 1979, pp. 1085–92.

Robison, L. J., and P. J. Barry. "Portfolio Adjustments: An Application to Rural Banking," *American Journal of Agricultural Economics,* Vol. 59, 1977, pp. 311–20.

Robison, L. J., and J. R. Brake. "Inflation, Cash Flows and Growth: Some Implications for the Farm Firm," *Southern Journal of Agricultural Economics,* Vol. 12, Dec. 1980, pp. 131–38.

————. "Applications of Portfolio Theory to Farmer and Lender Behavior," *American Journal of Agricultural Economics*, Vol. 61, 1979, pp. 158–64.

Tobin, J. "Liquidity Preference as a Behavior Toward Risk," *Review of Economic Studies*, Vol. 25, 1968, pp. 65–86.

Tsiang, S. C. "The Rationale of the Mean-Standard Deviation Analysis, Skewness Preference, and the Demand for Money," *American Economic Review*, Vol. 62, 1972, pp. 354–71.

Tweeten, L. "Macroeconomics in Crisis: Agriculture in an Underachieving Economy," *American Journal of Agricultural Economics*, Vol. 62, 1980, pp. 853–65.

University of Illinois, Department of Agricultural Economics. *Risk Analysis in Agriculture: Research and Educational Developments.* AE-4492. 1980.

Van Horne, J. C. *Financial Management and Policy.* 5th ed. Englewood Cliffs, N.J.: Prentice-Hall, Inc., 1980.

Washington State University, College of Agriculture. *Economic Growth of the Agricultural Firm.* TB-86. 1977.

Weingartner, H. M. "Capital Rationing: N Authors in Search of a Plot," *Journal of Finance*, Vol. 32, 1977, pp. 1403–32.

Wood, J. H. "Interest Rates and Inflation," *Economic Perspectives*, Federal Reserve Bank of Chicago, May-June, 1981, pp. 3–12.

7. Tax and Other Legal Considerations in the Organization of the Farm Firm

J. W. Looney

Abstract A variety of legal and tax considerations are involved in the selection either of the partnership or the corporation as a form of business organization for the firm. These factors are also important in the operation of the business. Therefore, the ultimate effect of choosing a particular organizational form is crucial in evaluating firm-level decisionmaking and related models.

Keywords Farm corporations, farm partnerships

The farm firm may be organized in a variety of ways. It may be operated as a sole proprietorship, a simple partnership, or a small business corporation; it may be operated as a limited partnership, multicapitalized corporation, or in a form that is a combination of the basic organization forms, such as a partnership of corporations. When leasing and trust arrangements are also used as business organizational devices, the farm firm can become quite complex.

When models are developed for effective farm-level analysis, the choice of organizational structure is a crucial element. In addition, the operator's reasons for selecting a particular form of business organization are important. Consideration must be given not only to the reasons for the choice of a particular business organizational structure but also the effect of that choice on behavior and economic activity once it has been made. This will be particularly true when an elaborate combination of business organizational forms is used as part of a business and estate plan.

The author is dean of the University of Arkansas School of Law and a member of the Arkansas, Missouri, and Virginia Bars.

Problems and Perspectives

The organization of the farm firm is closely related to estate planning. Farm business planning and farm estate planning are interrelated, and in many families the selection of a particular form of business organization may be dictated by the family's estate planning objectives. The farm family will have a number of varied objectives related to farm estate and business planning. A major, though not necessarily the only, objective will be that of minimizing taxes associated with the transfer of assets to the next generation. Most families involved in operating a farm also have as an objective the continuity of the business into subsequent generations. They are concerned with an equitable distribution of these assets and with security of income during their retirement years.

There are a number of major problems encountered in meeting these objectives in the farm firm. Of foremost concern is the potential erosion of capital that can occur in order to pay taxes and estate settlement costs and to provide for the farm's continuity in the hands of the on-farm heirs. A second major problem is the illiquidity of the assets, particularly land. A related problem is the indivisibility of assets and the sometimes disastrous consequence of dividing the farm into uneconomic units. The problems of illiquidity, and to some extent, indivisibility, have been addressed by recent tax legislation. Both the 1976 Tax Reform Act and the 1981 Economic Recovery Tax, Act addressed the problem of illiquidity and provided some relief for many farm firms by providing for special valuation of farmland for estate tax purposes and for the optional deferral of estate tax payments.

Other major problems in meeting the objectives of the farm family are those arising from the possible inequitable treatment of off-farm heirs if a decision is made to continue the farm business. This consideration must be balanced with possible inequitable treatment of on-farm heirs in the absence of well-conceived plans for the continuity of the business. Furthermore, many farm operations will be unable to produce sufficient income for 2 or more families. Other families face the problem of an inability to resolve questions of decisionmaking authority. Because of the difficulty in resolving some of these issues, these problems must be considered at the earliest stage in organizing the farm business to avoid inappropriate organizational structures.

The Partnership as a Business Organizational Structure

Many farm families find it advantageous to participate in a partnership arrangement. Typically, partnerships evolve from an employer-employee arrangement or a profit-sharing arrangement between family members.

The partnership provides a way of combining resources for a more efficient operation. It also serves as a way to plan business continuity and is a very attractive estate planning device for many families.

Partnerships are used as estate business planning devices to accomplish a variety of goals. The partnership may initially be used as a trial arrangement to determine if the operation can sustain the involvement of more than one family and to determine if the parties are compatible. Since the Uniform Partnership Act, in effect in most states, permits considerable flexibility in forming, operating, and dissolving the partnership, the partnership is an attractive vehicle as a trial organization.

The partnership may be chosen as a means of allocating income to younger family members in lower tax brackets and as a means whereby younger family members can accumulate capital to become involved in the farming operation. It can serve as a vehicle for the gradual transfer of farm assets to the younger generation and as a way of freezing the value of the interest of the parents. Appreciation in value may be shifted to the younger generation while reducing transfer taxes.

Since a primary objective in using the partnership is to provide for the continuation of the business following retirement or death of parents, a continuation clause should be a part of every partnership agreement. This is usually in the form of a buy-sell agreement, which will set the terms for transfer of the business. The continuation clause may be either mandatory or optional and may be funded either by insurance or through installment payments built into the buy-sell agreement. Such agreements may set a price for the business interest or contain a formula for determining the price or provide for an appraisal.

A major problem involved in business continuity arrangements is the valuation of partners' interest at death. The price stated in the buy-sell agreement is not necessarily determinitive for estate tax purposes; thus, it may become necessary to make an evaluation on an item-by-item basis.

Continuity arrangements may provide security for a retiring partner. For example, a buy-out agreement with installment payments can be an important source of retirement income. The agreement may also provide that a retiring partner is to receive a share of profits in liquidation of the parent's interest. In either case, the existence of the arrangement affects the income tax status of the partnership and of the individual involved.

A number of major legal and tax considerations are involved when the partnership is chosen as the organizational structure for the farm firm. Many of these come about during the planning stage. First is the question of determining the proportions of contributions to the partnership to be made by each partner. This determination involves the question of whether the property is to be contributed to or used by the partnership. A related question is how payments for capital contributions, labor, and management are to be determined. Often, the business continuity objectives of the family firm will be defeated if the arrangement fails to permit younger family members to acquire an interest in appreciating assets, especially land. Therefore, consideration must be given to the question of determining how property is to be owned by the partnership and what property is to be involved in the partnership arrangement.

Second, several tax problems may exist when the partnership is formed. For example, contribution of property to the partnership may generally be made without the realization of taxable gain or loss by partners or the partnership. However, special problems may be involved if a transfer is made of assets subject to liabilities or when a partner's capital interest in the partnership differs from the value of the contribution of property made.

Basis considerations are also crucial at formation. The partnership's tax basis is essentially the owner's adjusted investment in the property and is important in determining gain or loss upon sale of the asset. The partnership's tax basis in the property is the same as the basis of the property in the hands of the partners. Once the partnership is formed, each partner has a basis in his partnership interest equal to the original basis on contributions adjusted by subsequent contributions, distributions of property or money, partnership losses, partnership liabilities and other items. A record of the basis of each partner in the partnership and of the partnership in each asset is important in order to determine the tax effect of transfer of the partnership interest by a partner. It is also important upon transfer of an asset of the partnership. The tax basis is particularly crucial in determining the tax effect at dissolution of the partnership. Although dissolution is not necessarily anticipated by the parties, at least some plans should be made at the time of formation and should be considered during the operational life of the arrangement in case dissolution should occur at a future time.

Effect of Selection of the Partnership Form

The selection of the partnership form of organizational structure affects a number of firm-level decisions. These should be taken into account when modeling the farm firm. Because the partnership is not a tax-paying entity, models must be able to trace income and related deductions, tax credits, and depreciation (or cost recovery as it is known under the 1981 Tax Act) to each of the individual partners to accurately determine the tax status of the business. This procedure is complicated by the number of partners involved, their individual tax brackets, and the allocation of income to each.

Income allocation in the partnership does not necessarily depend on the proportion of contribution of property to the business by the partner receiving the income. Often, partners receive a share of profits based on contributions of labor and management rather than on contributions of property. Information on both the relative contributions of each partner and their rights to share income allocation is important to accurately analyze the tax status of the arrangement. This analysis is further complicated by the presence of a buy-sell agreement in which a retired partner or the surviving spouse of a retired partner is receiving a share of profits in liquidation of a prior interest.

Limited Partnerships

Some farm families have turned to the limited partnership as an alternative form of business organization. This has occurred because of some of the problems that arise from the use of the partnership entity related to the unlimited nature of the liability of general partners who, under partnership law, are individually liable for all debts and obligations of the partnership. The limited partnership is designed to protect partners whose only involvement is investment of capital and who do not expect to take an active role in managing the business. Limited partnerships are potentially useful in family situations involving retired landowners who no longer wish to have an active role in the business operation, but who wish to be compensated for the use of the land. The limited partnership is also useful as a business continuity device. For example, in those families in which there are off-farm heirs, the limited partnership may be used as a property-holding device in which the off-farm heirs hold a limited partnership interest and the on-farm heirs, or the parents, are general partners and are actively involved in managing the business.

When the limited partnership is chosen as an organizational structure, firm decisionmaking is altered. In the simple partnership, decisionmaking authority belongs equally to all partners unless the agreement provides for specific authority in certain partners. In the limited partnership, the general partners must have management control, so decisionmaking is shifted away from those who have only an investment interest in the business. Additionally, the limited partnership may provide for a sharing of profits which gives preferred treatment to the holders of the limited partnership interest. These decisions affect the income and tax status of the individuals involved and must be accurately reflected in farm-level models.

The Corporation as a Business Organizational Structure

A decision to incorporate the farm firm is often related to the ease of transferring interests in property and to the possibility of achieving continuity in management and ownership. Also, some families have found that they can achieve income tax savings by using the corporate form. Limited liability, possible improvement in credit status, and employee benefits for owners are further considerations. Although the number of farms that have been incorporated remains relatively small, there is a growing interest in their use, particularly for estate and business planning purposes and for achieving income tax flexibility.

Income Tax Considerations in Incorporating

By careful planning, individuals can realize income tax savings in the regular corporation ("Subchapter C corporations"). This is true particularly under the provisions of the Economic Recovery Tax Act of 1981 in which the income tax rates for regular corporations were lowered. For 1982, the federal income tax rates imposed on the Subchapter C corporation were as follows: for the first $25,000 of taxable income, 16 percent; for the second $25,000 in taxable income, 19 percent; for the third $25,000 in taxable income, 30 percent; for the fourth $25,000 in taxable income, 40 percent; and all taxable income over $100,000, 46 percent. For 1983 and thereafter, the tax on the first $25,000 will be lowered to 15 percent and on the second $25,000 to 18 percent. All other rates remain the same. Given these rates, some families may find that additional income tax savings can be realized. In the regular corporation, the tax rates are imposed after deducting all business expenses. Reasonable salaries paid for work actually performed are an expense deduction to the corporation. The same rule applies to reasonable rental payments on land or equipment and to interest payments on debts of the corporation. This means that with proper planning of these deductions, the family may be able to limit the income taxation to the lower ranges of the corporate brackets.

Residual earned income distributed to shareholders as dividends is taxable at the shareholder level—the so-called double taxation of corporate profits. Subchapter C corporations are permitted to accumulate a sizable reserve of retained earnings without the imposition of the accumulated earnings tax. This tax is a special tax on amounts that are not distributed as dividends. This reserve may permit some protection of income from the imposition of either taxes on dividends or taxes on accumulated earnings. The 1981 act increased the amount that may be accumulated from $150,000 to $250,000 without imposition of the accumulated earnings tax. Once that level is reached, additional accumulations may be justified by the reasonable needs of the business. Again, some tax savings may be achieved through the proper utilization of the accumulated earnings tax exemption.

In addition to the potential tax planning flexibility in using the regular corporation, one advantage of using the corporate form for the family farm business is that there is an optional tax status available by election. This permits the small family corporation (Subchapter S corporation) to be taxed as if it were a partnership. Thus, income, deductions, and other items can be passed through to the individual partners without the corporation paying tax. The corporation pays salaries, interest, rent, and other expenses as usual and then, for tax purposes, passes through to the shareholder a pro rata share of undistributed taxable income, long-term capital gains, operating losses, and a number of other tax items. This option has been expanded by the 1982 subchapters and is available for corporations that have 35 or fewer shareholders. Under prior law, the corporation had to have 15 or fewer shareholders to qualify. All shareholders must be individuals or the estates of individuals and must consent to this election. To elect this tax treatment, the corporation may have only one class of stock outstanding and at least 80 percent of the corporation's gross income must be from actual operation of the business. The 1981 act made it possible for certain types of trusts to hold stock in the Subchapter S corporation, producing additional flexibility when the tax option status is used.

Estate Planning Considerations in Incorporating

The corporation may be used as an attractive estate-planning tool. It may serve as a vehicle for gifts, as a means of maintaining the economic unit intact, and as a way of providing employee benefits to owners. The transfer of assets to a corporation with a simple capital structure permits transfers of stock to younger family members, including minors, yet allows retention of control of the corporation by the donor. In more complex arrangements, the use of debt securities such as notes, bonds, and debentures, as a part of the corporate capitalization can serve as gift instruments to off-farm heirs or can provide income to parents during retirement. Such techniques can provide a cap or freeze on the value of the estate and thereby mitigate appreciation of asset values. Capitalizing the corporation with both preferred and common stock may provide for continuity and shift of control and may assist in the implementation of buy-sell agreements.

These techniques allow the transfer of specific assets. Some protection for off-farm heirs and retired shareholders is also achieved; the use of debt securities or dual capitalization techniques provides a means whereby off-farm heirs or retired shareholders may hold a valuable interest in the corporation, while management authority and operating control are left with the on-farm heirs. Buy-sell and first-option agreements can be included in the bylaws or in a separate agreement to create a market for the stock of the off-farm heirs. These protections are sometimes necessary because minority shareholders may find that they hold shares of stock in a corporation that have little market value. The stock may be unmarketable because of transfer restrictions, lack of management rights, or because dividends are seldom declared in a small closely held corporation.

Additional opportunities for retirement planning are created by the fact that employee status for the owners permits the use of benefit programs such as health and accident plans, deferred compensation plans, group life insurance programs, and possibly even greater social security benefits, although costs will usually increase.

Considerations at Incorporation

Incorporation of the farm firm can normally be made tax-free. This means that assets can be transferred to the corporation in return for shares of stock and that no tax will be incurred by either the owners making the transfer or by the corporation itself. To achieve such an exchange, property must be transferred solely in exchange for stock or securities in the corporation and the transferors must be in control of the corporation immediately after the exchange. This means that they must obtain at least 80 percent of the combined voting power.

An additional tax consideration involves the basis of the stock received and of the property transferred in exchange for stock. The basis of stock received in exchange is the basis of the property transferred less any additional cash payment received plus gain recognized, if any. The corporation's tax basis for property received is the transferor's basis plus the amount of gain recognized, if any. The basis considerations are extremely important if the property or the corporate stock is ever transferred. These considerations may become particularly important at the dissolution of the corporation. The income tax consequences of dissolution will vary, depending on the exact procedure chosen; however, a good record of basis for both the property and the corporate stock is extremely important.

A number of special considerations are involved in determining whether or not a farm should be incorporated. For example, the question of whether or not the land should be transferred to the corporation is crucial. Land represents a major portion of the estate and is the asset with most rapid appreciation. Land is important as collateral for production credit. However, more flexibility exists if land is retained outside the corporation as it is easier to get spendable dollars to the owners (in the form of rent) and double taxation can be avoided upon sale of the land.

A second major question is whether or not the farmstead and residences should be owned by the corporation. If an occupant-shareholder lives in the residence, as they usually do, some additional tax effect may result. In some cases, maintenance expenses may be deductible and the taxpayer may not be required to consider the use of the property as income. This could occur in those circumstances where the corporate employer requires the presence of the employee on the premises at all times and in other limited circumstances.

A related question arises concerning whether or not automobiles should be owned by the corporation. There is the possibility of corporate liability for claims arising out of the personal use of the corporate automobile. Also, personal use of the business automobile is not a business deduction and may be considered taxable income to the employee.

There is also the question of how credit status will be affected by incorporation. Credit through Federal Land Bank, Production Credit Association, and the Farmers Home Administration is not prohibited to farm corporations. However, there are certain restrictions that must be met and the credit status must be thoroughly considered before incorporating. For example, production credit associations and federal land banks limit loans to legal entities in which 50 percent or more of the voting stock or equity is controlled by an individual conducting an agricultural operation or to one in which 50 percent or more of the value of its assets relates to agricultural production or 50 percent or more of its income originates from agricultural production.

Effect of Selection of the Corporation

The selection of the corporate form affects firm-level behavior and decisionmaking in several significant ways. First, the corporation is a legal entity which exists separate and apart from the owners. This means that final decisionmaking authority is in the hands of the directors of that corporation and individuals who are not active in the farming operation may have direct input into farm-level decisions particularly if retired shareholders are on the board. But, day-to-day operating control may rest with those who are only part owners of the business—those shareholders who are also employees.

The selection of the corporate form as a means of assuring business continuity, although done for purely nontax reasons, has an effect on both the business tax rate and the potential profits. The overall tax impact will be affected by the number of shareholders, the respective tax brackets of the shareholders and of the corporation itself, and the amount of income that is passed through to the shareholders in the form of salaries or dividends. Although it may be easy to determine that a business is incorporated and, therefore, to make certain assumptions regarding its tax status, the modeler must keep in mind that the ultimate financial impact will depend on the extent to which assets are transferred through the use of the corporate form.

If the model is only specified to assume that an incorporated business is a regular Subchapter C corporation, incorrect results may be obtained. For example, the dominant shareholder may be an employee and receive a major part of the total corporate net income as a salary. This situation would leave the tax liability very similar to the situation prior to incorporation. Or there may be a real effort to shift income to other shareholders. In this case dividend payments would play an important role in the individual's income situation and the taxes paid by the corporation.

The constructive dividend problem must also be considered. In those situations in which dollars are being transferred to shareholder-employees in the form of salaries, and if it is determined that such payments are unreasonable or for work that has not been performed, these payments may be considered dividends. The payments will be taxed both to the corporation (because the deduction is lost) and then to the shareholder as dividends.

Capitalization is another key element in determining the ultimate tax effect of the form of corporation. If the corporation has both common and preferred stock, a shift of income as dividends to preferred stockholders may occur. The same situation may occur when debt securities are used in the corporate financial structure. In these cases, the income of off-farm heirs or inactive shareholders may be an important factor when evaluating the ultimate tax effect of using the corporate form.

Multi-Entity Arrangements

In sophisticated estate and business planning arrangements, the farm family may find it desirable to make use of multiple entities. For example, some farm families use the corporation in combination with an irrevocable trust to which they make gifts of corporate stock for the benefit of minor children. This combination may shift income from the grantors to either the trust or the beneficiaries. The imposition of the trust arrangement may also significantly affect the tax status of the firm.

Some families use multiple corporations—one for landholding and another for business operations—or a partnership-corporation combination. When multiple entities are involved, evaluation of firm-level decisionmaking is greatly complicated. Both legal and tax constraints may be imposed on the parties and the typical model used to evaluate decisionmaking for basic farm operations will likely not be appropriate.

Summary and Conclusions

These examples illustrate the fact that realistic and accurate results may not be obtained when simulating and evaluating a farm firm's financial behavior if only simple assumptions regarding the organization structure are made. Scrutiny must be given to both the tax and nontax reasons for the selection of a particular entity, the problems associated with organizing in the particular form, and the ultimate results of that choice.

Discussion of Part 4

Dealing with Institutional
and Legal Information
in Models of the Farm Firm

Steve A. Halbrook

Institutional and legal information creates unique problems for the economist who builds models. The papers by Barry and Looney should be reviewed in terms of how they help to solve these unique problems. Neither paper offers any specific suggestion on this point, but both provide a background for this purpose. They bring three issues to my mind which I call the economist-lawyer conflict, the data availability problem, and the theory in a world of changing institutions problem. The model builder must resolve, or at least consider, these issues before using institutional or legal information in any model.

The economist-lawyer conflict is more than an identity crisis for those of us with some training in both fields. It is a conflict of approaches to problems that makes it difficult for lawyers to have significant input into economic model building.

Economists observe the economy and the behavior of firms and households with the objective of discovering a few general rules that explain the workings of the whole and its units. Theirs is an attempt to gain a general understanding from the observation of specific facts and to explain the complex by simple relationships.

In contrast, lawyers are constantly reviewing the general rules of law, looking for ways to solve a client's specific problems. In the extreme, the lawyer views every problem as unique. The Barry and Looney papers are (without attempting to be) perfect examples of this difference.

Peter Barry's paper is an excellent review of many recent attempts to develop and test hypotheses relating to the reaction of farm firms to changing economic conditions. In each case, he attempts to explain the complex by a series of simple economic relationships. Jake Looney's paper, by contrast, briefly explains the variety of forms of business organization that are available to farm firms and discusses the advantages and disadvantages of each. I am left with the impression that the

The author is an agricultural economist in the Dairy Division of the Agricultural Marketing Service, U.S. Department of Agriculture, and a member of the District of Columbia Bar.

decision to incorporate or form a partnership or execute a trust agreement is unique for each firm.

If we are to successfully integrate legal information into models of the farm firm, we must attempt to reconcile this conflict between inductive and deductive reasoning. Both the lawyer and the economist must evaluate the nature of the information in question and the capability desired in the model being constructed. Economists must realize that the legal information they wish to analyze or use as an explanatory variable is difficult to quantify. Consequently, economists may not be able to construct measurement variables that meet all the assumptions necessary to insure the integrity of the statistical technique being contemplated. Less sophisticated techniques may have to be accepted in order to accommodate the additional information. The implications for usefulness of the model must be carefully evaluated by the researchers.

The lawyer must be willing to sacrifice specificity. An economic model cannot include all legal implications of a decision. Computers presently available could not handle a model that large. Not even the Census of Agriculture contains enough data for testing purposes.

The lawyer must serve 2 functions in the model-building process. During the methodological development of the model, the lawyer must set priorities by identifying events and situations that have legal significance. Furthermore, the lawyer should help the economist construct variables that accurately reflect the impact of a statute or regulation, but that are general enough to offer the economist increased explanatory value in the model. For example, you cannot easily integrate the choice of organizational form into models of risk taking or portfolio analysis if you attempt to include every circumstance discussed by Mr. Looney. The treatment of off-farm heirs may not be important to the analysis, but the limited liability feature of a corporation may have great significance in economic, as well as in legal, terms. Since both estate planning and limited financial liability are motivations for incorporating a farm firm, it may take some careful reasoning to establish the appropriate way to integrate the legal information into models of the farm firm.

The second problem that comes to mind when one reads these papers is a very practical one, data availability. How do we quantify legal concepts and how do we collect data consistent with these concepts? There are at least 3 aspects of this problem: data that are impossible to obtain, data that can be obtained only at great expense, and data that are available, but not in usable form.

Foreign investment in U.S. farm land illustrates the problems of data availability. It is possible to discover the identity of the owner of record through a search of county land records. This process is slow and expensive. However, even this process will yield only the owner of record, which could be a corporation, partnership, or trust. Determining the ownership of an equitable interest in any of the above may be impossible without some statutory obligation to disclose that information. Even this information may not be meaningful in some instances because use of the land may be determined by a long-term contract that is not public information. The economist and the lawyer must work together to determine the relevant question and the data needed to answer the question.

My final thoughts are directed at the impact of institutional and legal changes on economic theories of behavior. The studies reviewed by Peter Barry hypothesize about how economic units react to change. Like all economic models, their construction involves making explicit and implicit assumptions that certain factors remain constant or change in a predetermined way. My question is: Do these assumptions remain valid when institutional and legal factors change, or do the models need to be constructed so that the relationships in the model adjust to be consistent with the institutional and legal factors?

The agricultural economy has undergone some dramatic legal and institutional shocks over the past decade. Many tax laws have been rewritten. Financial institutions are undergoing extensive change. Certain segments of the transportation sector have been deregulated. Do the old observed patterns of economic behavior work in this new legal world? I don't think this question can be answered by simply adding some dummy variables to "represent" these changes in a pre-existing model because changes in the institutional framework in which a market exists can alter economic relationships in the market. The shock effect of deregulation in the financial and transportation sectors may change decisionmaking criteria at the firm level which may force the model builder to add information to the model and retest the relationships among variables already included in the model. Therefore, one aspect of integrating institutional and legal information into models of the farm firm is discovering the extent to which our hypothesized relationships are altered by changes in legal institutions.

I continue to believe that the integration of legal and other institutional information into micromodels of the farm firm will make these efforts more useful to policymakers and farm operators, but the integration process is not without hazards. The problems I have raised will continue to plague the theoretical underpinning of our models, and the availability of data for their construction. However, if the lawyer and the economist can utilize each other's talents, the resulting models can provide lawyers, economists, and policymakers with improved insight into the impact of legal and other institutional changes on farm firms and the agricultural sector of the economy.

Discussion of Part 4

The Role of Financial and Legal
Concepts in Micromodel Building

Duane G. Harris

The purposes of this session, as implied by its title, are to identify institutional and legal variables that play a role in farm decisionmaking and to determine appropriate ways to incorporate those variables into micromodels. The papers presented by Barry and Looney offer useful summaries of relevant concepts in finance, taxation, and business organization. There are, however, some aspects of the session in general and of the papers in particular that could be strengthened. My remarks are divided into 3 categories: (1) comments on the organization of the session, (2) comments on the individual papers, and (3) additional observations regarding the role of financial and legal aspects in micromodel building.

Organization of the Session

If the purpose of the session is indeed to cover institutional and legal variables that are important in farm decisionmaking, then the selection of the papers falls short of accomplishing that goal. Or if the purpose is to examine financial models of the farm firm, then the title of the session is misleading.

Specifically, the makeup of the session implies that financial decisionmaking is "institutional" in nature. But, financial decisions are no more dominated by institutions than are marketing or production decisions. In fact, financial, marketing, and production decisions are made in an environment that is heavily dominated by institutional and legal forces.

Looney's paper certainly is in keeping with an examination of the institutional environment, whereas Barry's paper is less so. Looney outlines tax and legal considerations that affect farm firm decisions. Barry examines important analytical constructs that can be used to make farm firm decisions. Looney helps identify important institutional vari-

---

The author, formerly a professor of economics at Iowa State University, Ames, is currently manager of business analysis, General Mills, Inc., Minneapolis.

ables to be included in micromodels. Barry discusses the structure and form of equations that incorporate certain institutional variables and are consistent with some type of optimization process. The papers are quite different.

For the session to more completely identify important institutional and legal factors bearing on farm decisionmaking, it would require a consideration of additional topics. It might include the following topics: usury ceilings, pollution control regulations, tax incentives for investment in real capital, legislative programs to promote conservation, provisions of food and agricultural programs, or one of many others. If the session were intended to investigate building financial models of the farm firm, then it could have been more accurately titled. As the session is currently titled, Barry's paper stands out as a fine piece of work that appears misplaced.

Evaluation of the Papers

In spite of the fact that the 2 papers do not form a cohesive unit for addressing the issues of the session, they do make substantial individual contributions. However, both papers could be improved to assist in the task of model building of the farm firm.

Barry's Paper

A substantial portion of Barry's paper deals with portfolio theory and its application to farm firm decisionmaking. Although Barry presents a well-written summary of portfolio theory concepts, such a summary is already available in virtually every intermediate finance or investment textbook. Also, Barry devotes most of his efforts to only one aspect of the portfolio problem—that of generating an efficient frontier in mean-variance space. Little attention is paid to the problem of helping farmers choose among many points on that efficient frontier.

Barry is not alone in his relatively disproportionate attention to the generation of efficient frontiers. In fact, most work in agricultural economics that uses the mean-variance paradigm does just that. One reason for this emphasis is that generating efficient frontiers is easier than tackling the difficult problem of determining investors' risk-return tradeoffs. Those risk-return tradeoffs must be determined, either directly or indirectly, from knowledge about the investors' utility functions. These functions are much more difficult to estimate empirically than are efficient frontiers. But progress in modeling farm firm decisionmaking will be slow until we tackle head-on the utility function problem. Barry's paper could be improved if he were to give greater balance to the efficient frontier and utility function aspects involved with portfolio choices.

Looney's Paper

Looney's paper presents a useful review of legal forms of ownership, one that is especially helpful to a nonlawyer. Looney demonstrates that different forms of ownership imply different tax environments. Unfortunately, he stops at that stage, leaving us to figure out how to incorporate those tax environments into models. Looney does not address the issue of how tax policy and its changes bear on the day-to-day operating decisions of the farm firm. He emphasizes primarily organizational structure. But, decisions about organizational structure may be confronted only once in several decades. Looney could have improved the paper by paying more attention to tax policies and other legal restraints that affect operating decisions of the farm.

Observations on Financial Modeling

Although there have been attempts to develop general equilibrium approaches to financial modeling of the farm firm, the outlook for large-scale financial models is not promising. Much of the useful future research will be single-issue oriented, as has been the case traditionally, and it will deal with concepts such as capital budgeting, optimal capital structure, asset valuation, and portfolio selection. Since the agricultural finance literature lags the general finance literature by about 10 years, the researcher can make much progress in modeling agricultural firm decisionmaking by simply drawing from the work that has already been done in the general finance profession.

There is also substantial room for work that incorporates financial and legal considerations into micromodels, when decision variables are not financial variables. For example, cash flow and credit availability constraints may significantly affect the acquisition of factors of production. Income and capital gains taxes play a dominant role in firm-level decisions regarding resource allocation. And, interest rate levels and their volatility can substantially affect the behavior of the farm firm. These financial and legal variables must be correctly incorporated into micromodels to give meaningful positive and normative results.

More research is needed on the control of factors of production and on innovative methods for adapting our traditional institutions to exercise that control. The day when a farmer controlled all factors of production through ownership is clearly past. Owner and family labor have been supplemented by hired labor to accommodate larger scale farms. Owned land has been combined with rented land to take advantage of economies of scale in production. And, owned machinery has been combined with or replaced by leased machinery to afford the timeliness and flexibility necessary for modern crop and livestock operations. Problems related to maintenance of balance between labor, land, and capital and to reduction of the risk of loss of control of those factors require investigation of methods of long-term control other than by ownership. For example, research needs to be extended to analyses of land rental agreements, labor acquisition and retention programs, and machinery lease arrangements. Models need to be developed that will examine the profit and cash-flow implications of different ownership-leasing combinations.

Finally, we should not place too much reliance on our models to generate policy prescriptions in a rapidly changing world. First, it is difficult to build models with the capacity to handle massive structural shifts caused by such things as decontrol of the financial sector or dramatic shifts in tax policy. Furthermore, if parameters in our models are estimated from historical data, those parameters may be meaningless after structural shifts have taken place. If we insist on relying on models in such an environment, we always will be tempted to tinker with the models to produce "reasonable" results. We must recognize the limitations of model building.

Discussion of Part 4

Institutional Aspects of Firm-Level Models

George D. Irwin

My plan is to review for you some lessons from the previous generation of studies using firm-level models and to discuss the institutional aspects of some of the points raised by Barry and by Looney.

It was only about 20 years ago that the aggregate-supply response studies were in vogue in USDA and at the major land-grant universities. The general approach was to define a series of representative farm firms and use linear programming on each with various combinations of commodity prices to obtain synthetic supply response functions. These were then weighted by numbers of firms to build up aggregate supply response functions on each for various combinations of commodity prices. Thus, one phase of the studies was intended to address directly a policy issue of supply response and adjustment. This phase was of special interest to USDA and to university policy specialists. The results for individual firms were also interpreted as the types of price response adjustments that take place in typical firms. The representative firm analysis was intended to address the kinds of farm management-adjustment questions that were relevant to farm management extension (see Colyer and Irwin, for example). The postmortems on these studies were largely accomplished by the late 1960s. A major difficulty was the enormous staff-resource commitment required. Another was difficulty in obtaining credible aggregate estimates due to aggregation problems.

This group of studies yielded to another in which models were either focused more directly on growth of the individual farm firm through attempts to create multiperiod models or else they were strictly aggregate models on policy. No attempt was made to link the two. Initial efforts in the firm modeling area were with multiperiod linear programming, but these soon gave way to various sorts of simulations. Here, too, the interest was relatively short lived; it was also strictly an interest at the micro- or firm-level.

One residual effort from the earlier work was a study of how to modify farm accounting projects (microdata) to make the information useful in addressing aggregate questions (Irwin and Havlicek). The

The author is the director of the Economic Analysis Division of the Farm Credit Administration.

suggestions were to record data in ways that would permit purpose-specific identification of like-behaving farm firms and to attach census referents on farm type, which would produce a method of weighting from representative firms (to obtain aggregates). A similar technique might be a useful addition to micro-approaches to policy proposed by speakers at this conference.

The shock waves created by the volatile environment of the 1970s now seem to have moved us back toward the types of interests we had in the 1960s. We are again being asked at this conference whether micro-level models are needed alongside (more aggregate) macro-level models to address questions about policy. One difference is that no direct aggregation of results is implied. Why? The lesson in the Irwin and Havlicek article for recording data and for developing weights for types of farms might add a useful dimension. At a minimum, the need is to identify a model that is applicable to a series of firms, in the sense that they will behave alike over the range of policy parameters in which we are interested. This can be useful information, even if it is not possible to derive an estimate of the total or aggregate type of response to a change in that policy variable.

Both papers presented in this session provide a consolidation and exposition of knowledge, rather than an innovative breakthrough in modeling. They address the manner in which the current institutional and economic environment will force changes in the types of specifications of models that are appropriate to the future. Looney sticks closely to this agenda, whereas Barry occasionally slips into the use of micromodels for policy prescriptions.

Barry adopts the portfolio approach, considering the level and variance of return as the criterion for a decision on a static mix of assets and liabilities. His paper is more elegant than Looney's systematic review of estate planning and tax factors affecting the legal type of business organization, and it addresses a much more broadly based set of economic environments. However, Looney's paper reminds us of the many factors that may be overlooked in a portfolio approach, which are essential to the treatment of the full question. To Barry's list of production, credit, and financing decisions, we might add organizational-choice decisions. To Looney's paper, I would add emphasis on estate creation to parallel that on estate disposition, as well as something on linkages between the two.

The concentration on a one-period or static model, in Barry's case, subsumes interperiod differences in expected values and variances. The exception is his interesting section on land-price analysis. Barry focuses narrowly on a portfolio of risky farm-operating assets and negative holdings of riskless borrowing assets or risky variable-rate borrowing.

The diversification model needs to be extended to include nonfarm risky assets. Recent data show considerable acquisition of nonfarm risky assets by many farmers in addition to the more familiar form of nonfarm diversification—nonfarm employment. Low, and perhaps even negative, income covariances (in the short run) often provide a happy combination. Modeling reality must take this into account as a risk-response phenomenon.

Looney's paper makes a major point on multiple objectives in estate planning: maximize size of estate, minimize taxes, ensure business continuity, treat heirs equitably, and provide security of income for heirs or senior members. They are often incompatible with each other. His analysis considers the conflicts and tradeoffs involved, and he suggests that the choice of organizational structure makes a critical difference in the models of the firm. Looney also considers the legal aspects of continuing operations on multi-person operations. Here, too, there is surely an opportunity for goal conflicts. One wonders how Barry's neat theoretical approach could be brought to bear on continuity of operations of multi-person units. Such situations seem to have the neat, repetitive characteristics of his one-period model. I would also prefer that his model be expanded to think of adjustment in an encompassing sense, involving entry, growth, consolidation, decline, exit, and intergenerational transfer. Barry concentrates on the prime operating lifetime of full-time individual-proprietorship commercial farm operators—the growth phases. He focuses on those with high levels of financial leverage. Would generalizing Barry's conceptual framework to encompass other circumstances require shifts in utility functions over time, a change of emphasis on liquidity, or a consideration of feedback conditions affecting behavior?

I question Barry's implication that current adjustment strategies are just "bridging the gap" to a more favorable future environment. Both interest rates and farm prices are likely to remain volatile for the indefinite future. However, I do agree that a part of the current environment may be transitional because the level of relative indebtedness (financial leverage) may decrease. If we move to a more volatile environment and permanently lowered government "safety net," farmers must make transitional adjustments to lower leverage positions. Thus, I conclude that modeling will need to include more extensive risk parameters, both in the immediate future and in the long run. A consequence may be slower optimal growth paths for firms that are expanding.

Much casual attention has been paid to the subject of risk for borrowers paying variable interest rates to the Farm Credit System and, increasingly, to all other lenders as well. But, I would emphasize Barry's point that the "credit risks" may or may not add to financial risk. Barry notes that the effect on rate-of-return to equity depends on whether the covariance term between asset returns and interest rates is positive or negative, and large or small. If the forces move in the same direction, the result is stabilizing. As he notes, the "lender response" component of such risk is likely to be more predictable than the market-risk component. Some of our recent work indicates that the market-risk component of the variable rate added to fluctuations in income for certain farmers, but not for certain cooperatives.

We can hypothesize rather complex and differing relationships among these variables in the short and the long run. Increasingly, risks involve relative interest rates among countries, relative commodity prices, and the exchange value of the dollar. To the extent that there is an inflation premium in interest rates at certain times and to the extent that inflation is driven by food prices, the correlation tends to be positive. But there is a negative correlation if good weather gives us bountiful crops and inflation continues. Adding capital-gain returns on farmland, as Barry does, simply adds more complexity to the analysis. One could extend Barry's analysis one step further by adding futures markets—as forward contracts in either financial instruments or agricultural commodities. Another set of variances and programs would be introduced, as well as a third participant in the risk-sharing environment. There may be a potential for shifting at least a portion of the credit risk to third parties.

I am not comfortable with Barry's proposal to modify loan-payment policies as an answer to the small share of land returns accruing as cash, provided there is a compensating adjustment for an opposite-risk transfer caused by the inflation. I simply have no evidence on that point. How would the lender be able to assume the risk, if the risk component in the interest rate being charged is now quite small? What happens if the expectations of inflation or of capital gains were to decline, which may be the difference between Barry's 14(a) and 14(b) equations? Does one extract only "real gain" in equity or the entire nominal gain? Who now has the risk of land-price decline?

An inflation-adjusted repayment scheme also creates complexity for lenders who are issuing fixed-payment securities (bonds) to finance their operations. Certainly, some borrowers have attempted to utilize capital gains by extracting a cash flow stream from them by refinancing. But, it seems possible that pyramiding debt can occur under specific circumstances. This pyramiding can again create a cash flow problem as payments come due. What happens when the capital gains don't continue? I would ask Barry to conduct 2 additional exercises to help clarify this uncertainty:

1. Analyze those circumstances under which lending against inflated equity would produce unfavorable results for farmers. (Is it bad advice for farmers to borrow if they experience a series of years in which "luck of the draw" produced outcomes on the negative side of the variance of expected returns? Or when $g < i_f$?)

2. Consider another set of E-V models from the lender's viewpoint, and analyze the bargaining situation where lender and borrower optima differ.

As Barry's paradox about attempts to increase earnings in agriculture suggests, any widespread action of this type by farmers may be self-defeating. If so, the lesson from micromodels is indeed useful.

References

Colyer, Dale, and George D. Irwin. *Beef, Pork and Feed Grains in the Corn Belt: Supply Response and Resource Adjustments.* Missouri Agr. Exp. Sta. Research Bulletin No. 921. Aug. 1967.

Irwin, George D., and Joseph Havlicek, Jr. "Tailoring Farm Account Projects to Answer Aggregate Questions," *American Journal of Agricultural Economics,* Vol. 48, No. 5, Dec. 1966, pp. 1623–28.

PART 5

Risk Management in Models of the Farm

8. Cash Flow, Price Risk, and Production Uncertainty Considerations

Vernon R. Eidman

Abstract Models of agricultural production units that consider risk typically do not base the analysis on appropriate estimates of the distribution of the performance indicator. This paper argues the change in net worth plus additional consumption is the appropriate performance indicator. With this performance indicator it is necessary to consider financial risk as well as business risk and to estimate the distribution on an after-tax basis. The efficient set of operator actions can be selected for either normative or positive studies using stochastic dominance with respect to a function.

Keywords Microeconomic models, risk, stochastic dominance

This paper is concerned with modeling agricultural production situations where the decisionmaker views the uncertain environment in probabilistic terms.[1] The paper emphasizes considerations in developing micromodels either to provide normative recommendations for farmers or to predict the firm-level response to changes in the economic, technical, and institutional environment. The development of conditionally normative recommendations (that is, conditional on the expressed goals and the information available) for use in extension education programs will also continue to be an important use of such models.

The author is a professor of agricultural and applied economics, University of Minnesota, St. Paul. Minnesota Agricultural Experiment Station Miscellaneous Journal Series Paper No. 1842-0.

An extensive body of research has developed over the past 2 decades on the development of micro or firm growth models for normative and positive applications.[2] This work demonstrates the importance of including the major pricing, production, and financial alternatives for the type of firm and the problem being analyzed. This literature has also stressed the incorporation of institutional constraints and behavioral relationships in firm-level models.

In predictive applications the model is usually constructed for the typical or representative firm and then used to draw inferences about the population of such individuals. Evaluating the effects of changing economic conditions reflected through the prices of inputs and outputs and the availability of new technology on farm firms are examples of predictive applications. Other areas of predictive applications include evaluating farmers' likely response to changes in environmental restrictions, income tax regulations, income support programs, and the availability of crop insurance.

Modeling producers' response to risk is a behavioral relationship that has received a great deal of attention both in firm growth studies and in other applications of agricultural decisionmaking. Recent survey articles discussing alternative methods of incorporating risk into models for microanalyses are available elsewhere (Anderson, Dillon, Hardaker, Chapters 6 and 7; Anderson; Boussard). The detailed structure of several programming and simulation models will be discussed in other papers presented at this conference. For these reasons I will not attempt to present another survey of alternative methods to incorporate risk into firm-level models.

Apparently few research and extension workers doubt the importance of considering risk in microlevel models, but the wide range of approaches used indicates general agreement is lacking on the best method of handling risk. The expected utility hypothesis is cited as the basis for many of these models. This hypothesis provides a general decision rule—that is, expected utility maximization—which integrates data on both the decisionmaker's preferences and the risk associated with alternative actions to identify preferred choices. In spite of its wide acceptance, the successful application of this hypothesis has been limited by an inability to develop data on producer risk preferences. Without single-valued utility functions it has been difficult to develop an operational efficiency criterion to use in selecting preferred choices for the wide range of probability distributions associated with agricultural decisions. The assumption of normal distributions of outcomes associated with mean-variance analysis appears to be questionable in most applications.

The development of a more flexible efficiency criterion, referred to as stochastic dominance with respect to a function, permits more realistic applications of the expected utility hypothesis to a wide range of normative and positive applications. This criterion can provide a partial ordering of alternatives without estimating the utility function. Perhaps a characteristic of greater importance in dealing with a range of applications is the ability to evaluate either discrete, continuous, or mixed distributions. This characteristic allows more complete specification of the distribution of outcomes.

This paper discusses several of the underlying issues that are important in modeling agricultural production units, considering the appropriate subset of production, marketing, and financial alternatives for the problem being analyzed. The first section discusses the selection of the performance indicator, or consequence. The following section notes that financial risk has been ignored in most studies and that we should be preparing the estimates on an after-tax basis. A brief review of the procedures to estimate risk preferences and the empirical estimates available is provided in the third section. The following section introduces the use of stochastic dominance with respect to a function to select efficient choices. The final part discusses how the estimates of risk on an after-tax basis and stochastic dominance with respect to a function can be used in micro-modeling and policy analysis.

Selecting the Performance Indicator and Defining Risk

Previous studies have used a number of alternative performance indicators or consequences, including the maximization of gross margins, the maximization of after-tax net income, the maximization of ending net worth and minimizing the probability of being forced out of business. The corresponding measure of risk would be somewhat different for each, making selection of the consequence important.

Utility theory suggests using change in wealth (or consumption over time, which is related to current wealth and additions thereto) as the major consequence. Much of the empirical work on firm growth also documents the importance of providing desired levels of consumption and increasing net worth as goals of farm operators.

The change in wealth can be interpreted as the residual return to the owners of equity capital and, in the case of family businesses, the management ability of the operator(s). All withdrawals, including interest on debt capital, income taxes, social security taxes, and the minimum consumption requirement for family living, have been removed. The amount remaining can be used for additional consumption and addition to equity capital. The minimum consumption level is typically defined based on the family composition and the financial situation at the start of the period, while the additional consumption is based on the current period's income.[3] Thus, I will proceed assuming the consequence of interest is the change in net worth plus the additional consumption. Working with this objective over one time period is relatively straightforward, but the appropriate method of combining results over multiple time periods is less clear. Ideally, current period choices should be made in the context of a lifetime utility of wealth function. More conceptual development is required to specify a decisionmaker's currently held lifetime utility function in a way it can be made operational. Dillon suggests the following alternatives:

> In a world of perfect financial markets, liquidity, etc., a utility of present value approach could be argued as a guide to choice. Because of imperfections in the financial markets, however, present value will often be an inadequate standard and choice will be influenced by the possible patterns of time flows with their intra- and inter-period variations in payoffs. As outlined by Anderson, Dillon, and Hardaker; Bell; and Keeney and Raiffa; normative utility analysis then requires the treatment of each period as a separate dimension of utility in a multidimensional utility framework. (Dillon, p. 31.)

This paper does not suggest either single- or multiple-period utility functions should be estimated for individual producers. However, it should be possible to evaluate distributions for one or more time periods using stochastic dominance procedures and evaluate them using a multidimensional utility approach. I will return to this issue in the final section of the paper.

One of two general approaches is commonly used by applied researchers to quantify risk. Measures of dispersion, such as the variance, standard deviation, and other moments represent one approach. Expected utility is potentially a function of all statistical moments of the consequence. Casual empiricism suggests that farmers do not necessarily associate high levels of variability with high levels of risk, implying they also consider more moments than the mean and variance in their assessment of risk.[4] These observations suggest we should be concerned with more completely specifying the probability distribution of outcomes when dispersion is used as the measure of risk, rather than relying on the mean and variance.

The second approach to quantifying risk commonly used is based on the safety principle or, more formally, the probability (α) that the random outcome (c) will fall below some critical level (Z).[5] Applying this rule may involve minimizing the probability (α) that the consequence

(c) falls below a specified disaster level (Z), or it may involve maximizing the disaster level (Z) subject to a constraint that the probability of $c \leq Z = a$. A closely related formulation, the safety-first criterion, would maximize the expected consequence ($\bar{c}$) subject to a constraint that the probability of the outcome being less than a disaster level is less than or equal to a, or more formally, $P(c < Z) \geq a$. In each of these cases appealing to the Tchebychev inequality permits writing the constraint in terms of the mean and standard deviation of the consequence (Pyle and Turnovsky).

The above discussion relates the measure of risk to the consequence being evaluated in the objective function. In some applications, the variability of resource availability is the major source of uncertainty considered. In these cases, it may be difficult to link the source of uncertainty to variability of the objective function. Consider 2 examples. Selection of the machinery complement when the number of field days available is uncertain (due to uncertain weather conditions) has been analyzed using safety-first type restrictions on the field days available by period (Kletke and Griffin). Gabriel developed a model for use in studying firm growth that maximized the present value of after-tax net income subject to a safety-first type of liquidity constraint. In each of these examples the source of risk considered in the model applies to a constraint, rather than the objective function. This type of application will be dealt with in more detail later.

Measuring Risk

If an after-tax measure of the consequence is used in modeling, then the variation in that consequence must also be measured on an after-tax basis. This section explores some of the issues in estimating the variability of changes in net worth plus additional consumption.

Types of Risk

The finance literature divides the total risk of returns to equity capital into two parts, business risk and financial risk. Business risk is defined as the inherent uncertainty in the firm independent of the way it is financed (Weston and Brigham, p. 663). The major sources of business risk are input and product price uncertainty and production variation. It is common to evaluate risk either as the variability of net operating income or of net cash flows. Suppose the expected net operating income for a firm is defined as $\bar{c}$ and the standard deviation is σ_1. Business risk is commonly defined as the coefficient of variation $\sigma_1/\bar{c}$.

Financial risk is defined as the added variability of the net cash flows to owners' equity that results from the fixed financial obligation associated with debt financing (Weston and Brigham, p. 663). Defining σ_2 as the standard deviation of net cash flows with debt financing and I as the fixed debt servicing requirement, total risk can be written as

$$TR = \frac{\sigma_2}{\bar{c} - I}$$

Subtracting business risk results in a measure of financial risk, FR, where

$$FR = \frac{\sigma_2}{\bar{c} - I} - \frac{\sigma_1}{\bar{c}} \; .$$

Presumably, if 100 percent equity capital is used in financing, then $\sigma_2 = \sigma_1$ and financial risk as defined here is zero. However, when some debt financing is used ($I > 0$), financial risk will be positive if $\sigma_2 \geq \sigma_1$.

Most of the literature on risk and risk response treats only business risk, that is, production and price risk. Some authors have hypothesized that farmers adjust the production mix and pricing methods used to reduce business risk when they are faced with more financial risk and vice versa (Gabriel and Baker, p. 560). To the extent that financial risk is an important source of total risk and when such trade-offs are important to producers, consideration of financial risk will be important in modeling farm firms. Thus, the major concern is with estimating total risk as measured by σ_3^2 and other moments of the distribution and with incorporating the appropriate estimates in the usual programming and simulation models of farm firms.

An important issue for this discussion is how taxes are handled in the calculation. Conceptually both the net cash flows without debt financing used to calculate σ_1 and $\bar{c}$, as well as the net cash flows with debt financing used to calculate σ_2 and I, could be calculated on an after-tax basis. The additional calculation required would only warrant the additional effort when precise measures of business versus financial risk are required. More generally, we will be interested in the combination of the two. The following discussion assumes we are interested only in calculating σ_3^2 and other moments of the distribution on an after-tax basis.

As indicated, business and financial risk can be calculated based either on net operating income or net cash flows. The calculations for the first alternative can be written as equations (1) and (2).

$$EBIT = \sum_j (P_j - V_j) X_j - F - D - L \tag{1}$$

$$EAIT = EBIT - I - T \tag{2}$$

where:

P_j = average price per unit of product j;
X_j = amount of product j sold;
V_j = average variable cash cost per unit of product j;
F = fixed cash costs of operating the firm (considered constant);
D = the economic depreciation (considered constant);
L = the minimum withdrawal required for family consumption (considered constant);
$EBIT$ = operating earnings before interest and taxes;
I = interest on debt capital;
T = income tax payments; and
$EAIT$ = operating earnings after interest and taxes.

The variability of $EBIT$ depends on the variability of P_j, X_j, and V_j. These sources of price and production variability are commonly considered in programming and simulation models of farm firms. Subtraction of the constant minimum withdrawal required for family consumption at this point will not affect the size of the standard deviation. Handling consumption in this manner also simplifies some of the problems of dealing with sole proprietorships on a comparable basis with partnerships and corporations (where individuals draw salaries on a regular basis). Thus, the probability distribution of $EBIT$ can be considered a measure of business risk.

In the relatively simple world without changes in asset values that are unaccounted for by economic depreciation (such as land appreciation) and changes in inventories, $EAIT$ represents the money available for additional consumption (that is, additional to the minimum requirement L) and for addition to equity. If I and T are constants, the standard deviations of $EAIT$ and $EBIT$ will be equal. However, the predominance of variable interest rates and the recent experience with variation in nonprice costs of financing make the assumption of a constant I inappropriate. Furthermore, the importance of capital gains versus ordinary income, progressive tax rates, investment credit, and other features of the tax system suggest the standard deviation, and perhaps the form of the distribution of $EAIT$, is quite different than $EBIT$. These observations suggest that the variation of $EBIT$ is unlikely to be an accurate approximation of the variation in $EAIT$.

An alternative way to calculate business and financial risk is with data on net cash flows. The calculations are summarized in equations (3) and (4):

$$NBPT = \sum_j (P_j - V_j)X_j - F - L \tag{3}$$

$$NAPT = NBPT - I - P - T \tag{4}$$

where:

	=	net cash flow before debt servicing payments and taxes;
NBPT	=	net cash flow before debt servicing payments and
NAPT	=	taxes; and
P	=	required principal payments (considered constant).

In comparing the two approaches, it can be seen that the first removes depreciation, but does not remove principal payments. The opposite is true of the second approach. If depreciation is a constant, all moments of the distribution of business risk except the mean will be the same whether based on variation in *EBIT* or variation in *NBPT*. Similarly, if principal payments are constant, all moments of total risk except the mean will be the same whether based on variation in *EAIT* or *NAPT*. Thus, selection of a consistent method to calculate business risk and financial risk is important when these measures are expressed as a ratio of the standard deviation divided by the expected value. However, if the evaluation of alternative actions is based on the moments of the distribution or the probability distributions themselves, then either method should provide nearly equivalent results.

Issues in Estimating Business Risk

Many empirical studies of business risk for specific groups of agricultural producers have been published. Much of the literature appearing during the 1970s has been reviewed and critically appraised by Young (1980). This discussion is limited to a few summary comments concerning the source of data to calculate business risk and the selection of an expectations model.

One issue is the extent to which personal probabilities provided by decisionmakers should be used. Conceptually, it can be argued these are the only relevant probabilities for decisionmaking and that they should be used for both normative and positive applications. An evaluation of methods to elicit personal probabilities suggests that: (1) the elicitations must be performed very carefully to avoid biasing responses; (2) success in obtaining such estimates has been limited largely to providing estimates for one variable (as opposed to joint specification); and (3) such estimates are relatively time consuming and expensive to prepare (see Anderson, Dillon, and Hardaker, pp. 17–44). Given these problems, historic data will have to be used in many cases to estimate proxies for price and production risk.

The variability of interest in calculating risk is the variation about the decisionmaker's expectation. The criteria suggested by many authors for use in selecting an expectations model have been summarized by Young (1980, p. 5). He suggests that the expectation be based on a series of one-step-ahead forecasts, using only data available to decisionmakers at the time the expectation is formed, with more recent data given more weight. Furthermore, he suggests the estimates of the functional specification and the parameters of the expectation model, as well as the estimates of the expected component and the variability, should be subject to revision each period.

It is more difficult to meet these criteria in developing an expectations model for prices than for yields. Many commonly used methods of calculating the expectation for price, such as the simple mean, time trend polynomials, and the variate difference method, do not satisfy these criteria. Methods, such as a weighted autoregression reestimated each period, moving weighted average of several previous years, and relatively simple supply and demand models commonly used in extension outlook programs, are more consistent with these criteria.

Issues in Estimating Financial Risk

The two major areas of concern in estimating financial risk are estimating credit risks and the effect of income tax payments on risk. In a recent article, Barry and others note that risks associated with the cost and availability of credit are reflected partly in interest rates for loans and partly through nonprice sources. Their analysis suggests that data on changes in interest rates on Farm Credit System loans should indicate the part of farmers' credit risk that can be attributed to changes in financial market conditions. However, they caution that lenders nonprice responses (including differing loan limits, security requirements, and maturities among borrowers) are used to reflect the creditworthiness of borrowers and the availability of loan funds. These observations suggest that both estimates of the variation in interest rates and estimates of variability of loan availability should be included in models to reflect credit risk. The first can be reflected as price variability, while the second would be included as variation in a constraint in the typical programming and simulation models.

Income tax calculations are based on the whole-farm analysis, including the financing cost. Although income tax withdrawals for expected levels of taxable income have been computed in programming analyses, it does not appear that the effect on the after-tax distribution of the consequence can be calculated directly in programming models. However, the after-tax distribution can be calculated for alternative solutions on the pre-tax efficiency frontiers using a Monte Carlo simulation model, as applied by Lin, Dean, and Moore. Simulation models can be used to generate after-tax distributions of the consequences directly.

Measuring Risk Preferences

In addition to estimating the risk associated with alternative actions, information is needed on decisionmakers' risk attitudes. A decisionmaker whose preferences are consistent with the 3 axioms of ordering and transitivity, continuity, and independence is said to possess a well-defined utility function for risky prospects (Anderson, Dillon, and Hardaker, p. 67). Formal definitions of risk aversion are normally based on expected utility functions for income or wealth.

Given a utility of money function, $U(m)$, it is generally assumed that all decisionmakers prefer more money to less (that is, the first derivative of the utility function, $U'(m)$, is positive). Risk aversion is consistent with decreasing marginal utility of money (that is, the second derivative, $U''(m)$, is negative), while risk preference implies $U''(m) > 0$. Thus, the sign of the second derivative is one measure of risk preference. The difference between the expected monetary value (EMV) and the certainty equivalent (CE) is the risk premium. It follows from the signs on the derivatives that this difference, $EMV - CE$, is positive, zero, and negative for risk-averse, risk-neutral and risk-preferring decisionmakers, respectively. Measuring the amount an individual will pay to avoid participation in a fair bet (that is, one whose EMV is equal to the cost of participating) is one relatively simple and intuitively appealing method of measuring risk preferences.

The degree of risk aversion that a decisionmaker possesses can be uniquely measured by the coefficient of absolute risk aversion $r(m_i) = -U''(m_i)/U'(m_i)$, where m_i represents a particular monetary level. This coefficient is invariant to positive linear transformations of the utility function. As such, it can be used to compare local risk attitudes among decisionmakers. It follows from the definition that $r(m_i)$ is positive for risk-averse, zero for risk-neutral, and negative for risk-preferring decisionmakers. It should be noted that risk aversion is a local measure. Its magnitude and sign may vary as the magnitude of the monetary outcomes vary.

A more generalized concept of risk preferences is the absolute risk-aversion function. This function is defined for all values of monetary outcomes for which the utility-of-money function is defined and is twice differentiable. The absolute risk-aversion function corresponds uniquely to a decisionmaker's preferences among risky prospects (Pratt, Arrow).

As such, the absolute risk-aversion function represents an alternative (but not independent) means for ordering risky prospects.

Alternative Methods of Measuring Risk Preferences

Three measurement methods have been used to obtain much of the empirical evidence available on risk preferences of agricultural producers. These can be referred to as methods that use: (1) response to hypothetical choices, (2) experimental choices, and (3) actual economic decisions. A fourth method, referred to as the interval approach, also uses hypothetical choices. It is listed separately because it is relatively new and seems to avoid some problems of other methods using hypothetical choices.

Hypothetical Choices: The decisionmaker is asked to express preferences for monetary gains and losses in several simple and hypothetical lottery situations. The responses are used to empirically estimate the individual's utility function. The elicitation schemes are outlined elsewhere (Anderson, Dillon, and Hardaker, pp. 69–76). These methods have been criticized as subject to bias arising from misconceptions of the notion of probabilities, preferences for specific probabilities, attitudes toward gambling, and differences in interviewers. (See Young, 1979, for a review of the biases in measurement of risk preferences.)

Experimental Choices: The decisionmaker is asked to make a single selection among a specified set of gambles in which a significant level of actual financial compensation is made. The choice made is used to compute the individual's risk preferences (Binswanger). The advantage claimed for this approach is that it is not subject to the interview bias and many of the other errors of the hypothetical choice method. Binswanger compared the use of hypothetical choices and the experimental method in measuring risk preferences for more than 350 peasants in rural India. His field checks led him to conclude that results from the experimental method are more reliable and reproducible than those from the hypothetical choice method (Binswanger, p. 401). An obvious disadvantage of conducting a comparable experiment among commercial farmers in the United States is that substantial funding would be required for the prizes. Furthermore, Knowles (1980b) cites several studies in the psychological literature which indicate that changing the wealth position of individuals (by giving them money with which to play the game) and the use of games in which the participants average out as gainers rather than losers lead to different responses than individuals give when playing with money they consider to be their own. While Binswanger made efforts to avoid the effect of these problems, the actual choices are observed under somewhat artificial and contrived circumstances. Given these concerns and the lack of a test comparing predicted and actual business decisions, it is difficult to make strong inferences or generalizations about risk attitudes based on these experiments.

Actual Economic Choices: The problems with estimation of risk preferences based on hypothetical choices and experimental methods have increased interest in analyzing actual economic decisions to draw inferences about risk aversion. The approach is to develop a theoretical model of input demand or output supply and compare it with actual choices that were made to estimate the implied risk preference. For example, Anderson, Dillon, and Hardaker (p. 163) have shown that maximization of expected utility results in expression (5).

$$MFC_i = E(MVP_i) - \frac{dE(m)}{dV(m)} (MRI_i) \qquad (5)$$

which indicates that the optimum level of input *i* occurs when the marginal factor cost, *MFC_i* (the input price) equals the value of the expected marginal value product, $E(MVP_i)$, minus a marginal risk deduction. In this case the risk deduction is the product of the rate of marginal utility substitution between $E(m)$ and $V(m)$ and the marginal contribution to risk of additional input use (MRI_i). Assuming $MRI > 0$ and risk aversion (preference), then

$$\frac{dE(m)}{dV(m)} > 0 \quad (<0)$$

and utility-maximizing input use is less (more) than for a risk-neutral individual, where

$$\frac{dE(m)}{dV(m)} = 0$$

Thus, an expression such as (5) can be solved for

$$\frac{dE(m)}{dV(m)}$$

in terms of empirically observable magnitudes. (See Moscardi and deJanvry for an example of an econometric application.)

Brink and McCarl used a MOTAD-type linear programming model to derive indirect estimates of risk-aversion coefficients of 38 Corn Belt farmers. The value of the risk-aversion coefficient was varied parametrically, and the actual plan was compared with those selected by the programming model. The value of the risk-aversion coefficient that minimized the difference between the actual and the predicted plan was selected to represent the farmer's risk preference.

Risk preference estimates based on actual economic behavior appear to avoid the charge of being irrelevant for use in predicting actual decisions. However, the entire difference between the profit-maximizing solution and the actual solution is attributed to risk aversion. Thus, incorrect or inaccurate specifications of technical data, market data, resource availabilities, different subjective probability assessments, and different objective functions could be at least partially responsible for the risk aversion inferred.

Interval Approach: The use of an interval approach is a recent addition to the literature on methods of estimating risk preferences (King and Robison, 1981a, b). The procedure is based on the premise that a choice between two distributions defined over a relatively narrow range of monetary levels divides absolute risk-aversion space over that range into one region consistent with the choice and another region that is not consistent with the choice. Stochastic dominance with respect to a function (see the following section) is used to evaluate the two distributions. The evaluation determines the division between the consistent and inconsistent regions in absolute risk-aversion space (Meyer, 1977b). This boundary level of absolute risk aversion is calculated for carefully selected pairs of distributions at each of several ranges in monetary outcomes.

The respondent is asked to choose between pairs of distributions. The sequence of pairwise comparisons is arranged so that each successive choice narrows the interval in risk-aversion space. The process can be continued until the desired precision (narrowness of the risk-aversion interval) is attained.

The application of this procedure to estimate the absolute risk-aversion intervals for 17 farmers participating in a series of extension workshops is reported by King and Robison (1981a). The authors report that the respondents had little difficulty completing the questionnaire. The empirical estimates for most individuals exhibited increasing absolute risk aversion over lower income levels and decreasing absolute risk aversion at higher income levels. Furthermore, the interval measurement of absolute risk aversion for most individuals included negative as well as positive values at some income levels. Only four had lower-level absolute risk-aversion functions that were everywhere nonnegative (p. 519).

Empirical Evidence of Producers' Risk Preferences

Young (1979) summarizes much of the empirical evidence of agricultural producers' risk preferences reported from 1965 through 1979. The 11 studies summarized represent a very small number of agricultural producers and, in general, are not a representative sample of producers in the area or country of origin. The empirical evidence suggests the predominance of producers are risk averse, with some producers exhibiting risk neutrality and others risk preference. A comparison of the studies indicates developed countries have a somewhat higher proportion of producers with risk-neutral and risk-preferring attitudes than the developing countries. As Young notes, this is consistent with the hypothesis of decreasing absolute risk aversion with increasing wealth.

A study by Knowles (1980a) completed since the above survey was published reported estimates of utility functions for 5 commercial farmers in Minnesota. Although some responses reflected risk neutrality for questions with smaller stakes, all respondents eventually displayed risk aversion as the stakes were raised. The extent to which raising the stakes might have changed the results in other studies reporting some risk preferences is unclear.

The direct application of risk-preference data in microanalyses is limited by the methodological problems and the small number of empirical estimates that have been made. However, existing research provides some guidelines to follow in both normative and positive analyses. Although some empirical evidence to the contrary exists, it is reasonable to continue using the general hypothesis that agricultural producers are risk averse and that they display decreasing absolute risk aversion as wealth increases. Furthermore, existing empirical studies indicate the coefficient of absolute risk aversion falls within a relatively narrow range of approximately -0.0002 to $+0.0006$ for a relatively wide range of current wealth and income levels. The following section indicates how this information on risk preferences can be used in both normative and positive analyses. Attempts have been made to generalize the results by estimating the relationship between producers' attributes and risk preferences. Young (1979) summarizes four of these studies and concludes they have been of limited success. "Overall, few consistent relationships emerged over all four studies. The one exception was a positive correlation between education and the willingness to bear risk" (p. 1067).

Selecting the Preferred Choice

The traditional decision-theoretic approach to maximization of expected utility is based on the assumption that the preferences of the decisionmaker are known or easily obtained and quantified. The brief review in the previous section indicates detailed knowledge of producers' utility functions will typically not be available for agricultural research, extension, and policy problems. Fortunately, an alternative approach for efficiency analysis, stochastic dominance, has been developed for situations in which little is known about the decisionmakers' preferences. This approach is based on the expected utility-maximization framework, but does not require complete specification of the decisionmakers' utility function.[6]

One risky prospect is said to dominate another if the expected utility of the first prospect exceeds the expected utility of the second for all possible utility functions within a defined class. The approach is implemented by defining the class of utility functions to be considered, and then identifying the stochastically efficient set by making pairwise comparisons of the cumulative distribution functions (CDF) for the risky prospect. The risky prospects which are dominated are inefficient in the sense they would not be preferred to members of the dominant set by any expected utility maximizer whose preferences are characterized by the behavioral assumptions. Those which are not dominated are members of the stochastically efficient set. This set will contain the risky prospect that maximizes expected utility for decisionmakers whose preferences belong to the defined class of utility functions.

Three stochastic dominance rules have been routinely applied. They place increasingly restrictive assumptions on the class of underlying utility functions. First-degree stochastic dominance (FSD) orders risky prospects assuming only that the underlying utility functions are monotonically increasing, that is, $U'(m) > 0$. As such FSD places no restriction on the nature of the risk preferences, and consequently, the risk aversion functions of the decisionmakers.

Second-degree stochastic dominance (SSD) assumes both risk aversion and monotonically increasing utility functions. These assumptions imply that the underlying utility functions belong to the class that exhibit positive but decreasing marginal utility, that is, $U'(m) > 0$, $U''(m) < 0$. The absolute risk-aversion functions for these individuals are restricted to be greater than zero for all values of monetary outcomes in the defined domain, excluding risk-neutral and risk-preferring decisionmakers.

In addition to the previous assumptions, third-degree stochastic dominance (TSD) assumes the third derivative of the underlying utility function is positive, that is, $U'(m) > 0$, $U''(m) < 0$, $U'''(m) > 0$. This additional assumption is a necessary condition for decreasing absolute risk aversion.

Relatively few agricultural decision problems have been analyzed with the three ordinary stochastic dominance rules. A review of these applications (Zentner and others) suggests the FSD rules were of rather limited effectiveness in reducing the size of the efficient set of risky prospects. The SSD rules were generally much more effective in reducing the size of the stochastically efficient set, while the TSD rules did little to further reduce its size. Compared with the expected value-variance $(E - V)$ efficient set, the SSD efficient set was similar in size but often contained different risky prospects. The $E - V$ efficient set generally included some elements that were eliminated by the SSD rules. Also, the SSD criterion was generally more difficult to apply than the $E - V$ criterion.

Meyer's Criterion

The above discussion notes that FSD imposes no restrictions on the decisionmaker's (or a set of decisionmakers') absolute risk-aversion function interval; that is, the lower bound is $r_1(m) = -\infty$ and the upper bound is $r_2(m) = +\infty$. Both SSD and TSD restrict the group of decisionmakers to those who are risk averse, which implies the corresponding absolute risk-aversion function interval is $r_1(m) > 0$ and $r_2(m) = +\infty$. Meyer (1977b) generalized the ordinary stochastic dominance rules. He developed a procedure that chooses between 2 distributions when it is known only that the decisionmaker's risk-aversion function lies between an arbitrarily set $r_1(m)$ and $r_2(m)$ defined over a given domain of monetary outcomes. The procedure requires the identification of a utility function, $U_0(m)$ which minimizes:

$$\int_{-\infty}^{+\infty} [G(m) - F(m)]U'(m)dm \qquad (6)$$

subject to the constraint:

$$r_1(m) \leqslant -U_0''(m)/U_0'(m) \leqslant r_2(m) \qquad (7)$$

where $F(m)$ and $G(m)$ are cumulative distribution functions. Equation (6) equals the expected utility associated with $F(m)$ minus the expected utility corresponding to $G(m)$, and the minimization process requires the difference between the expected utilities of the two choices to be as small as possible. Therefore, all other decisionmakers with risk-aversion functions within the defined absolute risk-aversion interval would have a difference in expected utility greater than the amount calculated.

Given this condition, the following preference orderings can be specified for the particular set of decisonmakers defined by $S[r_1(m), r_2(m)]$:

1. If the minimum of this difference is positive, then $F(m)$ is unanimously preferred to $G(m)$ for the particular set of decisionmakers.
2. If the minimum is zero, then the decisionmakers in the particular set are said to be indifferent between the 2 activities and they cannot be ordered.
3. If the minimum is negative, then the particular set of decisionmakers does not unanimously prefer $F(m)$ to $G(m)$.

If the last situation occurs, equation (6) is changed to:

$$\int_{-\infty}^{+\infty} [F(m) - G(m)] U'(m) dm \qquad (8)$$

and minimized subject to equation (7) to determine whether $G(m)$ is unanimously preferred to $F(m)$ for the same set of decisionmakers. If $G(m)$ is shown not to be unanimously preferred or indifferent to $F(m)$, that is, equation (8) is also negative, then the 2 activities cannot be ordered. Thus, a complete ordering is not ensured by this procedure.

Meyer (1977a, pp. 331–33) reformulated the above problem into an optimal control framework using the absolute risk-aversion function, $r_0(m)$ as the control variable and $U'(m)$ as a state variable. (King and Robison, 1981a, and Zentner and others present examples of this solution procedure.) The optimal control procedure makes it possible to apply Meyer's criterion to determine the preference ordering between a pair of risky prospects for a particular set of decisionmakers defined by $[r_1(m), r_2(m)]$.

One major advantage of this criterion is that it does not place restrictions on the width of the relevant risk-aversion interval. King and Robison (1981a) demonstrate that the width and level can vary across the relevant range in monetary values. Furthermore, the procedure (like FSD, SSD, and TSD) can evaluate discrete, continuous, and mixed probability distributions. The empirical results of an application are presented by King and Oamek in these proceedings.

Implications for Micromodeling and Policy Analysis

Much of this discussion is based on the premise that agricultural producers make decisions based on their assessment of after-tax returns. Although I know of no formal test of this hypothesis, the emphasis on developing models that calculate expected returns on an after-tax basis suggests many extension and research workers share this premise. It seems clear that normative recommendations should be made on an after-tax basis. Furthermore, I know of no evidence that basing prediction on pretax calculations is more accurate. The measure proposed is the change in net worth plus additional consumption, which facilitates comparison of operations that withdraw salaries regularly with those that withdraw money for family living as needed. The change in net worth (which could include unrealized capital gains as well as ordinary income and realized capital gains) plus total consumption also could be used because minimum consumption is considered a constant and would not affect the ranking of alternatives.

If an after-tax criterion is appropriate, then the distribution of outcomes relating to that criterion (whether it is used with a dispersion or safety-first measure of risk) should also be prepared on an after-tax basis. It might be argued that farmers cannot estimate the appropriate after-tax distribution subjectively. While that is probably correct, it seems reasonable to assume they are aware the after-tax distribution is frequently quite different than the before-tax distribution. Whether their subjective assessment is closer to an after-tax or a before-tax distribution is a matter to test empirically. In any event normative recommendations should be based on after-tax assessments. Consideration of risk on an after-tax basis may also be important in predicting the difference in supply response and input use for farmers in different tax situations. Related to this point is the more complete assessment of the relevant risk. Recent changes in the availability and cost of credit have highlighted the importance of considering these sources of risk and their interaction with business risk in our analyses.

Additional and carefully conducted methodological research is needed on assessing risk preferences. The premise that decisions are made on an after-tax basis suggests that some of the difficulty in developing estimates for commercial producers may result from failure to express alternatives on an after-tax basis. (See Lin, Dean, and Moore for an exception.) Furthermore, testing the predictive accuracy of the assessments should be emphasized. In the meantime, we can work with the hypothesis that farmers tend to be somewhat risk averse to risk preferring, with coefficients of absolute risk aversion ranging from -0.0002 to 0.0006. At least one study has used this approach. Danok, McCarl, and White argue the available data indicate a value of the absolute risk-aversion function ranging from 0.000001 to 0.1 is appropriate for commercial farmers in Indiana. While they used a somewhat different range in the absolute risk-aversion coefficient, their study illustrates that existing estimates may allow us to reduce the size of the efficient set for predictive as well as normative analyses.

Stochastic dominance is an effective means to obtain the efficient set, but the alternative actions to be evaluated must be generated before applying stochastic dominance. I suggest using existing programming and simulation techniques to select alternative actions for consideration. Quadratic programming and MOTAD models that maximize an appropriate after-tax objective function subject to before-tax risk can be used to estimate an efficiency frontier. The after-tax variability of plans on the efficiency frontier can be estimated, and stochastic dominance with respect to a function can be used to select the efficient set. Likewise the after-tax distribution of returns for alternative actions estimated with simulation models can be evaluated.

An earlier section of the paper noted that some applications naturally lead to incorporation of risk in the constraints. The inclusion of field day constraints in machinery selection models and the detailed consideration of a liquidity constraint were cited as examples. Danok, McCarl, and White estimate the distribution of returns for each machinery set considered over a range in field days available. The resulting distributions of net returns were evaluated using stochastic dominance to select the efficient set of machinery complements. This approach may be appropriate to other problems which logically incorporate the major sources of risk in the constraint set.

As noted earlier, multiperiod problems pose a unique challenge. If the computational burden can be managed, it should be possible to estimate the distribution of outcomes for the alternatives by period and select the efficient set at the end of each period. Any alternatives remaining in the efficient set across time presumably would be quite attractive to the decisionmaker. To reduce the computational burden, the efficient set could be selected at one point in time, such as the end of the planning horizon, subject to appropriate constraints on liquidity, consumption, and so forth for each period.

The procedures discussed appear to be particularly useful in identifying policy options that are likely to be selected by farmers. For example, micromodeling might be used to evaluate farmers' response to USDA's all-risk crop insurance program. Representative farms could be defined based on the financial situation as well as land and labor resources by geographic area. Alternative farm plans with and without crop insurance can be specified for each representative farm being analyzed. The after-tax distributions of returns can be calculated for each alternative plan. Stochastic dominance with respect to a function could be applied to determine which plans are included in the efficient set for producers with an absolute risk-aversion coefficient in the specified range. One or more of the plans included in the efficient set would be preferred to any plan not included by producers with an absolute risk-aversion function in the specified range and that perceive the probability distribution of net returns as modeled. Thus, the insurance options included (not included) in the efficient set are likely (unlikely) to be selected by operators having a similar resource situation.

The consideration of both business and financial risk as well as after-tax estimates of the distributions also appears to be important, but is even more difficult in macromodels. Perhaps firm-level studies that estimate the difference in distributions for size and type of farm by region of the country will facilitate considering these issues in macromodels as well.

Finally, I suggest that risk only be considered explicitly in analyses when the benefits are expected to justify the cost of preparing valid estimates of the distributions. Preparing estimates of business and financial risk on an after-tax basis is both time consuming and expensive. However, it is not clear that the before-tax estimates of price or production risk prepared with some conveniently available statistical procedures are reasonable proxies of the after-tax distributions advocated in the paper. Unless the returns distributions are reasonable approximations of producers' subjective estimates, the consideration of risk is unlikely to improve our predictive ability. When resources are not available to prepare such estimates, relevant deterministic studies with appropriate sensitivity analyses are likely to be more valuable in evaluating policy alternatives.

Notes

1. The terms *risk* and *uncertainty* are used interchangeably in this paper. Modern decision theory is concerned with the choice among alternatives, some or all of which have consequences that are not certain. The current decision theory literature refers to all situations having consequences that are not certain as *uncertainty*. However, some of the applied work uses the terms *risk* and *uncertainty* interchangeably. The two terms are used interchangeably here because the conference organizers chose to include both words in the title of the paper. However, the use of risk does not necessarily imply that the decisionmaker has any more information on the likelihood of the consequences.

2. A comprehensive summary of this research is provided by Baker and others.

3. Vandeputte and Baker discuss basing consumption withdrawals on both current and past levels of ordinary income and capital gains.

4. Whether including higher moments is worth the effort is an empirical question.

5. This approach to quantifying risk is consistent with the cautious behavior described by Richard Day earlier in this conference.

6. A more detailed explanation of the concepts discussed in this section is presented elsewhere (Zentner and others). The publication also includes a more complete bibliography of the stochastic dominance literature.

References

Anderson, J. R. "Perspectives on Models of Uncertain Decisions," *Risk, Uncertainty and Agricultural Development*. New York: Agricultural Development Council, 1979, pp. 39–62.

Anderson, J., J. Dillon, and J. Hardaker. *Agricultural Decision Analysis*. Ames: Iowa State Univ. Press, 1977.

Arrow, K. J. *Aspects of the Theory of Risk Bearing*. Helsinki: Academic Book Store, 1964.

Baker, C. B., P. Barry, W. F. Lee, C. E. Olson, E. Hochman, G. Rausser, and M. Kottke. *Economic Growth of the Agricultural Firm*. College of Agriculture

Research Center Technical Bulletin 86. Washington State Univ., Pullman, Feb. 1977.

Barry, P. J., C. B. Baker, and L. R. Sanint. "Farmers' Credit Risks and Liquidity Management," *American Journal of Agricultural Economics*, Vol. 63, No. 2, May 1981, pp. 216–27.

Bell, D. E. *A Utility Function for Time Streams Having Interperiod Dependencies.* International Institute for Applied Systems Analysis RR-75-43. Laxenburg, Austria, 1975.

Binswanger, H. P. "Attitudes Toward Risk: Experimental Measurement in Rural India," *American Journal of Agricultural Economics*, Vol. 62, No. 3, Aug. 1980, pp. 395–407.

Boussard, Jean-Marc. "Risk and Uncertainty on Programming Models: A Review." *Risk, Uncertainty and Agricultural Development.* New York: Agricultural Development Council, 1979, pp. 63–84.

Brink, L., and B. McCarl. "The Trade Off Between Expected Return and Risk Among Corn Belt Farmers," *American Journal of Agricultural Economics*, Vol. 60, No. 2, May 1978, pp. 259–63.

Danok, A., B. A. McCarl, and T. K. White. "Machinery Selection Modeling: Incorporation of Weather Variability," *American Journal of Agricultural Economics*, Vol. 62, No. 4, Nov. 1980, pp. 700–08.

Dillon, J. L. "Bernouillian Decision Theory: Outline and Problems." *Risk, Uncertainty and Agricultural Development.* New York: Agricultural Development Council, 1979, pp. 23–38.

Gabriel, S. C. "Financial Responses to Changes in Business Risk on Central Illinois Cash Grain Farms." Ph.D. dissertation, Univ. of Illinois, 1979.

————, and C. B. Baker. "Concepts of Business and Financial Risk," *American Journal of Agricultural Economics*, Vol. 62, No. 3, Aug. 1980, pp. 560–64.

Keeney, R. L., and H. Raiffa. *Decisions with Multiple Objectives: Preferences and Value Trade-offs.* New York: John Wiley and Sons, 1976.

King, R. P., and L. J. Robison. "An Interval Approach to Measuring Decision Maker Preferences," *American Journal of Agricultural Economics*, Vol. 63, No. 3, Aug. 1981a, pp. 510–20.,

————. *Implementation of the Interval Approach to the Measurement of Decision Maker Preferences.* Michigan State Univ. Agr. Expt. Sta. Res. Report 418, 1981b.

Kletke, D., and S. Griffin. "Machinery Complement Selection in a Changing Environment." Oklahoma State Univ., Dept. of Agricultural Economics, 1976.

Knowles, G. J. "Estimating Utility of Gain Functions for Southwest Minnesota Farmers." Ph.D. dissertation, Univ. of Minnesota, Aug. 1980a.

————. "Estimating Utility Functions," *Risk Analysis in Agriculture: Research and Educational Developments.* AE-4492. Univ. of Illinois, Dept. of Agricultural Economics, June 1980b, pp. 186–216.

Lin, W., G. Dean, and C. Moore. "An Empirical Test of Utility vs. Profit Maximization in Agricultural Production," *American Journal of Agricultural Economics*, Vol. 56, No. 3, Aug. 1974, pp. 497–508.

Magnusson, G. *Production Under Risk: A Theoretical Study.* Uppsala, Sweden: Almquist and Wicksells, 1969.

Meyer, J. "Choice Among Distributions," *Journal of Economic Theory*, Vol. 14, 1977a, pp. 326–36.

―――――. "Second Degree Stochastic Dominance with Respect to a Function," *International Economic Review*, Vol. 18, No. 2, June 1977b, pp. 477–87.

Moscardi, E., and A. deJanvry. "Attitudes Toward Risk Among Peasants: An Econometric Approach," *American Journal of Agricultural Economics*, Vol. 59, No. 4, Nov. 1977, pp. 710–16.

Pratt, J. "Risk Aversion in the Small and in the Large," *Econometrica*, Vol. 32, No. 1, Jan.-Apr. 1964, pp. 122–36.

Pyle, D., and S. Turnovsky. "Safety-First and Expected Utility Maximization in Mean-Standard Deviation Portfolio Analysis," *Review of Economics and Statistics*, Vol. 59, No. 1, 1970, pp. 75–81.

Roumasset, J. A. "Introduction and State of the Arts," *Risk Uncertainty and Agricultural Development*. New York: Agricultural Development Council, 1977, pp. 3–21.

Vandeputte, J. M., and C. B. Baker. "Specifying the Allocation of Income Among Taxes, Consumption and Savings in Linear Programming Models," *American Journal of Agricultural Economics*, Vol. 52, No. 2, May 1970, pp. 521–27.

Weston, J. F., and E. F. Brigham. *Managerial Finance*. Hinsdale, Ill.: Dryden Press, 1978.

Young, D. L. "Evaluating Procedures for Computing Objective Risk from Historical Time Series," *Risk Analysis in Agriculture: Research and Educational Developments*. AE-4492. Univ. of Illinois-Urbana, Dept. of Agricultural Economics, June 1980, pp. 1–21.

―――――. "Risk Preferences of Agricultural Producers: Their Use in Extension and Research," *American Journal of Agricultural Economics*, Vol. 61, No. 5, Dec. 1979, pp. 1063–70.

Young, D., W. Lin, R. Pope, L. Robison, and R. Selley. "Risk Preferences of Agricultural Producers: Their Measurement and Use," *Risk Management in Agriculture: Behavioral, Managerial and Policy Issues*. AE-4478. Univ. of Illinois Urbana, Dept. of Agricultural Economics, July 1979, pp. 1–28.

Zentner, R. P., D. D. Greene, T. L. Hickenbotham, and V. R. Eidman. *Ordinary and Generalized Stochastic Dominance: A Primer*. Staff Paper P81-27. Univ. of Minnesota—St. Paul, Dept. of Agricultural and Applied Economics, Sept. 1981.

9. Hedging Procedures and Other Marketing Considerations

Wayne D. Purcell

Abstract Price risk for storable commodities has increased significantly in the past 10 years. Models of farm-level decision processes generally pay only casual attention to price risk. To incorporate this risk factor, more attention needs to be paid to the futures markets. Strategies to more effectively manage price risk are hypothesized and tested. Alternatives include a target price strategy, the use of econometric models, and the use of technical indicators of price direction to guide decisions on when to seek price protection and when to speculate in the cash market.

Keywords Fundamental, futures, hedging, income, mean, modeling, price risk, technical, variance

Most intra- or inter-year models of the farm have paid only casual attention to price risk. A typical approach is that recently followed by Richardson and Condra. In their simulation analysis from 1960 to 1977, prices were drawn at random from a user-specified probability distribution. However, this period spans significant changes in price variability of soybeans, one of the crops in the study. Changes in the price variability for other storable commodities can be shown to be similar through a visual examination of their monthly futures price charts.

A partial explanation for the failure to focus more attention on price risk is that price variability, at the level observed in the early 1980s for the grains and fibers, is a new phenomenon. Before the 1973–74 crop year, production controls and direct price supports were in place for most storable agricultural commodities. Partly as a result of the Russian grain purchases in the early 1970s, government programs were

---

The author is a professor of agricultural economics, Virginia Polytechnic Institute and State University.

changed and domestic markets became more directly exposed to the vagaries of world markets. Increased levels of price variability and concurrent price risk have resulted.

But, there are other and perhaps more important reasons for the failure to treat price risk more effectively in models. It is difficult to effectively discuss and analyze the topic of price risk[1] without also considering the futures markets and related pricing strategies. Many agricultural economists have tried to ignore the futures markets, particularly in analyses of livestock commodities. This posture has been perpetuated by trade groups who go on record opposing trade in futures because of an assumed negative effect of trade in futures on cash prices. These often poorly informed positions have in turn been encouraged by the efforts of political entities such as the House Committee on Small Business. The Committee staff has published numerous and largely negative reports dealing with trade in live cattle futures. Other analysts have questioned both the conceptual base and analytical procedures in some of these staff reports, but the studies are apparently not subject to a rigorous review process before publication.[2] A tendency to ignore the futures markets is also supported by the attitudes of many producers and financial agencies. There is ample evidence that some producers and producer groups have not accepted futures markets as a legitimate marketing alternative with the potential to allow more effective management of price risk. Financial institutions, especially commercial banks, register their lack of understanding and expertise in this area by refusing to advance a credit line allowing producers to finance margins.[3] The use barriers thrown up by producers and their financial representatives mask an important point: Futures markets give producers an opportunity to forward price which is available over a longer time period than most cash-forward contracts and, therefore, allow producers more flexibility in pricing.

Agricultural research economists involved in modeling firm activity or measures of the output of that activity cannot afford to be eventually pulled into the futures analysis with a major time lag. They should be leading. It is clear as we watch the debt-equity ratio climb that production skills alone are no longer a sufficient condition for success of the farm firm and that incorporating production-related management skills is not sufficient when modeling the behavior of the farm firm. Complementary marketing strategies and marketing management skills must be included in these models, especially when the model results will be affected by intrayear price variability.

Futures markets are a reality. They will play an increasingly important role in price discovery and as an ingredient in marketing strategies. Anne Peck put it this way in 1976:

Futures markets exist for most of the major agricultural commodities; to ignore their presence in analyzing these markets is to present a partial analysis at best. Any attempt to discuss the incorporation of price risk firm models must deal with the futures markets. (p. 422.)

Some Conceptual Issues

Two broad dimensions of futures markets must be considered in relation to price risk and firm modeling. These relate to the long-established economic functions of futures markets to aid in the process of price discovery and to provide a hedging mechanism to facilitate the transfer of price risk.

Price Discovery

Most surveys indicate a small percentage of farmers actually trade futures. However, the same surveys do indicate a large majority of farmers watch the futures market. This tendency is leading to an increasingly apparent effect of futures prices on the supply of agricultural commodities.

A simple model to explain current feeder cattle prices is:

$$P_{FC} = f(P_{SC}, P_C)$$

where

P_{FC} = price of feeder cattle
P_{SC} = price of slaughter cattle
P_C = price of corn

At any point in time, using the current or spot quote for the live cattle futures maturing 5 to 6 months later as a proxy variable for P_{SC} will typically give a significantly better regression "fit" than using current slaughter cattle prices to predict the price of feeder cattle.[4] This result is not surprising. The current quote for distant live cattle futures is, as Peck points out, a market price. Whether it turns out to be an accurate prediction of price may not even be important when consideration is restricted to the role of futures prices in supply response. At the time an investor is considering buying feeder cattle and placing them on feed, the futures quote (properly adjusted for basis)[5] offers a market price and offers an opportunity to forward price. Clearly, the current cash price is not a price that allows forward pricing.[6]

In the work by Purcell, Flood, and Plaxico a causal relationship between current quotes on distant live cattle futures and current cash prices for feeder cattle was statistically confirmed. The concept of derived demand theory suggests such a causal relationship would exist. Thus, futures prices for fed cattle (and surely for corn as well) are apparently an important factor in the discovery of feeder cattle prices.

In both livestock grains and fibers, futures quotes are increasingly important factors in determining producers' intended supply. Examination of the relationship between distant live cattle futures, current cash prices for feeder cattle and current corn prices indicate that timing opportunities to place feeder cattle, buy corn, and forward price at a significant profit exist for only short periods. Since cattle feeders bid up the price of feeder cattle and sell distant live cattle futures, they also close the profit "window" which had opened. Essentially the same behavior happens in the hog markets. In this case the initial response is often to hold back gilts from market to build total hog numbers. Consequently, a longer period is involved before the supply response runs its course.

Planting decisions about corn and soybeans as competing enterprises are made at least partly on price relationships. A simple model of the planting decision is:

$$RA_{SC} = f(P_C, P_S)$$

where

RA_{SC} = the ratio of soybean to corn acreage
P_C = corn prices before planting
P_S = soybean prices before planting

Using March-April levels of December corn futures and November soybean futures versus preplanting cash prices for corn and soybeans typically gives a better statistical "fit" of this simple model.[7] Despite the rigidities imposed by land quality, herbicide treatment, and costs of putting in a crop (variable costs are less for soybeans), producers clearly use futures quotes in making their planting decisions.

Without pursuing an exhaustive consideration of alternatives, it should be accepted by agricultural economists that futures prices are considered by decisionmakers, supply can and will respond to changes in futures prices, and cash price levels and cash price changes are at least partly related to futures prices through the supply response mechanism. Yet, Peck notes that most firm models have assumed that producers make decisions on past prices. Gardner, one of the analysts who has investigated the use of futures prices in supply analysis, concludes (p. 83):

The use of futures prices seems a promising tool in supply analysis. In the soybean and cotton (supply) equations, futures prices gave reasonable results and performed at least as well as Nerlovian lagged price, lagged dependent variable specifications. . . . Futures prices may be even more helpful in modeling the complex relationships of the livestock sector.

Hedging Framework

The literature on hedging strategies is in the embryo stage but is growing rapidly. A stimulus for an expanded base of research and analysis came with the start of trade in live cattle, live hog, pork belly, and feeder cattle futures in the 1964–72 period. Trade was thus expanded to nonstorable or semistorable commodities which had experienced high levels of price variability during the 1960s.

Futures markets allow decisionmakers to transfer price risk via hedging. Technically, hedging involves taking equal and opposite positions in the cash and futures markets. Risk exposure is reduced to what is called "basis risk," the risk that basis will change in an unpredictable manner after the futures contract is sold and the hedge is set.

In the early 1970s, most research concentrated on measuring the impact of selected hedging strategies on the mean and variance of net returns to the farmer's production program (Holland and others). Routine hedging,[8] hedging strategies based on seasonal patterns in cash prices, hedging to secure preselected levels in cash prices, and hedging to secure preselected profit targets were tested for year-round cattle feeding operations. In general, the results suggested hedging would decrease the variance of net returns with some concurrent decrease in the mean of net returns. Decisionmakers were, therefore, required to choose a strategy with the mean-variance trade-off a function of their risk preference pattern.

A second development phase involved attempts to use econometric models to predict cash prices and then to make a yes or no decision on hedging based on the cash price prediction. The work by Brown and Purcell and by Purcell and Richardson illustrates these approaches. Results varied across commodities. In most of the strategies based on econometric models, there were significant decreases in mean net returns relative to speculation in the cash market and little improvement relative to routine hedging strategies. During this period, Heifner and Ikerd looked at the relationships between production risk and market-related risk. In general, the analyses suggested that strategies which decrease exposure to market-related risk would allow increased exposure to production risk in the form of expanded production, increased borrowing of capital, or both.

Later in the 1970s, increased attention was paid to technical analysis[9] as an additional guide to hedging decisions. Selective hedging, using fundamental analysis,[10] technical analysis, or both to select appropriate timing of the hedge and when to speculate in the cash market, became increasingly popular. Research by a number of analysts (Brown and Purcell; Purcell and Richardson; Shafer and others; Russell and Franzmann; Kenyon and Cooper) indicated it is possible to both decrease the variance of income streams and increase the mean level of those income streams by applying selective hedging strategies based on technical price direction indicators. These results strongly suggest that price fluctuations in futures markets are not a random walk. Thus, analysts who have summarily dismissed technical analyses as not being valid should further investigate selective hedging based on technical indicators. Most commercial advisory brokerage services use both technical and fundamental analyses when issuing recommendations designed to guide the decisionmaker through a selective hedging program. Moving averages, various types of oscillators, relative strength indexes, and point and figure charts are among the many technical tools employed with the basic bar chart in attempts to monitor and project market direction.[11]

The increased levels of price variability have also prompted renewed interest in cash-futures relationships measured by the basis. For example, in the Midwest, the "basis contract"[12] has become a popular means of marketing for progressive and well-informed grain producers. By learning normal seasonal patterns in market area basis, producers look for a basis contract when the basis is unusually favorable (the discount of cash to futures is significantly smaller than normal). An agreement should:

- Tie the price to a distant futures contract, thus securing the favorable basis;
- Provide immediate delivery of the grain to the elevator, thus eliminating carrying costs;
- Receive immediate payment for 50–75 percent of the crop value based on current cash prices; and
- Specify the final date the producer has to decide to accept the prevailing net price of the futures contract minus the fixed basis adjustment.

A strategy involving a basis contract can be extremely effective for the grain producer when carrying costs are high. Elevator managers buy futures when they enter the basis contract to protect their position. They are still exposed to basis risk, but this is discounted in their contractual offer. The basis contract works for both seller and buyer in markets when the elevator needs grain and bids up cash price (narrowing the basis) when the farmer is aware of the logic of tying down the favorable basis while it exists.

Implications for Firm Modeling

Most firm modeling efforts have not incorporated the latest developments in marketing and related hedging strategies. The marketing strategies analyzed by Lutgen and Helmers in a recent article included only such simplistic alternatives as selling all at harvest, spreading sales through the market year, hedging an arbitrary fraction of the crops at harvest, and spreading the lifting of the hedges throughout the market year. As the authors noted, the hedging strategies were not designed to reflect either costs or market price movements. Recalling Anne Peck's point, it appears that only partial firm-level analyses of futures markets, market movements, and the operator's marketing behavior are being incorporated into analytical modeling efforts.

Target Price Approach

Modeling highly sophisticated marketing strategies can be complex and time consuming. For example, most technical indicators require an extensive data set of daily futures prices to test or optimize the parameters which determine the sensitivity of the technical indicator to changes in market price direction. In comparison, the target price approach offers a decision criterion which can be easily incorporated. The concept of a target price can apply in intrayear storage decisions, intrayear production decisions, or interyear production-related decisions. Conceptually, the establishing of a target price could involve:

- Estimate of costs,
- Salary for management (if not included in costs),
- A return on investment (if not included in costs),
- Estimate of basis for the end point of the decision period, and
- Incorporation of a discount for exposure to basis risk.

A production decision (deciding on acreage to commit to soybeans, how many cattle to place on feed, whether to hold back gilts for breeding) can employ the target price approach before the production process is begun. If resource flexibility is sufficient, the resources can be committed to those areas in which the target price is met (or most closely approached) and diverted away from those areas where the available prices are less favorable.

The target planning approach would obviously affect specification of whole farm planning models. The share of expected production that should be target priced requires a concurrent analysis of the financial position of the firm and the inclination and ability of the decisionmaker to manage risk. The more sophisticated modeling efforts will eventually also incorporate the probability of a resulting supply response and bring it into the decision on whether and how much to forward price. Most futures contracts are traded for at least a calendar year and allow adequate time to complete forward-pricing objectives.

When farm resource flexibility is restricted, an alternative marketing strategy can be incorporated into the models by assuming that whenever price exceeds average variable cost, the resources will be committed. In this situation, the decisionmaker monitors the market price offered and establishes forward-price protection whenever the price target is met.[13]

The financial dangers of using this simple target price approach are apparent. If the target price is not reached, forward pricing does not occur. Consequently, the manager is speculating in the cash markets and is totally exposed to price risk. But the use of such an approach, with reasonable assignments to manager's salary and return on the investment, will simulate businesslike behavior more closely and will typically allow forward pricing. For modeling purposes, this is an easy strategy to incorporate. More sophisticated marketing strategies can be combined with this basic approach as experience is accumulated. For example, a strategy based on a technical indicator of market direction could be used until the target price is reached and then deactivated. If the target price is not reached, the technical-based strategy occupies a fail-safe position.

Basis-Related Strategies

The cash-futures basis is the barometer of the market in the storable grain and fiber commodities. The important indicator is the level of the basis relative to the normal pattern, not the absolute level at any point in time. In general, the basis is interpreted as follows:

Unusually narrow: The market is saying "sell, don't hold." Anticipations are for a deteriorating basis as the crop year progresses. Logical approaches include selling or establishing a basis contract. Speculating in cash should be avoided.

Unusually wide: The market is saying "hold." Cash prices have been forced down by an excess supply relative to demand, and expectations are for an improving basis during the crop year. Logical alternatives include speculating in the cash market or holding cash and selling futures (hedging) to benefit from the expected improvement in basis.

The historical data pattern must be analyzed (but not back beyond 1970) to incorporate this type of strategy into a model. An unusual basis could be defined as a basis value outside the mean basis value plus or minus one standard deviation for a particular time period and futures contract. The basis-related strategies could be especially useful in combination with the target approach for storage-related decisions. For example, the decision to store could be "yes" even though the target price is not currently offered by the market, if the basis is unusually wide and offers the potential of basis improvement during the year. Storing the grain and waiting for the target price could, under these circumstances, be intelligent management of price risk.[14] Since a combination of strategies could be tied to the basis level, they could easily be incorporated into modeling efforts.

Fundamental Analysis Strategies

Fundamental analysis is extremely important in determining the probable direction of price trend for a particular commodity. Consequently, fundamental analysis can and should be used to supplement other strategies such as the basic target price approach. When the projected price trend is falling, the target price could be adjusted or some other type of strategy activated to protect the financial position of the firm. Whether formal econometric techniques or basic balance sheet decision criteria are used, awareness of the fundamental supply-demand balance in the market is a necessary condition for the intelligent management of marketing strategies.

Econometric models have not been highly effective as predictive indicators of when to have price protection in place by hedging or forward contracting. Most efforts discussed earlier concluded that the technical strategies produced better results in terms of mean and variance of income streams than did strategies tied to econometric price forecasting models (Kenyon and Cooper; Purcell and Riffe; and Brown and Purcell). This is a difficult conclusion to accept for many academic price analysts trained in, and predisposed to use, econometrics. This result is based on the lack of decision or data input flexibility in the fundamental models. The data are typically monthly or quarterly, are available with a time lag, and do not allow model updating to be completed to quickly discover turning points in the market price patterns.

Issues of conceptualization and specification are another problem with these models. There is surely some specification which would incorporate and anticipate the supply response and would improve the precision of estimates and decisions as well as the flexibility of the models. Perhaps it is our shortcoming as analysts. However, our continued failure to capture the behavioral dimensions and their related supply response is the key issue.

As our quantitative ability to capture the supply response in aggregate models improves, marketing models based on fundamental factors will become still more important and valuable. In addition, analyses which take account of developing and pending structural changes will have the capacity to project the trends in prices throughout the year or over longer time periods. In combination with strategies capable of capturing the shortrun movements in market prices, the choice of marketing strategy will be more nearly correct.

Technical Strategies

A single technical strategy does not exist that will work for all commodities or for the same commodity across time periods characterized by changes in the amplitude and frequency of price change. However, the conclusions in the literature are becoming increasingly clear with regard to important issues:

1. Price changes in futures markets do not follow the theoretical random walk. Hedging strategies based on technical indicators have been shown by many analysts to stabilize income streams and, on occasion, also increase the mean level of those income streams. This type of performance would be inconsistent with the concept of a random walk.

2. Optimum parameters for technical indicators have a significant degree of stability over time. For example, the choice of length on moving averages or the cell size and reversal requirements for point and figure charts perform well for most commodities for several years. Periodic reexamination is needed, but there appears to be sufficient stability to make the properly selected set of parameters functional.

Incorporation of selective hedging strategies into firm models is being blocked by many economists unfamiliar with these techniques, the often massive data sets required to optimize the parameters, and the conceptual problems in linking hedging strategies to the production and financial assets of firm models. Incorporating a selective hedging strategy would require:

1. Selection of the type of technical strategy criteria. It could be a technical oscillator, a point and figure chart approach tied to double bottoms and double tops, or a pair of moving averages using the crossover approach to signal when a hedge should be placed and lifted.

2. Selection of a strategy that is appropriate for the commodity in question. The literature offers some guidelines. Analysis designed to optimize parameters needs to be done. A technical indicator is needed which is sensitive to major price moves, but not so sensitive that it

whipsaws the user. One approach is to use computer programs to simulate results across a historic data set, select the best combination of averages, and then test them across the most recent years not included in the data set before making a final choice.

3. A selection process to examine the mean, variance, ranges, and the percent of profitable hedges (a net profit from placing and lifting the hedge). Calculation of the correlation coefficient between 30-day cash and 30-day futures flows would be very important. A large negative coefficient suggests success in offsetting the dips in the cash market while allowing gains in a rising cash market.

The technical strategies appear to offer the most potential in efforts to model risk. After investigating and selecting a technical system and related parameters, the additional step of incorporating the decision criteria into an appropriate model should not be overly difficult. The technical relationships are, or can be, totally objective without requiring subjective input. In firm models designed to simulate and to calculate cash intrayear flows, technical systems can be incorporated and used without the problems of frequent updating associated with the fundamental models. And these strategies can be used in conjunction with such approaches as a target price. If the target price is not reached, the properly designed technical system will be in place to protect the operation from unexpected or disastrous breaks in cash price. More research, continued refinement of the systems, more work to combine technical and fundamental systems, and a much broader base of awareness by analysts are still needed.

A Final Observation

With minor exceptions, the economic analyst as a model builder has not been able to effectively incorporate price risk and marketing strategies into firm models. The increased price variability and related risk at the level we saw during the 1970s is relatively new, and the time lag required for analysts to recognize its increasing importance to firm decisionmaking is not yet over. In addition, the subset of agricultural economists that has worked on marketing strategies, the use of futures markets, and ways to manage price risk more effectively have not really met, as yet, the subset that works on modeling agricultural firms and micro behavior. The result is the continued absence of a literature on modeling which deals effectively with fundamental, technical, and related management strategies in managing price risk. It is time to span this gap and move toward a more complete treatment in our analysis and modeling efforts.

Notes

1. Price risk will refer to intra-year price risk during the remainder of this paper.

2. An example is "A Report to the Honorable Neal Smith, Member of Congress, on the Systematic Downward Bias in Live Cattle Futures." When released in early 1981, the report received a great deal of publicity. The Chicago Mercantile Exchange responded that the report was poorly conceived and subject to criticism in terms of methodology and analytical procedure. The Commodity Futures Trading Commission (CFTC), the federal agency charged with the responsibility of overseeing trade in commodity futures, reviewed the report and found no reason for action or further investigation by its staff. When released from the U.S. Government Printing Office, most of the reports of the Committee on Small Business have on the front cover: "This staff report has not been approved or disapproved by members of the Committee on Small Business."

3. An initial margin is required for each contract traded. Amounts vary across brokers, but are usually 5–10 percent of the face value of the contract. If prices move higher after the producer sells futures to forward price, additional margin monies are required to protect the price which has been established.

4. For example, see Robert A. Brown and Wayne D. Purcell, *Price Prediction Models and Related Hedging Programs for Feeder Cattle*, Okla. Agr. Expt. Sta. Bull. B-734, Stillwater, Jan. 1978.

5. Basis is the difference between cash and futures for a particular futures contract in a specific market at a point in time.

6. Recognizing that forward cash contracts are always an alternative, I will restrict discussion to the use of futures as a means of forward pricing. When cash contracts are employed, the elevator or processor does the hedging for the producer and typically charges for the service in the margin or basis used to set the cash offer. I have no quarrel with this approach; but for many commodities, especially livestock, the producer cannot always get an opportunity to forward price via cash contracts.

7. This is a favorite exercise of students in my price analysis course. The R^2 is typically significantly higher for the model containing the futures quote.

8. Routine hedging strategies are those in which a hedge is set automatically at the time a position is taken in the cash market.

9. Technical analysis refers to the use of chart patterns, measures of volatility, levels of trading volume and open interest, and so forth to determine the direction of price trends and to guide the timing of hedges.

10. Fundamental analysis deals with the supply-demand forces in the market via econometric modeling, balance sheet approaches, and so forth.

11. Literature is sparse, but growing. Research by Franzmann at Oklahoma State, Price and McCoy at Kansas State, Shafer and his colleagues at Texas A&M, and Kenyon and Purcell at Virginia Tech is included in the efforts to test technical and fundamental analyses. A good source of reference material is *Commodities*, 219 Parkade, Cedar Falls, Iowa 50613.

12. A basis contract between producer and buyer sets the cash-futures basis. For example, corn at harvest might be priced at $0.20 under the May futures. Producers can select the day prior to May 1 on which they wish to price. Under this contractual arrangement, the producer is still exposed to price level risk but is not exposed to basis risk.

13. Examination of costs and trading level of futures in live cattle, for example, indicates only a small probability the decisionmaker can realize the target price when the cattle are being placed on feed. But there is a much higher probability that a profitable target price will be available some time during the production period. The same situation prevails in grains. Even in good crop years, periodic weather scares and announcement of export sales cause the market to bounce to levels which might meet the preestablished price objectives.

14. Assume the soybean basis at harvest is −$1.00 using the May soybean futures for the next year. If the normal basis pattern for May is −$0.20 and the "worst case" in recent years is −$0.30, the market is "offering" $0.70 (with high probability) and $0.80 or more (at a lower probability) if the producer stores and sells May futures. It is this offer that must be assessed against all the costs of carrying the beans.

References

Brown, Robert A., and Wayne Purcell. *Price Prediction Models and Related Hedging Programs for Feeder Cattle,* Bull. B-734. Oklahoma Agr. Expt. Sta., Stillwater, Jan. 1978.

Gardner, Bruce L. "Futures Prices in Supply Analysis," *American Journal of Agricultural Economics,* Vol. 58, No. 1, Feb. 1976.

Heifner, Richard G. "Optimal Hedging Levels and Hedging Effectiveness in Cattle Feeding," *Agricultural Economics Research,* Vol. 24, 1972.

Holland, David, W. Purcell, and T. Hague. "Mean-Variance Analysis of Alternative Hedging Strategies," *Southern Journal of Agricultural Economics,* Vol. 4, No. 1, July 1972.

Ikerd, John E. "Managing Total Market Risk in Feeding Operations," *Proceedings,* Oklahoma Cattle Feeders Seminar, Jan. 1977.

Kenyon, David, and Craig Cooper. "Selected Fundamental and Technical Pricing Strategies for Corn," *North Central Journal of Agricultural Economics,* Vol. 2, No. 2, July 1980.

Lutgen, Lynn H., and Glenn A. Helmers. "Simulation of Production-Marketing Alternatives for Grain Farms Under Uncertainty," *North Central Journal of Agricultural Economics,* Vol. 1, No. 1, Jan. 1979.

Mapp, Harry P. Jr., M. Hardin, O. Walker, and T. Persaud. "Analysis of Risk Management Strategies for Agricultural Producers," *American Journal of Agricultural Economics,* Vol. 61, No. 5, Dec. 1979.

Peck, Anne E. "Futures Markets, Supply Response and Price Stability," *The Quarterly Journal of Economics,* Vol. 90, No. 3, Aug. 1976.

Purcell, Wayne, D. Flood, and J. Plaxico. "Cash-Futures Interrelationships in Live Cattle: Causality, Variability and Pricing Processes," *Proceedings*, Livestock Futures Research Symposium, Chicago Mercantile Exchange, June 1978.

_____ , and Thomas Richardson. *Quantitative Models to Predict Quarterly Average Cash Corn Prices and Related Hedging Strategies*, Bull. B-731. Oklahoma Agr. Expt. Sta., Stillwater, Dec. 1977.

_____ , and Don Riffe. "The Impact of Selected Hedging Strategies on the Cash Flow Positions of Cattle Feeders," *Southern Journal of Agricultural Economics*, Vol. 12, No. 1, July 1980.

Richardson, James W., and Gary D. Condra. "Farm Size Evaluation in the El Paso Valley: A Survival/Success Approach," *American Journal of Agricultural Economics*, Vol. 63, No. 3, Aug. 1981.

Russell, James R., and J. Franzmann. "Oscillators as Decision Guides in Hedging Feeder Cattle: An Economic Evaluation," *Southern Journal of Agricultural Economics*, Vol. 11, No. 1, July 1979.

Shafer, Carol E., W. Griffin, and L. Johnston. "Integrated Cattle Feeding Strategies, 1972-76," *Southern Journal of Agricultural Economics*, Vol. 4, No. 1, July 1978.

Discussion of Part 5

Modeling Farm Decisionmaking to Account for Risk

Glenn A. Helmers

The 2 papers directed toward risk modeling differ considerably in their scope regarding risk management and its inclusion in firm models. The Eidman paper is of broad scope encompassing risk resulting from production, product prices, inputs, input prices, and finance. A narrower risk focus is presented in the Purcell paper examining product price risk and strategy in hedging using futures markets. Eidman is more directed toward elements of risk models while Purcell emphasizes the use of hedging in a profit-maximizing framework without explicit consideration of different risk-aversion levels of firm managers.

In recent years concepts of firm risk have broadened as a result of large changes in economic variables coming from outside agriculture. Traditionally, agricultural firm risk management has examined yield and product price variability with relatively less emphasis on input and input price variability and financial risk. However, increased international market activity, along with increased dependence on purchased inputs, has increased firm risk related to product prices, input prices, and input availability in recent periods. Similarly, changes in monetary variables as well as changes in tax laws have caused agricultural firm managers concern over credit restrictions, inflation, variable interest rates, and changing tax, inheritance, and retirement laws. Thus, national economic instability now appears to be a more important factor related to agricultural firm risk than in previous decades. Elements of these factors in Eidman's paper relate to financial risk resulting from debt financing and in Purcell's paper relate to apparent increased product price variability.

Interest in risk modeling arises because of the hypothesis that including behavioral aspects of risk management helps better explain decision processes of firm managers. If, for example, dairy cow replacement strategy is better explained by a stochastic model compared to a certainty model, our confidence in the model is increased. Unfortunately, much modeling research stops before the validation stage. The understanding

The author is a professor of agricultural economics, University of Nebraska.

of decision processes of managers is essential before micromodels can be constructed for normative use.

Risk Management Approaches

In a broad sense, risk management modeling approaches fall into 2 categories. The first is active incorporation of behavioral adjustments to risk. This category would include optimizing models (based on utility) using subjective or objective risk estimates with particular attention paid to the risk-aversion level of the manager. These models may be static or dynamic and adaptive or nonadaptive in response to the time element. The second approach does not explicitly consider risk adjustment in the model, but instead improved forecasting becomes the risk control factor. Such models are evaluated on predictive performance and may place little emphasis on the risk position of the decisionmaker. Often, farm operators suggest their method of minimizing risk is one of maximizing returns.

Modeling risk requires the incorporation of risk measurement to parallel concepts of risk. Here the issue becomes complex when considering production, prices, and finance risk. In general, variability of returns has been more widely accepted as a measure of risk relative to minimizing probabilities of disaster events (safety first) because of the theoretical relationship of income variability to the expected utility hypothesis. Risk is considered to refer to departure from expectations. Thus high levels of variability and instability, if predicted well, may not necessarily be synonymous with high risk. Yet in some cases, the occurrence of disaster events, even though well predicted, causes extreme problems. Some producers may perceive the risk level of an event, such as the price level, as inseparable from its probability of occurrence in terms of risk conception. Another problem relates to producers who appear to have great disutility when not securing the highest price ex post facto. Risk strategies such as hedging with satisficing behavior in models of firm behavior may be unrealistic for such firms. Yet, such firms may appear to behave according to conventional risk models with respect to other aspects of their business.

The Present Status of Research on Risk

Eidman deals with the following issues:

- the basis of much risk research—the expected utility hypothesis— and the assumption of normality in the distribution of outcomes,
- the advocacy of after-tax wealth as the model objective function and basis from which to measure dispersion,

- the need to consider resource availability risk,
- the types of farm risk including both business risk and financial risk which result from debt financing,
- the problems of measuring risk preferences, and
- the use of stochastic dominance in selecting efficient choices of risky alternatives.

This paper is an excellent summary of the status of risk modeling issues. I would support the use of a multiperiod after-tax wealth function in modeling firm behavior. At the same time, its use adds a complicating modeling feature which for even simple risk analyses requires considerably more research effort. A major difficulty in incorporating the behavioral aspects of this function lies in the decisionmaker's perception of the meaning of after-tax wealth and changes or differences in its level, particularly under increasingly complex forms of business organization. Thus, the choice of a $1,000 difference in after-tax, long-term wealth among alternative decisions when each decision is accompanied by its respective probability is more difficult to assess than short-term income differences. Under more complex forms of business organization, the financial adviser is already often required to interpret cash flows. Further, tax consequences are often difficult to model, particularly the revisional aspects which are dependent upon subsequent incomes.

Another point related to multiyear models is changing risk-aversion levels by producers over time in response to various factors. The direct measurement of risk preferences has been shown to entail considerable problems. Temporal aspects to such measurements add even more complexity, but are essential to time-related risk choices where growth and investments form the focus of choices.

In firm research related to financial risk, it is likely that a substantial amount of sensitivity research will be attempted examining alternative monetary and tax changes. Sensitivity analysis is usually attempted in the absence of strong convictions over the magnitudes of economic parameters. In the firm financial area, sensitivity analysis involves alternative risk scenarios of price movements, credit restraints, and so forth. I would like to emphasize using caution toward such activities. Little is generally learned from such modeling activity.

Varying the economic parameters individually can result in analyses of disequilibria which are largely irrelevant because changes in some parameters are interrelated and are not independent. Recently this activity has been most evident in models where price movements resulting from inflation are hypothesized without specifying equilibrium-based relationships between inflation, interest rates, and product and resource prices. Also scenario research cannot instill confidence in either a micro-

or macromodel; confidence can only be secured in well thought out research validation and testing efforts with consistent sets of assumptions about exogenous variable levels.

Including Hedging and the Futures Market

Purcell makes the following points:

- agricultural economists have tended to ignore the futures markets;
- commodity futures act both as a price discovery tool and as a transfer of risk mechanism;
- much risk reduction can be secured by selected hedging without a necessary trade-off in returns;
- target prices can be effectively hedged; and
- hedging strategies can be developed from both technical and basic market information.

This paper is directed toward one specific strategy among alternative strategies for reducing product price risk and is devoted almost entirely to hedging and hedging research. It summarizes much of the marketing research which became focused on product-price selling strategies after the occurrence of increased price variability during the past decade. Purcell's paper is not directed to a discussion of detailed firm modeling aspects except for the selective hedging mechanism of target prices and selective strategies based on basic and technical aspects.

A few points discussed in the paper need further research, and I list them here. The causal relationship between futures prices and current prices is not always obvious, particularly for distant futures prices. The continued existence of price cycles suggests that producers still project current prices rather than basing production plans on forecasted or futures prices. The argument over the random walk of the futures market continues. The use of target prices as a basis to hedge upon is a satisficing criterion and somewhat subjective. Finally, it is not clear that a futures market is a more effective price-forecasting mechanism than updated econometric models. Perhaps neither tool is accurate enough to reduce price risk dramatically in the absence of more active risk strategies. Thus, as Purcell suggests, hedging may be the best possible risk-reducing mechanism we have to minimize shortrun product price risk.

I see 2 problems that relate to the use of hedging in modeling risk decisions. One is related to the issue of longrun investment decisions. Longer run knowledge of prices is necessary to these investment decisions. For such longer run modeling, econometric forecasting may be useful in developing price expectations when futures prices are not available. A second problem lies in distinguishing where hedging leaves off and speculation begins. The implementation of selective hedging as a risk-reducing strategy requires such a distinction.

More research is necessary related to the concept of price risk and the theoretical processes of risk transfer through hedging. Finally, selling decisions must be made in a broad concept of firm risk. For example, because of income tax considerations, a lower price received at a point in time may be preferable to a higher one in the next period. Consequently, after-tax firm models may be preferable to analyze behavior decisions. Nevertheless, the major issue is that production, selling, investment, and financial decisions are not totally separate, and much may be gained by combining them in comprehensive risk and planning models.

Discussion of Part 5

Thoughts on Modeling Risk Management on the Farm

Steven T. Sonka

One criterion for evaluation of a presented paper is whether that paper was thought provoking. Did the paper provide insights or raise questions that expose weaknesses or strengths in our conventional wisdom? For me both Eidman's and Purcell's papers met that criterion. Further, I commend both authors for not just reviewing the standard literature but also for attempting to extend our thinking about modeling the risk management process. This is not to say that there aren't several areas of concern or clarification which warrant further discussion. In these comments, I will not attempt to exhaustively critique the Eidman and Purcell papers. Rather I propose to share some thoughts which came to my mind after studying them.

Wealth Measure of Risk

I will discuss the Eidman paper first. His suggestion for using a wealth measure of risk should be commended. Some past empirical applications have used variability of annual cash flows in an inadequate manner to model risk. Focusing on wealth, as properly defined by Eidman, forces us to recognize that risk is a longrun concern relating to personal decisionmaking at the farm firm level. Although there may be instances where variability of annual cash flows is an appropriate proxy for wealth, we should not automatically use that variable without first assessing its appropriateness.

Eidman is correct in his concern about how we have incorporated the effect of income taxes on farm decisionmaking. Income tax considerations may have as important an effect on variability of actual returns as they do on expected returns. His suggested modeling approach may be theoretically unsatisfactory. However, it appears to be a heuristic approach which could lead to better identification of the decision setting

---

The author is an associate professor of agricultural economics, University of Illinois, Urbana-Champaign.

facing the farmer. It certainly deserves empirical testing and analysis as a potential means to solve this problem.

Eidman's discussion of portfolio analysis and the E-V frontier raises several issues. The E-V frontier is an attractive conceptual framework. The concept of a trade-off between risk and expected income is intuitive. I am concerned, however, with our extensive empirical use of this framework. We can become frustrated evaluating the positive contributions of those empirical applications focusing on tracing out efficient frontiers. But as an increasing number of efficient frontiers are estimated, it is not clear that an improved understanding of the risk management problem results.

There are 3 problems associated with empirical use of the E-V frontier. The first is that we are really interested in farmer decisionmaking, not just the efficient frontier. The key question is: How will or should farmers make decisions in an environment where risk is a major issue? Many empirical applications trace the possibility set and evaluate which are most efficient. However, we have not spent sufficient time concentrating on the other half of the choice question: How do the farmer's utility function and other decision criteria affect decisions actually made?

A second concern is the use of a risk-aversion coefficient which is constant throughout the range of alternatives facing the farmer. It is not intuitively clear, and probably false, to assume that the farmer's risk-aversion coefficient is the same for a production decision (what varieties of corn to plant) as it is for an investment decision (how much to pay for an acre of land). We need to develop and use techniques which will allow risk preferences to vary, based on the magnitude of the decision.

A third concern is the way outcome variability is measured. In many of our applications, we tend to regard variability as bad. Variability in a positive direction is often presumed to have the same effect on the farmer as variability in a negative direction. Of course, use of the normal distribution, whether it fits the data or not, may be a strong impetus for this practice.

Some farmers may not regard upside and downside risks as being equivalent. Some agricultural economists might argue that farmers are only concerned with downside risk. I would disagree. From the standpoint of the farmer, positive variability represents the potential for making wrong decisions. For example, a decision to not purchase land implies a foregone opportunity to increase production and capture capital gains. In the short run, not buying land could be a safe decision from the standpoint of current cash flows. But, from the longer run view of the farmer's expected future net worth, that shortrun decision may be risky.

The farm firm is, to a large degree, a small closely held firm. The decisions of the farmer directly affect the personal wealth of that individual. Therefore, decisions causing foregone wealth increasing opportunities are often of major concern to particular operators. Again, we do not know enough about how farmers actually make decisions. Our concentration on risk as a source of non–profit-maximizing behavior may have led us off the main trail. Several factors including production constraints, multiple goals, and lack of information could cause farmers to make non-profit-maximizing decisions. We should be concerned about assessing the impact of all these factors—not just the presence of risk.

Modeling a Particular Source of Risk

The Purcell paper takes a different approach to modeling the risk question than does Eidman's. I initially had a difficult time assessing this paper because its approach to modeling risks is somewhat different than I had expected. It does point out that there are really 2 possible approaches. To illustrate, consider an assessment of the withdrawal of a certain kind of pesticide. Information could be gained from the study, on an experimental plot basis, of plant growth with and without that pesticide. An output of this study would be estimates of the effects of using that pesticide on mean yields and fluctuations in yields. A second kind of study would assess the impact of withdrawal of this pesticide on the whole farm system. The second study would draw upon the parameters estimated in the first. The Purcell paper concentrates on the first approach, that of attempting to model a particular source of price risk. This attempt, of course, is a useful and necessary activity to pursue.

Another major question is: How do people attempting to model the second type of analysis incorporate price risk in an appropriate and sophisticated manner? For example, Purcell notes the increasing use of basis contracts for sale of grain commodities. For a researcher attempting to assess the larger farm decisionmaking process, how does the emergence of basis contracts affect the measures of average income and fluctuations in income now available to the farmer? This also is an important area of analysis.

Purcell does document an important reason why more sophisticated analyses of price risks have not yet been completed. Changes in the marketing structure brought about by the sudden increase in grain exports during the 1970s have limited the number of observations relative to the new market structure that farmers now face. A related question is: In the context of the whole farm model, how does the researcher handle the marketing information that the farmer actually has? For production risk, it may be appropriate to assume that the farmer only has historical information to use to estimate next year's weather conditions.

When the farmer makes marketing decisions, however, that farmer has a considerable amount of information about the future. For example, the farmer knows that the supply-demand conditions for corn in 1981 are greatly different than they were in the fall of 1980. Certainly the farmer's expected price would be different in those 2 years. But aren't the farmer's expectations of potential price fluctuations (the price range within which the farmer thinks trading could occur) also going to be different? Assessment and guidance in incorporating this kind of information into whole farm models is needed.

In summary, both papers would be valuable reading for economists concerned with modeling the risk management process on farm firms. Both papers imply that considerable work at both the conceptual and empirical levels is needed if we are to gain a better understanding of the decisionmaking process on farm firms.

Discussion of Part 5

The Need for Prescriptive Risk-Management Models: A Suggested Methodological Approach

James N. Trapp

In reviewing Eidman's and Purcell's papers, one quickly sees that they are focused in different directions. Eidman's focus is positivistic, pointing out the need to model producer's perceptions and response to risk. Purcell's focus is normative. He builds a strong case that normative firm models have used naive risk-management strategies and have, therefore, failed to develop effective, normative firm decision models.

Eidman makes a valid point that we must be able to determine how producers perceive risk before we can incorporate the effects of risk into our models, or, for that matter, before we can develop strategies for producers to use in managing risk. Likewise, if we are ever to evaluate changing the producer's risk environment, we must be able to ascertain how producers respond under different risk situations. Hence, I see Eidman's work as fundamental to risk modeling, but not actually involved in modeling risk management; however, it lays the groundwork for such activity.

Purcell, in contrast, leaps over the problems that concern Eidman and assumes that producers do face risks and that, given sound, demonstrated risk-management models, they will use them to their own benefit and presumably to the benefit of society. The latter point is never raised, however. Indeed, Purcell could be criticized on the grounds that he has not stopped to consider whether his suggested approaches to risk management are related to perceived risk situations held by producers and, hence, whether producers will ever use these approaches, if developed. The risk-management strategies suggested by Purcell are largely related to the futures market. Indeed, many producers seem to have very inhibiting perceptions about this market. These inhibitions make risk-management models based on any type of use of the futures market incompatible with many producers' management philosophies. However, just as Purcell's work can be criticized as having gone too

The author is an associate professor of agricultural economics, Department of Agricultural Economics, Oklahoma State University, Stillwater.

far too quickly, Eidman's work may suffer from the opposite problem. He has placed so much concern on defining and measuring risk that he has made no progress in presenting ways to cope with risk or in designing policies to alleviate risk.

With these caveats, I shall now discuss and expand on the work of Purcell. In doing so, I shall join him in attempting to solve problems which Eidman justly contends we don't fully understand.

The Need for Normative Risk-Management Models

Purcell builds the case that recent increases in price variability have increased the need for normative models of firm decisionmaking that incorporate price variability and for decision criteria that recognize the existence of uncertain prices. Purcell's evidence for increased price variability is demonstrated by only one graphic illustration. Although intuitively appealing, this brief presentation lacks convincing rigor. Ray and Trapp give a broader and more rigorous measure of changing price variation in which coefficients of price variation for pork, fed beef, feeder calves, corn, and soybean meal are all shown to have had statistically significant increases in magnitude over the 1949–76 period.

Purcell contends that the profession has failed to effectively treat price risk in its modeling efforts because it has virtually ignored the futures market in its modeling development research. I believe this contention is somewhat overstated. I contend that the primary reason the profession has not done a better job of modeling normative risk-management strategy is because the methodology has not yet been developed for adequate risk-management strategy models. Our previous modeling methodologies have led us astray and caused inept normative risk-management models to be developed and then quickly discarded. I must agree with Purcell that some of the efforts in developing models to cope with price risk which use futures market technical analysis tools should be considered for their potential. But more important, I think we can learn why these tools work and which methodological requirements are necessary for normative risk-management modeling. Before discussing the contributions of technical analytical methods to those requirements, I want to touch on several more points Purcell makes with regard to using the futures market to manage risk.

Conceptual Issues Regarding Risk Management Strategies

Purcell's paper leaves the impression that the only type of risk-management strategies that can be used to cope with price variation are futures market related. Although the futures market may be a primary tool one can use in coping with price variation, other types or classes of risk management methods exist—for example, government program participation strategies, storage decisions, price forecast application, diversification and portfolio theory, and so forth. In using the futures market as a risk management tool, Purcell makes three points I would like to expand on.

First, Purcell makes a critical point that in dealing with risk management, researchers must consider both the production and marketing aspects of risk management simultaneously. Strategies must integrate both production and marketing decisions because one affects the other; that is, one cannot choose to contract for future delivery dates of grain or cattle without considering the resource allocation effects within the firm. Our profession has often drawn a distinction between production/ farm management research and marketing research. This distinction has served as a barrier to "firm decision modeling" which in reality must span both production and marketing decisions.

Second, Purcell presents a strong case that producers do make decisions based on expectations and that these expectations are more accurately reflected by futures market prices than by some set of weighted historical prices. This aspect of Purcell's paper relates to one of Eidman's issues, that of the value of considering the expectation model used by producers. Purcell cites several cases in which futures prices have been statistically shown to be better explanatory variables of producers' actions than lagged market prices. Recent work by Firch indicates that futures prices may actually serve as "inverse forecasters" of market price because of producers' responses to them or, perhaps more correctly, because of the feedback actions which follow from the expectations the futures market reflects. Firch estimated that high harvest period futures prices for crops at planting are correlated with increased production efforts and hence with subsequent declines in price over the crop year. Therefore, futures prices for the harvest period that existed at planting are inversely correlated to eventual harvest prices and to the level of the harvest period contract price on its expiration.

The third, and perhaps the most important, point Purcell makes is that risk management models using technical strategies have performed better than econometric/fundamental-based models. Technical strategies referred to include such techniques as moving averages, oscillators, and point and figures chart formations. Purcell fails, however, to ascertain why technical strategy-based models work better than econometric-based models.

Technical, strategy-based risk-management models perform in a superior manner to fundamental, strategy-based risk-management models because they are specified with direct objective functions instead of indirect objective functions. To elaborate and illustrate this point of view, let us consider that the parameters selected for technical, futures market trading models are selected on the basis of those parameters which will directly optimize some form of trading profits. The parameters of an econometric model are selected to minimize error squared of a set of forecasted values. The forecasted values are then used in a decision model framework to make decisions aimed at maximizing some management objective, generally profit or some variation of profit. Hence, the parameter selection criteria of the econometric model contributes only indirectly to maximization of the management objective because accuracy of the variable used in the decision model enhances the ability of management to obtain its objectives.

There is no reason to believe that the criteria of minimizing the error squared of forecasted values of prices used by management in making decisions will result in the maximization of management's objective. One must recognize that management will make mistakes in its decision process due to imperfect information. The cost of these mistakes in relation to management's objective criteria will only by chance be weighted the same as the error squared criteria constituting the econometric model objective function. Consider the case where a forecasted wheat price above $4 per bushel would result in a decision to store wheat, while a forecast below $4 will result in a decision to sell. A forecast of $4.02 versus an actual price of $3.97 will result in a wrong decision and loss. But a forecast of $4.30 versus an actual price of $4.20 would result in a correct decision and no loss, just lower profits than expected. The latter error would be weighted four times more heavily by the error squared objective criteria. However, its negative impact upon management's objective may have been much less than that of the former 5-cent error. Hence, the point is, econometric-based decision models tend to maximize an unrelated or indirectly related objective function while technical strategy decision models maximize objectives functions much more closely related to management's objectives. The technical strategy model tends to directly consider the decision process and decisionmakers' objectives. The econometric model does not. Given that a performance criterion related to the management's objective is used to evaluate a model's performance, it is not surprising that the technical strategy decision model outperforms the econometric-based model. The evidence cited by Purcell confirms this.

Comments Toward the Development of a Decision Modeling Methodology

A comprehensive firm decision model must: (1) describe the environment and information context within which the decision is made; (2) forecast the environment and the consequences of alternative actions;

and (3) prescribe an action in light of the decisionmakers' objective and the degree of risk associated with imperfect forecast. I believe the profession has concentrated too much on forecasting the general environment that decisionmakers face. It has not placed enough emphasis on describing the specific environment of the decisionmaker and the nature of that person's objectives. Our models have tended to be static and discrete, as opposed to dynamic and continuous. Decisionmaking is a time-dynamic phenomenon. Little effort has been placed on expanding the information/data base the decisionmaker has to use. Rather more emphasis has been placed on maximizing the extraction of information from limited available data. Finally, the most neglected area may be that of making decision prescriptions. Our emphasis on forecasting has caused us to separate the forecasting objective from the decision objective. All too often this results in a model where decision prescriptions are an afterthought of the modeling process instead of an integral part of the process.

In light of the criteria cited above, let us consider the merits of the technical analysis tools Purcell discusses. I believe the strength of technical analysis tools is that they are inherently prescriptive and were designed to provide a futures trading prescription. The profession has tended to degrade technical analysis tools because they are not forecasting models in the traditional sense and because their parameters have generally been derived with highly subjective methods.

The refusal of the profession to accept technical analysis models reflects its tendency to exclude consideration of the decisionmaker's environment and objective in its market forecasting models. We have continued to insist that we must first accurately forecast a specific price and then use that price to make a prescription. This is an oblique method of prescriptive decision modeling. Technical analysis models integrate the decision prescription into the model and directly consider the decisionmaker's objective, i.e., maximize trading profits. As such, technical analysis models do not forecast, they prescribe. Larger profits can be generated from hedging strategies guided by technical models which simply call turning points in the market than from those guided by econometric model price forecasts. The key is that the structure and parameters of the technical strategy model were predicated upon maximizing trading profits. The objective is not to minimize forecasting error, but to directly maximize profits through correct prescriptions.

The profession has rigorous procedures for estimating model parameters which optimize criteria other than squared error. I believe it is time we started to use them in our decision modeling efforts. Indeed, they can be used to improve upon and lend rigor to the parameter selection process for existing technical models such as moving averages and oscillators. Recent research at Oklahoma State by Franzmann and Shields has already used such optimizing methods in the study of moving average models. Once we begin to specify our models with prescriptive objective functions instead of forecasting objective functions, progress in firm decision modeling will be made.

The most prevalent discussions of generalized optimization procedures that are capable of optimizing prescriptive objectives are in the literature related to optimal control theory. Actually the formulation of a risk-management decision strategy is really a search for a stochastic control function. The parameters sought are parameters of the control function. By stretching our imaginations, I believe we can view technical analysis models as "sequential open loop" control functions. This assertion, if correct, points out the tremendous gap in communication and conceptualization in our profession between the traditional marketing economist and the new generation of quantitative modelers. Indeed, I believe the optimal control approach to modeling has much to offer in terms of developing prescriptive decision models instead of forecasting models which are only indirectly prescriptive.

In summary, several comments that reemphasize previous points are warranted. Models describing the decisionmaking environment should be dynamic because I believe the decisionmaking environment itself is dynamic. Hence, the framework of optimal control and simulation modeling is well oriented to the firm decision modeling problem. Second, the firm simulation model must encompass both the production and marketing aspects of the firm's environment. Finally, the model must be prescriptive in nature and specified to maximize what has been referred to here as a "prescriptive objective function."

References

Firch, R. S. "Futures Markets as Inverse Forecasters." *Proceedings, Conference on Applied Commodity Price Analysis and Forecasting.* Ames, Iowa, Oct. 1981.

Franzmann, John R., and Mike E. Shields. *Managing Feedlot Price Risks: Fed Cattle, Feeder Cattle and Corn.* Bull. B-759. Oklahoma Agr. Expt. Sta., Stillwater, Oct. 1981.

Franzmann, John R., and Mike E. Shields. *Multiple Hedging Slaughter Cattle Using Moving Averages.* Bull. B-753. Oklahoma Agr. Expt. Sta., Stillwater, Jan. 1981.

Ray, Daryll E., and James N. Trapp. "Grain-Livestock Interrelations and Trade-offs with Implications for the Structure of the Livestock Industry," *Proceedings, Symposium on Agricultural Policy.* Kansas City, Mo., Feb. 1977.

PART 6

Simulation Models

10. FLIPCOM: Farm-Level Continuous Optimization Models for Integrated Policy Analysis

Kenneth H. Baum and James W. Richardson

Abstract Two alternative recursive programming-simulation (hybrid) models, FLIPSIM and FLIPRIP, were developed with the objective of providing decisionmakers with economic intelligence relating to the dynamic effects of policies and economic conditions on various representative farm situations. The user-specified data sets must include information on farm size, behavioral parameters, costs of production, other technical coefficients, tenure, equity, and types .of crop and livestock activities. The differential growth and adjustment behavior of Illinois, Texas, and Mississippi representative farms in the same policy and economic environment was stochastically simulated from 1980 to 1989 to illustrate the models' usefulness.

Keywords Microeconomic policy analysis, Monte-Carlo techniques, recursive programming-simulation models, representative farms

The farm-level continuous optimization models for integrated policy analysis (FLIPCOM) consist of 2 FORTRAN computer programs, FLIPSIM and FLIPRIP. These models were developed to describe and predict the effects of alternative agricultural policies and economic conditions on the income flows, resource use, and financial characteristics of typical farms over a 10-year horizon.

The authors are, respectively, an agricultural economist, National Economics Division, Economic Research Service, U.S. Department of Agriculture, and an associate professor, Department of Agricultural Economics, Texas A&M University.

FLIPSIM and FLIPRIP are recursive and incorporate elements of mathematical programming and simulation techniques. FLIPSIM and FLIPRIP are similar in operational structure, but differ in their respective use of mathematical programming solution algorithms. Neither model is farm or location specific, and both are capable of using data sets describing different sizes, tenure, equity, and types of crop and livestock farms in different production regions. Each data set also describes agricultural policies including commodity income and price-support programs, the farm's behavioral parameters, and the macroeconomic environment. Consequently, the models are analytical tools capable of providing decisionmakers with economic intelligence relating to the effects of a large number of farm programs and policies on particular types of farms or a particular farm in an extension setting.

FLIPSIM was developed at Texas A & M University (Richardson and Nixon) under a cooperative agreement with USDA's Economic Research Service (ERS). FLIPRIP was developed in ERS by modifying FLIPSIM in order to provide an alternative modeling approach to various methodological problems, to provide a closer link with other computer software available to ERS, and to provide greater policy analysis flexibility by ERS staff (Jensen and others; T. Hatch; Baum and others).

Review of Previous Research

Aggregate sector simulation or programming models have often been used to analyze the effects of agricultural policies and commodities on the farm sector (Heady and others; Ray and Richardson; Wade and Heady). Typically, these models estimate the levels and changes in macroeconomic variables such as net farm income, program benefits, crop and livestock prices, and production patterns. However, the analysis of the dynamic impacts of agricultural policies at the individual farm level has not received the same level of research attention, even though actual production and resource decisions are made by individual producers. An implicit assumption often made by users of these macromodels is that an average farmer's behavior and decisionmaking process can be sufficiently well estimated with aggregated data to answer microeconomic questions about the farm sector. This assumption may not be entirely correct. There is undoubtedly a compromise perspective that will allow the researcher and policymaker to simultaneously consider the farm sector as a whole, while also considering crop and livestock farms as component parts.

Research studying the dynamic farm or micro-level impacts of macroeconomic scenarios or agricultural policies has occurred sporadically during the last 15 years. It started perhaps a decade earlier with multiple-period linear programming optimization models. Although the methodological approaches have varied widely, the whole-farm specification problem seems to have been uniformly recognized and accepted, partially as a result of farm management research from earlier decades. Perhaps this approach has been due to the widespread recognition that modeling farm and producer behavior is extremely complex and that empirical results can be greatly influenced through unintentionally biased assumptions and the exclusion of important behavioral details and constraints. These models have typically been used once, having been constructed as an integral part of a thesis or dissertation studying a particular aspect of financial planning, income taxes, estate taxes, growth strategies, investment behavior, inflation, alternative accounting procedures, or commodity programs. Whole-farm dynamic modeling has been sporadic and noncontinuous, and few of these models seem to be currently utilized in ongoing research. In addition, very few of these models are constructed in a manner that would permit analysis of representative farms in different regions of the country.

Three categories of whole-farm models are reported in the agricultural economics literature. In order of appearance they are optimization models, simulation models, and hybrid models incorporating both simulation and multiple-constraint mathematical programming techniques. Optimization models are generally multiperiod, deterministic, and use linear programming techniques (Loftsgard and Heady; Walker and Martin; Boehlje and White; Heidhues; and Lins). Reference to these models as "models" is somewhat misleading, since the data are in a very real sense "the model." The data describing the farm and its technical relationships are used to solve an objective function with a linear programming algorithm. Thus, if the data are changed, the model is changed because the constraint and objective function structures are mutually dependent. The input-output technical coefficients also tend to be farm specific.

Whole-farm simulation models can be deterministic or stochastic. They differ from optimization models because a multiple-constraint optimization algorithm is not included. Consequently, these models do not have the internal problem-solving capability for determining an optimal growth strategy for a farm facing more than one resource constraint (Boehlje and Griffin; Halter and Dean; Patrick and Eisgruber; Patrick; Hutton and Hinman; Hardin; Held; Roush and others; Umberger; and Holland and Young). With the exception of the research studies done by Hardin and by Hutton and Hinman, simulation models have been developed primarily for analyses of farm- and region-specific problems. For example, Boehlje and Griffin studied the effect of alternative price-support programs on income and land values for different size farms in Iowa. Holland and Young analyzed the impacts of the 160-acre limitation on different size farms in Washington. The conceptual advantages of hybrid macro- or microeconomic decisionmaking and analytical models were suggested over a decade ago (Day, 1964; Dorfman). However, the empirical methodological problems and techniques involved with linking or interfacing simulation and programming elements with recursive interactive programming (RIP) and recursive adaptive programming (RAP) models have only been recently explored in any depth (Baum; Huang and others; Weisz and others).

Hybrid farm simulation models incorporate optimization techniques to select the annual crop mix or determine the optimal growth strategy (Armstrong, Connor, and Strickland; Chien and Bradford; Condra; Hardin and Lacewell; Harman; R. Hatch; Kingma; Richardson and Condra). In these studies, the optimization routines specified MOTAD, LP, or recursive programming (RP) problem formats. Again, these models are not readily applicable to farms in other regions for reasons discussed earlier. This limits their information input capability for aggregate analysis questions.

A comparison of selected whole-farm simulation and hybrid models is presented in Table 10.1. The Halter-Dean model is a landmark research effort since it was the first whole-farm simulation model. The Patrick-Eisgruber and Hutton-Hinman models made numerous improvements over the first whole-farm model by adding tax, balance sheet, cash flow, and debt use output information. Several models developed at Oklahoma State University during the early 1970s were modifications of the Hutton-Hinman model (Mapp and others; R. Hatch; Harman). The Hutton-Hinman type model has been recently replaced by a third generation of farm simulation models (Baum and others; Boehlje and Griffin; Hardin; Richardson and Condra), incorporating options including farm programs, farmland disinvestment, stochastic prices and yields, family consumption functions, tax laws, machinery replacement, flexible inflation rates, and optimization routines to determine the annual crop mix and resource use.

The Dynamic Optimization Problem

The primary problem from the point of view of the farm operator is to simultaneously allocate inputs or resources among crop, livestock, and off-farm income-producing activities while taking into account financial constraints. Inputs available to the producer include land, labor, capital, and management and are differentiated by quality, type, and quantity. The FLIPCOM models assume these resources must be optimized and allocated using a criteria function that maximizes expected utility from economic gains, π. Consequently, the models make production period, activity mix decisions by maximizing expected net revenue.[1]

Multiperiod optimizing decisionmaking and planning by individual producers has been characterized by Rausser and others, in terms of the putty-clay model originally formulated by Salter and by Day (1974), as myopic optimizing. The putty-clay model describes a situation where assets are fixed in the short run, limiting the choices of input mix and output. Over time, the quality and quantity of assets are more variable and output quantities can be changed. Myopic optimizing, on the other hand, describes a situation where the "best" decisionmaking is costly in terms of information and decisionmakers make choices based on imperfect information. Both paradigms strongly suggest that farm economic decisions should be described with a dynamic optimization methodology.[2]

The present FLIPCOM constraint structure reflects various assumptions about the producer's access and control over durable and nondurable inputs. The nondurable inputs, such as fertilizer and water, are assumed to be in completely elastic supply at a given price for the farm. However, the supply of durable inputs, such as land, machinery, capital, and information are more price inelastic from the producer's point of view, and act as the effective constraints within the system.[3] For example, assume the amount and type of land acreages owned or leased by the ith farmer, AC_i, can be represented by vectors $L_i = (L_{i1}, \ldots, L_{ij})$ and $Z_i = (Z_{i1}, \ldots, Z_{ij})$. The farmer may buy or sell parcels of land, ΔL_{ij}, or lease additional land, ΔZ_i, from or to other landowners. In each production period the acreage utilized by the ith farm must satisfy the following constraint:

$$0 \leqslant AC_i \leqslant L_i + Z_i + \Delta L_i + \Delta Z_i \tag{1}$$

Other durable input and resource constraints may also be important. For example, consider the distribution of various types of capital stock, where $S = 1, \ldots, s$ are technologies available to the farmer. Given the assumptions of the putty-clay model, the farmer may either continue operating with owned machinery, $K_i = (k_{i1}, \ldots, k_{is})$, or buy new equipment, ΔK_i. This investment cost is amortized in each production period, given associated depreciation, and other factors such as investment credits into $\lambda K^i + \lambda^0 K_i$, where λ is the cost per period of new machinery and λ^0 reflects the cost of old machinery. Machinery may also be rented to and from other farmers or be rented from a service sector at a cost of δK_i.

Each available technology is associated with a matrix of input-output coefficients, A_{sj}, where each element is defined as the amount of a

TABLE 10.1
Comparison of selected farm simulation models to FLIPCOM

Points of comparison	Halter and Dean (1965)	Patrick and Eisgruber (1968)	Hutton and Hinman (1971)	Hardin (1978)	Boehlje and Griffin (1979)	Richardson and Condra (1979)	FLIPSIM FLIPRIP (1981)
Applicable region	California	Indiana	general	general	Corn Belt	Texas	general
Enterprises	cattle ranch and feedlot	cash grains and livestock	general	cattle and wheat	cash grains	cotton, grains and vegetables	general
Stochastic prices/yields	no/yes	yes/yes	yes/yes	yes/yes	no/no	yes/yes	yes/yes
Optimization routine for crop mix	yes	yes	no	no	no	yes	optional
Production costs computed or inflated	1/	1/	computed	inflated	computed	inflated	computed
Cost production aggregated or disaggregated	aggregated	aggregated	aggregated	aggregated	aggregated	aggregated	disaggregated
Flexible inflation rates for:							
Input costs	no	yes	yes	yes	yes	yes	yes
Crop yields	no	no	no	yes	yes	yes	yes
Output prices	no	no	no	yes	yes	yes	yes
Machinery replacement	1/	1/	exogenous	exogenous	no	exogenous	yes
Economic accounting	yes	yes	yes	yes	yes	yes	yes
Cash flow accounting	no	no	yes	yes	yes	yes	yes
Partial balance sheet	no	yes	yes	yes	yes	yes	yes
Family consumption function	no	yes	fixed	fixed	yes	fixed	yes

TABLE 10.1 (cont.)

Personal income taxes	yes	yes	yes	yes	yes	no
Self-employment taxes	yes	no	no	yes	no	no
Allows growth	yes	no	yes	no	yes	no
Growth through cash purchase/ mortgage/leasing	yes/yes/yes	no/no/no	yes/yes/no	no/no/no	yes/no/no	no/no/no
Debt repayment	yes	yes	yes	yes	yes	no
Borrowing for debt refinancing	yes	yes	yes	yes	yes	no
Interest paid on cash reserves	yes	yes	no	yes	yes	no
Test for solvency	yes	yes	no	yes	yes	no
Disinvestment in farmland	yes	no	no	no	no	no
Bid price for farmland	yes	no	yes	no	no	no
Land value related to returns	yes	no	no	no	no [1]	no
Property taxes	yes	yes	yes	yes	yes	no
Off-farm income	yes	yes	yes	yes	yes	no
Price-support program	yes	no	yes	no	no	no
Target price program	yes	yes	no	no	no	no
Farmer-owned reserve program	yes	no	no	no	no	no
Acreage set-aside program	yes	no	no	no	no	no
Acreage diversion program	yes	no	no	no	no	no
Disaster programs	yes	no	no	no	no	no
All-risk crop insurance	yes	no	no	no	no	no
Payment limitations	yes	no	no	no	no	no
Marketing quota program	yes	no	no	no	no	no
Acreage allotment program	yes	no	no	no	no	no
Cost effectiveness	yes	yes	yes	yes	yes	?
Transferrable to other regions	yes	no	no	yes	no	no

[1] Not discussed in the publication cited.

particular input required per acre of type j land using technology s. Each technology is also associated with an expected output vector, Y, where each element y_{sj} is the yield per acre or quantity of livestock produced. Finally, use of each technology may be associated with a linear capacity constraint:

$$c_s AC_i \leqslant b_s \qquad (2)$$

where c_s is the maximum proportion of land acreage available for particular uses given a financial risk, resource, or technological capacity constraint.

To maximize expected utility by increasing net revenues, the producer must be able to calculate revenues and costs for alternative activities, including land and capital (dis)investments. If the producer faces competitive markets, then input, output, and rental prices (expenses) are determined exogenously for the producer. Total revenue is the sum of $PY_s AC_s$, where P is a vector of output prices. The vector of variable (cash) costs of production per acre is $f_s AC_s$, where f_s is a vector of average costs per acre. Finally, if $W = (w_1, \ldots, w_j)$ and $R = (r_1, \ldots, r_j)$ are competitive price vectors for land by type, new investment in land then is $W\Delta L_i$, and net rental expense is RZ_i. Capital appreciation on land holdings can then be written as

$$[W_i^* - (1+\Delta)W_i](L_i + \Delta L_i)$$

where θ is the effective interest rate on land investments, and W_i^* is the vector of expected prices at the end of the production period.[4]

The final set of constraints reflects the fact that investment in alternative technologies must always satisfy the availability of funds for investment, m_i:

$$K_i + W\Delta L_i \leqslant m_i \qquad (3)$$

Investment funds at any particular time depend on cash on hand, the value of durable assets, and outstanding debt. A farm's credit availability is endogenized within the model because credit becomes a function of the farm's debt-equity position and of its ability to maintain a cash flow sufficient to cover debt amortization. Consequently, specification of tax payments, payment of outstanding principle, renegotiated loans, a minimum debt-equity ratio or minimum living expenses could be included as additional constraints. These constraints could be solved simultaneously with or following the optimization decision problem. In the current version of FLIPCOM, these constraints are not specified as simultaneous constraints in the programming problem, but are solved after determining optimal enterprise combinations.

The farmer's objective function for maximizing total realized and unrealized financial gains for the production period can be expressed as:

$$\Pi = (Py_s - f_s)AC_i - RZ_i - \lambda K_i + \lambda^0 K_i - \delta^0 K_s + [W_i^* - (1+\theta)W_i](L + \Delta L_i) \quad (4)$$

Thus, the farmer's production decision is based on maximizing equation (4) subject to the constraints in equations (1), (2), and (3). The decision becomes a dynamic problem if production occurs over several periods, because the producer must choose among production technologies, make land portfolio adjustments, consider financial constraints, and choose the quantity of various inputs while simultaneously determining expected output levels and a set of expected net returns for each activity. The FLIPCOM approach hypothesizes that this economic problem can be solved by solving a linear programming problem within a simulation model.

Description of the FLIPCOM Computer Programs

In both models, the computer program calls a precise sequence of subroutines within a multiyear simulation loop and a stochastic iteration loop (see Figure 10.1).[5] The model is designed to recursively optimize a set of production activities over a 10-year period for a representative farm in a specified policy and economic environment. Table 10.2 illustrates a set of options chosen for a single simulation. The farm can be simulated deterministically or stochastically. At the completion of a deterministic simulation, an annual income statement, a balance sheet, and a cash flow statement are printed. At the end of a stochastic simulation, these tables are suppressed and probability distributions for up to 400 variables are statistically summarized.

The FLIPCOM models were constructed to use farm data sets representing alternative agricultural policy environments. Consequently, information must be provided by the user for variables such as:

1. Fixed costs of production;
2. Labor costs and input-output coefficients by crop and livestock;
3. Variable input costs per acre for each activity;
4. Annual inflation rates for production inputs;
5. Annual interest rates and financing terms for owned or purchased land and machinery and for cash-flow deficits;
6. The initial asset and liability position of the farm;
7. If government commodity programs are in effect; and

220

FIGURE 10.1
THE SUBROUTINE LINKAGE IN THE FLIPCOM COMPUTER PROGRAM

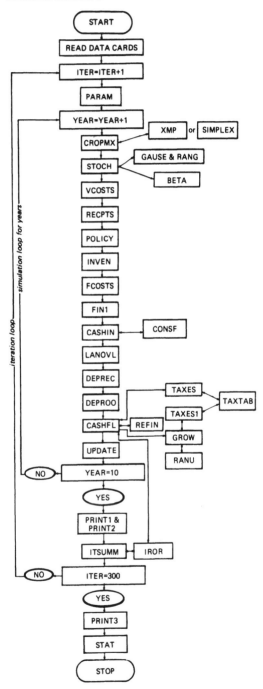

TABLE 10.2

Example of a Summary of Program Options Chosen as Exogenous

Parameters for a Representative Farm

SIMULATE THE REPRESENTATIVE FARM FOR 10 YEARS

FIRST YEAR TO BE SIMULATED IS 1980

THE SIMULATION WILL BE STOCHASTIC WITH 100 ITERATIONS

USER REQUESTED A STATISITCAL ANALYSIS OF ONLY 118 OUTPUT VARIABLES

CUMULATIVE PROBABILITY DISTRIBUTIONS WILL BE PRINTED FOR ONLY 18
OUTPUT VARIABLES

USER HAS SELECTED THE MULTIVARIATE NORMAL PROBABILITY DISTRIBUTION
FOR PRICES & YIELDS

THE REPRESENTATIVE FARM HAS 3 CROPS AND 0 LIVESTOCK ENTERPRISES

THE CROP MIX WILL VARY OVER TIME BASED ON EXPECTED NET RETURNS PER
ACRE

ANNUAL DEPRECIATION WILL BE CALCULATED USING THE DOUBLE DECLINING
BALANCE METHOD

THERE ARE 23 PIECES OF FARM MACHINERY TO BE DEPRECIATED

FULLY DEPRECIATED FARM MACHINERY TO BE DEPRECIATED

USER HAS SPECIFIED THE FAMILY CONSUMPTION FUNCTION FOR REGION 6

THE FARM MAY SELL CROPLAND TO AVOID INSOLVENCY

PER ACRE CASH LEASE COST FOR CROPLAND WILL BE CONSTANT OR ESCALATED
BY A USER-SPECIFIED RATE 0.086

ANNUAL INFLATION RATES FOR FARMLAND ARE PROVIDED BY THE USER

THE FARM WILL BE ALLOWED TO GROW THROUGH PURCHASE AND LEASE OF
CROPLAND; AVAILABILITY OF CROPLAND IS PREDETERMINED AND LEASED LAND
WILL BE RELEASED WHEN LAND IS PURCHASED

INFORMATION FOR 0 ALTERNATIVE FARMS IS PROVIDED BY THE USER

USE THE 1980 SCHEDULE Y TO CALCULATE PERSONAL INCOME TAXES

A PRICE SUPPORT PROGRAM WILL BE IN EFFECT
INTEREST ON FHR LOANS WILL BE WAIVED IN ALL YEARS

A TARGET PRICE PROGRAM WILL BE IN EFFECT AND WILl USE THE "HIGH" TARGET
PRICE

AN ALL-RISK CROP INSURANCE PROGRAM IS IN EFFECT

PAYMENT LIMITATIONS ARE IN EFFECT FOR DEFICIENCY PAYMENTS, DIVERSION
PAYMENTS AND DISASTER PAYMENTS

ALL FARMS ARE ELIGIBLE FOR ALL FARM PROGRAM BENEFITS

8. Prior estimated multivariate probability distributions for crop and livestock prices and yields.

Farm program options for cotton, wheat, sorghum, corn, barley, oats, soybeans, rice, and peanuts must also be selected from among the following:[6]

1. Target price program for both high and low target prices,
2. Price-support program with a 1-year Commodity Credit Corporation (CCC) nonrecourse loan,
3. Variable-length farmer-owner reserve (FOR) program which follows a 1-year nonrecourse CCC loan,
4. Variable-length direct (or immediate entry) FOR which bypasses the 1-year nonrecourse CCC loan,
5. FOR programs (items 3 and 4 above) that release stocks at either the "release price" or the "call price,"
6. Acreage set-aside,
7. Acreage diversion,
8. Disaster payments for low yields,
9. Disaster payments for prevented plantings,
10. Federal Crop Insurance Corporation all-risk crop insurance,
11. Farm program payment limitations,
12. Marketing quota for peanuts,
13. Acreage allotment for rice,
14. Scale farm program benefits based on acres farmed, and
15. Scale farm program benefits based on crop receipts.

The optimizing subroutines (CROPMX, XMP, or SIMPLEX) first calculate the objective function coefficients for each crop activity by using expected yields and by choosing among loan, target, and expected market prices. The expected unit market price and yield per acre for each activity are calculated with a Koyck distributed lag model. Expected market prices are compared with government support and target prices in effect, and the highest price is selected. Expected gross returns are estimated by multiplying expected prices by expected yields. Finally, expected net returns per acre are calculated by subtracting variable costs per acre from expected gross returns.

The constraint system consists of land available for crop activities, 12 monthly labor supplies, an irrigation acreage capacity restraint if needed, and minimum and maximum constraints for each crop each year. Crop rotation activities are defined on one land class. The exogenously specified monthly availability of operator plus hired labor is used to develop the labor resource constraints.

The technical input-output coefficient matrix of monthly labor needs and acreage use for each production activity are provided initially by the analyst. The producer and any hired employees were specified to be full-time. The producer was also allowed to hire part-time labor as needed, at $3.50 per hour. The exogenous data specifying the representative farms was partially developed from unpublished data available from the Department of Agricultural Economics, Texas A&M University, and census and FEDS survey information available to research personnel at ERS.[7]

The actual net returns are then estimated using random prices and yields generated from user-provided probability distributions (STOCH), updated budgets (VCOSTS), and estimated cash receipts (RECPTS). If the simulation is deterministic, annual prices and yields are set equal to their respective exogenous values. When a stochastic simulation is specified, random prices and yields are calculated from either multivariate or independent normal, triangular, and beta distributions using procedures suggested by Clements, Mapp, and Eidman and by King.

Cash receipts are revised to reflect the farm's participation in the various government commodity programs (POLICY). For example, if a set-aside or voluntary diversion program is specified, cash receipts are recalculated taking into consideration the required set-aside, slippage rate, and diversion payment rate. If a target price program is in effect, the deficiency payment per acre is added to cash receipts. The value of disaster or indemnity payments is added to cash receipts if actual yields are low enough to trigger the disaster program or FCIC insurance program, if either is in effect. The net cash position of the farm at the end of the year is calculated after the following expenses have been paid (CASHIN): (1) property taxes and other fixed costs (FCOSTS), (2) existing debts and current year debt amortization (FINI), and (3) family consumption expenditures (CONSF).

After the production period's financial picture has been calculated by the program, attention is turned to capital flows on the farm. The market value of machinery and land is calculated (LANDVL). Depreciation on machinery and buildings is estimated using straight line or double declining balance methods (DEPREC or DEPRDD). This information is used with the farm's cash balance, adjusted for year-end purchases, to estimate year-end cash flows. Year-end purchases of machinery and land, and state and federal income taxes paid (if applicable) further adjust the farm's cash balance. If the year-end cash reserve is negative, assets must be refinanced in the following priority (REFIN): (1) a second mortgage on long-term assets, (2) a second mortgage on intermediate term assets, and (3) the sale of cropland. Land is sold in order of acquisition only if the farm has a cash flow deficit and cannot refinance additional debt.

If the farm meets certain financial conditions and the user allows the farm to grow, cropland is purchased or leased (GROW). The farm's maximum bid price for cropland is estimated using a net present value formula described by Lins, Harl, and Frey. If the farm's maximum bid price per acre is less than the market value per acre, cropland is not purchased. If cropland is not available for purchase or the farm does not meet the financial conditions necessary to purchase the parcel, the availability of cropland for leasing is examined. Leased cropland may then be acquired if certain cash flow requirements are met.

Finally, the farm's federal and state income tax liabilities are computed using three different methods (TAXES, TAXES1, and TAXTAB): income averaging, standard, and maximum tax. The minimum tax liability method is used, based on these calculations. The program then moves either to a new production period (UPDATE), begins a new simulation (IT-SUMM), or ends (PRINT 1 and 2, PRINT 3, STAT, and so forth).

Simulation Results

The growth and adjustment behavior of three farms was stochastically simulated with FLIPRIP over 100 iterations for a 10-year time horizon, 1980 to 1989, using ERS baseline price data.[8] The data sets for each farm differed only in the initial crop mix, machinery complements, and total cropland farmed:

1. *Illinois corn-soybean farm* (East-Central) has 360 acres of cropland (180 acres each of corn and soybeans) and 20 acres of pasture or other land. The operator owns 280 acres of cropland, cash rents 80 acres, and owns the pastureland. The operator must depend on family labor plus an equivalent amount of part-time hourly labor. The market value of owned cropland and buildings is $788,000 and that of machinery is $113,000.

2. *Mississippi cotton-soybean farm* (Delta) has 1,040 acres of cropland (480 acres of cotton and 560 acres of soybeans) and 240 acres of pasture or woodland. The operator owns the pasture and woodland and 780 cropland acres, and cash rents 260 additional acres. The labor supply consists of the operator, two full-time hired workers, and an equivalent amount of part-time hourly help. The market value of owned cropland and buildings is $664,000 and that of machinery is $281,000.

3. *Texas cotton-sorghum farm* (High Plains) has 1,100 acres of cropland of which 360 acres are irrigated. The initial crop mix includes 342 acres of irrigated cotton, 630 acres of dryland cotton, and 73 acres of dryland sorghum. The operator owns 825 acres, and rents an additional 275

acres. The labor supply consists of the owner, a full-time laborer, and twice as much part-time hourly labor. The market value of owned cropland and buildings is $479,000 and that of machinery is $250,000.

The economic and policy scenario assumed that a full complement of federal programs was in effect, including target prices, a farmer-held reserve program, a price-support program with a 1-year CCC nonrecourse loan, and the all-risk crop insurance program (Johnson and Ericksen; U.S. Department of Agriculture). If the loan rate minus 9 months' storage costs for a crop is less than its market price, the farmer's share of the crop is put under CCC loan and cash receipts are adjusted. If the market price in the following year exceeds the current year's loan rate plus 9 months' interest, the crop is redeemed and cash receipts are recalculated. Otherwise, the crop is forfeited. The corn and sorghum crops could be placed in a multiple-year farmer-owned reserve (FOR) program. A 3-year-maximum-length grain reserve FOR program was in effect. When the market price is greater than the "release price" (145 percent of the loan rate for feed grains), stocks are redeemed by farmers and sold after repaying the CCC loan plus interest charges.

The target price program serves as an additional form of income support policy through the use of deficiency payments. The deficiency payment rate is the lesser of (1) the target price minus the 5-month average market price or (2) the target price minus the loan rate. The deficiency payment is the product of the deficiency payment rate and the farm program yield, times the harvested acreage. The price and yield covariance-variance matrix was estimated from unpublished data collected in the respective Crop Reporting Districts. Finally, crop price and yield distributions were considered to be independent in each district due to each district's relatively insignificant share of national production.

The loan rates increased from $0.48 in 1980 to $0.90 in 1989 per pound for cotton lint, $2.28 to $4.62 per bushel of grain sorghum, $2.40 to $4.87 per bushel of corn, and $5.02 to $10.48 per bushel of soybeans.[9] The interest rate for CCC loans was set at 14.5 percent. The target prices for cotton, sorghum, and corn increased from $0.58 to $1.23 per pound, $2.50 to $5.67 per bushel, and $2.35 to $5.09 per bushel, respectively. The deficiency payment limitation was held at $50,000.

The simulation results illustrate the type of information that can be obtained for various policy analyses. Selected summary statistics for the three simulations are presented in Table 10.3. The Texas, Illinois, and Mississippi farms remained solvent for all iterations. The relatively high market prices in the baseline created financial conditions leading to a very high probability of economic well-being and survival. A set of lower annual market prices would have undoubtedly caused the income

TABLE 10.3
Summary statistics for selected variables for all solvent iterations with support price, all-risk crop insurance, target price and farmer-owned reserve programs using FLIPRIP

Variable	Unit	Mean 1/	Standard Deviation 1/	Minimum	Maximum	Coefficient of variation
Net present value:						
Mississippi	Dollars	132,540	174,173	-229,019	597,805	131.41
Illinois	do.	390,653	49,089	247,901	478,503	12.57
Texas	do.	461,925	96,862	223,141	683,501	20.97
Present value of ending net worth:						
Mississippi	do.	869,564	147,081	525,137	1,222,633	16.91
Illinois	do.	960,956	36,624	861,826	1,041,704	3.81
Texas	do.	828,819	84,843	616,095	1,046,025	10.23
Total cropland:	Initial Acres					
Mississippi	1,040	0	0	0	0	0.00
Illinois	360	7,533	996	0	9,400	13.18
Texas	1,100	665	938	0	2,332	141.10
Cropland owned:	Initial Acres					
Mississippi	780	802	41	780	940	5.09
Illinois	290	302	20	280	320	6.64
Texas	825	964	113	825	1,185	11.76

TABLE 10.3 (cont.)

	Initial Acres					
Acres of cropland cash rented:						
Mississippi	260	333	106	100	620	31.80
Illinois	80	328	86	120	440	26.11
Texas	275	273	197	0	635	72.04
Ending cash revenue:						
Mississippi	Dollars	−166,424	$149,712	−488,999	479,493	−89.96
Illinois	do.	111,768	91,442	−11,235	400,343	81.80
Texas	do.	183,260	207,578	−140,925	1,152,446	113.27
Total long—term debt:						
Mississippi	do.	770,301	303,209	71,452	1,449,309	39.36
Illinois	do.	244,167	74,063	135,638	361,737	30.33
Texas	do.	203,440	93,402	59,663	445,485	45.91
Total intermediate— term debt:						
Mississippi	do.	54,420	39,472	0	203,853	72.53
Illinois	do.	15,390	1,518	14,996	23,521	9.86
Texas	do.	42,382	4,675	28,223	61,964	11.03

TABLE 10.3 (cont.)

Summary statistics for selected variables for all solvent interations with support price, all-risk crop insurance, target price and farmer-owned reserve programs using FLIPRIP

Variable	Unit	Mean 1/	Standard Deviation 1/	Minimum	Maximum	Coefficient of variation
Ending net worth:						
Mississippi	do.	2,255,434	381,492	1,362,072	3,171,196	16.91
Illinois	do.	2,492,479	94,995	2,235,356	2,701,913	3.81
Texas	do.	2,149,753	219,982	1,597,991	2,713,120	10.23
Leverage ratio:						
Mississippi		.567	.225	.178	1.414	39.70
Illinois		.238	.030	.178	.300	12.45
Texas		.277	.051	.168	.398	18.33
Equity-to-asset ratio:						
Mississippi		.650	.086	.414	.849	13.28
Illinois		.808	.019	.769	.849	2.40
Texas		.784	.031	.716	.857	3.97
Van Horn profit index:						
Mississippi		1.156	.204	.731	1.702	17.69
Illinois		1.555	.070	1.352	1.679	4.48
Texas		1.805	.169	1.389	2.191	9.35
Maximum bid price for farmland:						
Mississippi	Dollars	1,135	84	1,040	1,400	7.30
Illinois	do.	630	85	400	720	13.50
Texas	do.	1,237	102	1,100	1,460	8.27

N.A. = Not applicable.

1/ Values in year 10.

and price-support programs to become far more important to farm survival. Nevertheless, without further simulations to test the sensitivity of the results to lower market prices, any conclusions drawn with regard to the relative effect of these programs could not be clearly supported.

The mean present value of ending net worth for all 3 farms was quite similar, about $850,000. The variation in mean values was significantly larger for the cotton-producing farms. The respective Van Horn profit index values also indicate the much larger relative yearly variation in profitability of the Mississippi and Texas farms.

None of the farms was forced to sell cropland to maintain solvency during the simulations. The Mississippi and Texas farms showed a slightly larger inclination to purchase cropland, while the Illinois farm grew primarily through leasing additional cropland. Since 40 acres of land were available for purchase or lease in each year, the data set could be characterized as overemphasizing the rapidity and availability of land parcels for acquisition. Nevertheless, the mean total cropland farmed in the last year was significantly larger as a percentage of the original cropland base for the Illinois farm and slightly larger for the Texas farm. A related statistic is the mean maximum bid price for cropland in the last year. The Illinois and Texas farms' maximum bid price ranged from $0.00 to $9,400 and from $0.00 to $2,332, respectively, but the Mississippi farm's maximum bid was always $0.00.

The remaining financial variables also showed significant differences and variability among the 3 farms. The ending mean cash revenue in the last year was about −$166,000 on the Mississippi farm, but was almost $112,000 for the Illinois corn-soybean farm, and was even higher at $183,000 for the Texas cotton-sorghum farm. The deviation in this variable was much larger for the Mississippi farm, ranging between plus and minus $500,000, while the Illinois farm's minimum ending cash revenue was only $11,000.

The Mississippi farm also tended to increase long-term and intermediate-term debts from their initial levels. On average, the Mississippi farm owed 3 times as much as the Illinois and Texas farms. It is interesting to note, however, that the Illinois farm on average ended with a higher ending net worth by almost $250,000 and with a much smaller standard deviation than the Mississippi and Texas farms, perhaps as a result of its land-leasing, rather than its land-purchasing behavior. The leverage and equity-asset ratios also demonstrated the stronger ending financial position of the Illinois and Texas farms, on average. The Mississippi farm's mean leverage ratio was over twice the mean value of the Illinois and Texas farms. In addition, each farm began each iteration with a long-term debt-equity ratio of 0.25 (and an intermediate-term ratio of 0.20). On average, the Illinois and Texas farms' overall equity to asset ratios remained near 0.80, while the Mississippi farm's ratio fell to 0.65. One suggested explanation of the fall in the Mississippi farm's equity to asset ratio is a larger variation in yields and resulting net cash flows, together with the cropland purchases. Another explanation might be lower baseline cotton and sorghum prices relative to production costs compared with their respective production costs in other regions. In any case, both of these suggestions remain tentative.

The separation of various influences on selected financial and structural variables is extremely complicated due to the number of variables in the model. The sensitivity of the FLIPCOM models to variations in exogenous input data is still being determined. Nevertheless, a future analysis could maintain the farms at their respective original size with the same data sets and could again stochastically simulate the financial variables. This controlled experiment would provide additional information to help lend credence to the latter conclusion.

Another tentative hypothesis is that although the Mississippi farm is able to have enough very good income years in a stochastic environment to periodically buy land, it also has a number of very poor years. Obviously, since the Mississippi farm does buy cropland during the simulation period, it must be able to bid at least the market price. However, after the land purchase has occurred (most likely subsequent to a very good year), the likelihood of a poor income year or series of poor income years is large and the farm is then unable to grow further and will experience cash flow problems. On the other hand, the Illinois and Texas farms exhibited apparently steady growth behavior through the acquisition of leased land and purchased land, respectively. Evidently, the Illinois farm only has a good enough income year or series of years to build up the cash and equity needed for only a sporadic cropland purchase, but is able to almost always meet the criteria for leasing additional cropland. The Texas farm, however, is able to maintain its favorable financial condition while maintaining its leased land base and increasing its owned cropland.

These conclusions and hypotheses need to be examined further in more detail in future research. If cropland acquisition behavior by existing farms is related to income (price) variation as has been tentatively identified, then government policies designed to affect price variation can have a direct impact on farm structure and resource ownership patterns without the impact being immediately apparent.

Summary and Conclusions

The analysis of commercial farms' economic behavior can provide important and relevant information for agricultural policy analysis. The development of responsive and cost-effective farm programs directed toward specific needs and economic problems has always been of concern. Aggregate sector models and farm-level micromodels can provide complementary quantitative and qualitative information about the effects of agricultural programs and provide a means for testing new ideas, economic theories, and program strategies. Participatory behavior, distributional effects, resource ownership, and other economic responses to government programs and market price become important and interdependent from this research viewpoint. However, the effectiveness of representative farm models depends on their ability to blend observed behavior with economic theory. These models must be able to portray both the dynamic and adaptive aspects of microeconomic behavior and the resource allocation behavior of producers.

The research results reported in this paper suggest that micromodels of the farm enterprise incorporating programming and simulation techniques are useful tools that can provide additional microlevel information to that provided by aggregate sector econometric models. Although the results reported in this paper are limited in scope and the conclusions will remain tentative until further research is completed, the results are interesting and conform to economic logic and observed producer behavior. The authors anticipate that development of additional representative farm data sets and further research with these analytical techniques will provide an additional dimension to farm policy research.

Notes

1. See Richardson and Condra for a hybrid modeling approach to the same question.

2. The authors recognize that other types of objective functions could also have been used.

3. See Rausser and others and Day and others for added discussion.

4. The effective interest rate can be thought of as the vector of time-weighted interest rates during the fiscal year that reflect length of ownership.

5. For a lengthier description of the various subroutines, see Baum and others and Richardson and Nixon. Particular subroutines in Figure 10.1 are referenced throughout this section with parentheses.

6. The models reported on here incorporate decision rules for participation in government farm programs based only on the Food and Agricultural Act of 1977 (Johnson and Ericksen) and on the Agricultural Adjustment Act of 1980.

7. Some of the exogenous data used in the model, such as interest or inflation rates, were subjectively determined by the authors and do not necessarily represent official or unofficial institutional and government expectations or forecasts. The authors recognize that elements of this subjective data may have inadvertently biased the results.

8. The farm data sets use the same policy and economic environment and the same initial percentage debt and landownership conditions.

9. The loan and target prices from 1986 through 1989 were derived by using the average percentage increase in each commodity program price from 1983 to 1985.

References

Armstrong, D. L., L. J. Connor, and R. P. Strickland. "Combining Simulation and Linear Programming in Studying Farm Firm Growth," in *Simulation Uses in Agricultural Economics* (ed. D. L. Armstrong and R. E. Epp). East Lansing: Michigan State Univ., 1970.

Baum, K. "A National Recursive Simulation and Linear Programming Model of Some Major Crops in U.S. Agriculture." Ph.D. thesis, Iowa State Univ., 1978.

———, J. Richardson, and L. Schertz. "FLIPRIP: A Stochastic Recursive Interactive Programming Model of Farm Firm Growth," in *Proceedings of the Third Symposium on Mathematical Programming with Data Perturbations* (ed. Anthony V. Fiacco). Washington: George Washington Univ., 1982.

Bellman, R. *Adaptive Control Processes: A Guided Tour.* Princeton, N.J.: Princeton Univ. Press, 1961.

Boehlje, Michael, and S. Griffin. "Financial Impacts of Price Supports," *American Journal of Agricultural Economics*, Vol. 51, No. 2, May 1979.

———, and T. K. White. "A Production-Investment Decision Model of Farm Firm Growth," *American Journal of Agricultural Economics*, Vol. 58, No. 3, Aug. 1969.

Chien, Y. I., and G. L. Bradford. "A Sequential Model of the Farm Firm Growth Processes," *American Journal of Agricultural Economics*, Vol. 58, No. 4, Aug. 1976.

Clements, A. M., Jr., H. P. Mapp, Jr., and V. R. Eidman. *A Procedure for Correlating Events in Farm Firm Simulation Models.* Oklahoma Exp. Sta. Tech. Bull. T-131. Oklahoma State Univ., 1971.

Condra, G. D. "An Economic Feasibility Study of Irrigated Crop Production in the Pecos Valley of Texas." Ph.D. thesis, Texas A&M Univ., 1978.

Day, R. H. "Dynamic Coupling, Optimizing and Regional Interdependence," *Journal of Farm Economics*, Vol. 46, No. 3, Aug. 1964.

———, "Cautious Suboptimizing." Modeling Research Group Paper 7603. Univ. of Southern California at Los Angeles, Dept. of Economics, 1976.

———, S. Morley and K. R. Smith. "Myopic Optimizing and Rules of Thumb in a Micro Model of Industrial Growth," *American Economic Review*, Vol. 72, No. 1, Mar. 1974.

Dorfman, R. "Formal Models in the Design of Water Resource Systems," *Water Resources Research*, Vol. 1, No. 2, May 1965.

Halter, A. N., and G. W. Dean. "Use of Simulation in Evaluating Management Policies Under Uncertainty: Application to a Large Scale Ranch," *Journal of Farm Economics*, Vol. 47, No. 3, Aug. 1965.

Hardin, D. C. and R. D. Lacewell. "Implication of Improved Irrigation Pumping Efficiency on Farmer Profit and Energy Use." TA-14654. Texas Agr. Expt. Sta., 1979.

Hardin, M. L. "A Simulation Model for Analyzing Farm Capital Investment Alternatives." Ph.D. thesis, Oklahoma State Univ., 1978.

Harman, W. L. "Evaluating the Growth Potential of Irrigated Farms with Diminishing Water Supplies: A Multiple Goals Approach to Decision Making." Ph.D. thesis, Oklahoma State Univ., 1973.

Hatch, R. E. "Growth Potential and Survival Capability of Southern Plains Dryland Farms: A Simulation Analysis Incorporating Multiple-Goal Decision Making." Ph.D. thesis, Oklahoma State Univ., 1973.

Hatch, T. "The Census Typical Farm Program User Guide and Documentation." Staff Report No. AGES811230, U.S. Dept. Agr., Econ. Res. Serv., Feb. 1982.

Heidhues, T. "A Recursive Programming Model of Farm Growth in Northern Germany," *Journal of Farm Economics*, Vol. 48, No. 3, Aug. 1966.

Heady, Earl O., T. Reynolds, and K. Baum. "Tax Policies to Increase Farm Prices and Income: A Quantitative Simulation," *Canadian Journal of Agricultural Economics*, Vol. 24, No. 3, Nov. 1977.

Held, L. J. "Growth and Survival of a Wheat Farm Under Selected Financial Conditions: A Simulation Analysis." Ph.D. thesis, Univ. of Nebraska, 1977.

Holland, D., and D. Young. "An Examination of Cash Flow Conventional Net Income Measures for Evaluating the Economic Viability of Farms of Varying Size," *Western Journal of Agricultural Economics*, Vol. 5, No. 1, July 1980.

Huang, Wen-Yuan, and others. "The Recursive Adaptive Hybrid Programming Model: A Tool for Analyzing Agricultural Policies," *Agricultural Economics Research*, Vol. 2, No. 1, Jan. 1980.

Hutton, R. F., and H. R. Hinman. "Mechanics of Operating the General Agricultural Firm Simulator," *Agricultural Production Systems Simulation* (ed. V. R. Eidman). Stillwater: Oklahoma State Univ. Press, 1971.

Jensen, H. R., T. Hatch, and D. Harrington. *Economic Well-Being of Farms: Third Annual Report to Congress on the Status of Family Farms.* AER-469. U.S. Dept. of Agr., Econ. Res. Serv., 1981.

Johnson, J., and M. H. Ericksen. *Commodity Program Provisions Under the Food and Agriculture Act of 1977.* AER-389. U.S. Dept. Agr., Econ. Res. Serv., 1977.

King, R. P. "Operational Techniques for Applied Decision Analysis Under Uncertainty." Ph.D. thesis, Michigan State Univ., 1979.

Kingma, Onka T. "A Recursive Stochastic Programming Model of Farm Growth," in *Modelling Economic Change: The Recursive Programming Approach* (ed. R. H. Day and A. Cigno). Amsterdam: North-Holland Publishing Co., 1978.

Lins, David A. "An Empirical Comparison of Simulation and Recursive Linear Programming Firm Growth Models," *Agricultural Economic Research*, Vol. 41, No. 1, Jan. 1969.

———, N. Harl, and T. Frey. *Farmland.* Skokie, Ill.: Century Communications, 1982.

Loftsgard, L. D., and E. O. Heady. "Application of Dynamic Programming Models for Optimum Farm and Home Plans," *Journal of Farm Economics*, Vol. 41, No. 1, Feb. 1959.

Mapp, H. P., Jr., and others. "Simulating Soil Water and Atmospheric Stress-Crop Yield Relationships for Economic Analysis." Tech. Bull. T-141. Oklahoma Agr. Expt. Sta., 1975.

Marsten, Roy E. *The Design of the XMP Linear Programming Library.* Dept. of Management Information Systems Technical Report No. 80-2. Univ. of Arizona, 1980.

Patrick, G. F. "Some Impacts of Inflation on Farm Firm Growth," *Southern Journal of Agricultural Economics,* Vol. 10, No. 1, Jan. 1978.

_____, and L. M. Eisgruber. "The Impact of Managerial Ability and Capital Structure on Growth of the Farm Firm," *American Journal of Agricultural Economics,* Vol. 50, No. 3, Aug. 1968.

Rausser, Gordon C. "Active Learning Control Theory in Agricultural Policy," *American Journal of Agricultural Economics,* Vol. 60, No. 3, Aug. 1978.

_____, R. E. Just, and D. Zilberman. "Prospects and Limitations of Operations Research Applications in Agriculture and Agricultural Policy," *Operations Research in Agriculture and Water Resources* (ed. D. Yaron and C. Tapiero). Amsterdam: North-Holland Publishing Co., 1980.

Ray, D. E., and J. W. Richardson. "Detailed Description of POLYSIM." Tech. Bull. T-151. Oklahoma Agr. Expt. Sta., 1978.

Richardson, James, and C. Nixon. *FLIPSIM: The Farm Level Income and Policy Simulator.* Dept. of Agr. Econ. Tech. Rep. 81-2. Texas A&M Univ., 1980.

_____, and G. D. Condra. "Farm Size Evaluation in the El Paso Valley: A Survival/Success Approach," *American Journal of Agricultural Economics,* Vol. 63, No. 3, Aug. 1981.

Roush, C. E., H. P. Mapp, and C. D. Maynard. "Implications of the Tax Reform Act of 1976 for Farm Estate Planning," *Western Journal of Agricultural Economics,* Vol. 4, No. 2, Dec. 1979.

Salter, W. E. *Productivity and Technical Changes.* 2nd ed. Cambridge, England: Cambridge Univ. Press, 1966.

Umberger, D. E. "Factors Affecting Farm Growth Rates: A Simulation Analysis of Columbia Basin Irrigated Farms." Ph.D. thesis, Washington State Univ., 1972.

U.S. Department of Agriculture, Agricultural Stabilization and Conservation Service. *Implementing the Agricultural Adjustment Act of 1980.* PA-743, PA-744, and PA-745. Mar. 1980.

Wade, J., and E. O. Heady. "Controlling Non-Point Sediment Sources," *American Journal of Agricultural Economics,* Vol. 59, No. 1, Feb. 1977.

Walker, O. L., and J. R. Martin. "Firm Growth Research Opportunities and Techniques," *Journal of Farm Economics,* Vol. 48, No. 5, Dec. 1966.

Weisz, Reuben H., and others. *Six Alternative Models for Linking Normative and Positive National Agricultural Models.* VPI & SU Staff Paper No. SP-79-15. Virginia Polytechnic Inst. and State Univ., Dept. of Agr. Econ., 1979.

11. FLOSSIM: A Multiple-Farm Opportunity Set Simulation Model

Jerry R. Skees

Abstract Previous farm-level simulation models have ignored an essential component of the system environment—interaction among farmers for limited natural resources—namely, land. This paper demonstrates the importance of modeling multiple farms and allowing them to simultaneously compete to acquire the same parcel of land. A primary finding is that quite different results can be expected when farm growth is modeled with a single-farm model rather than a multiple-farm model. Single-farm models overestimate the growth potential of smaller farms. However, modeling a number of farms simultaneously presents several unique problems. This paper also discusses methodological approaches to overcome these problems.

Keywords Land competition, Monte Carlo, multiple-farm simulation

Reemerging concerns over the ownership and control of agricultural resources have been partly responsible for spawning a new generation of farm firm simulation models. A major objective of such models is to more accurately reproduce the economic and institutional environment of the farm firm. Although many theoretical and empirical problems remain, such models are providing insights into many important economic interactions. However, micromodels have ignored an extremely important dimension of the firm's environment—the interaction among farms as they compete for limited natural resources. Although this dimension is not essential for analysis of the well-being of farmers who are not growth-oriented, it is crucial for analysis of the possible impact of alternative policies on the ownership and control of agricultural natural resources.

The author is an assistant professor of agricultural economics at the University of Kentucky.

A multiple-farm, land competition simulation model, FLOSSIM, presented in this paper recognizes that farmer decisionmaking is interdependent not only with the economic and institutional environment, but also with decisions of other farmers. Previous farm-level simulation models have not attempted to include competitive behavior for limited resources at the micro-level. Few models have explicitly acknowledged that land availability is limited. Only Boehlje and Griffin have developed a model that simultaneously simulates the behavior of more than one farm. To compare different size or financially structured farms, previous studies have compared farms only after simulating each farm individually. FLOSSIM provides for annual competition among four farms for the same land parcel to be purchased or leased. The intent of this micro-modeling research is to address the methodological issues peculiar to a multiple-farm model and to demonstrate the advantages of providing for interaction among farms in a local land model. The following issues are also addressed: (1) the use of limited data to develop a local land market reflecting both the limited availability of land and the fact that different size parcels of land may become available in any given year, (2) the concept of designing farmer behavior that reflects multiple goals, (3) the development of useful measurement variables, (4) the concept of designing controlled experiments, (5) the techniques for testing the sensitivity of crucial assumptions, and (6) the use of FLOSSIM for policy analysis.

FLOSSIM: A Farm-Level Opportunity Set Simulator

FLOSSIM is a recursive Monte Carlo simulation model similar to many previous farm firm simulation models in its reliance on economic and cash accounting to simulate the financial well-being and decision-making process of farmers through time. Figure 11.1 presents the basic sequential information flows of the model.[1]

FLOSSIM is currently designed to simulate corn-soybean farms in Central Illinois because of the availability of reliable data and the homogenous farm structure in this region. Development of the model and availability of data were interdependent. Annual summaries of Illinois Farm Business Records were used to develop quadratic equations that relate farm size (tillable acres) to soil fertility, machinery, fuel and repair, seed and crop expenses, labor requirements, and so forth. Other items, such as yields and machinery investments, are also related to farm size using table look-up functions.[2] Cost and production items were disaggregated to allow for adjustments that reflect actual (varying) yields. Actual yields are a function of farm size, crop mix, level of machinery complement, and a random deviate. All farmers' expected yields are adjusted by the same random deviate.

A major cost saving in programming time occurred when a modified TELPLAN program developed by Black and Vitosh was used. This program substitutes for the need to use an optimization routine, such as linear programming (LP). It does this by calculating the expected net returns from growing alternative mixes of corn versus soybeans given the machinery complement. FLOSSIM then selects the crop mix that provides the highest expected net return above variable cost. A yield discount matrix that Black and Vitosh developed, by use of an LP model, accounts for tradeoffs associated with different levels of specialization in either corn or soybeans given different (discrete) levels of machinery.

Actual net returns per acre are calculated from an average storage value, actual (random) yields, random harvest prices, and tenure (share leasing is accounted for). Decisions on the percentage of the current year's crop to store are made based on cash flow needs and expected income tax payments. Less crop will be stored when cash flow needs are great or when income taxes would be below an exponentially smoothed average. Once storage decisions are made, the current year's taxes are calculated from either standard procedures or from income averaging. The tax calculations account for numerous opportunities provided to farmers: cash accounting, investment tax credits with carryover provisions, a variety of depreciation techniques, interest write-offs, and so forth.

A major innovation of the model involved the use of multivariate Beta process generator developed by King and modified by Hoskin. This generator was used to develop random corn and soybean prices and yields that were correlated among years as well as within years. Although it is common in many farm models to assume that there is no correlation between prices and farm-level yields, examination of Illinois farm data revealed that a negative correlation does exist in this area. The assumption that the correlation between prices and yields is zero would overestimate the variance in returns. In addition to the correlation matrix, the multivariate Beta process generator also requires means, standard deviations, and upper and lower bounds. Table 11.1 presents a summary of the data. A carefully selected sample (that is, only farms that did not undergo extreme adjustments such as major expansion) of farm-level yield data from 1972 to 1979 was used to develop yield distributions. Each farm's data was normalized by taking the mean yield of that farm and subtracting the yearly yield to obtain deviations. This sample was then used to obtain the appropriate parameters. Variation of price parameters were obtained by use of the

FIGURE 11.1a

Sequential Development of FLOSSIM

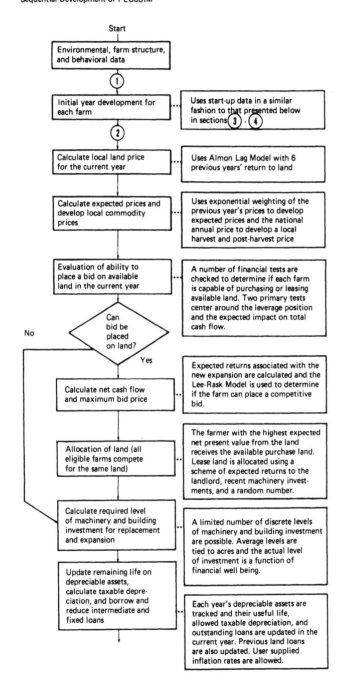

FIGURE 11.1b

Sequential Development of FLOSSIM

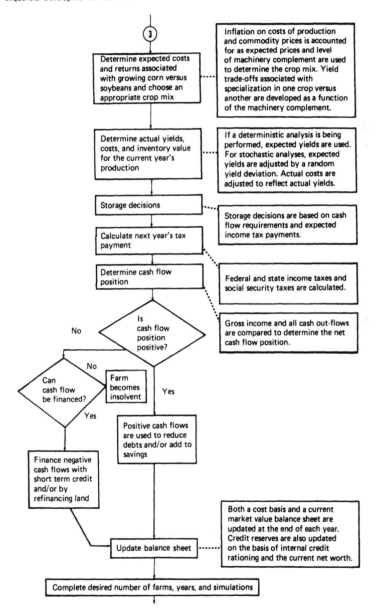

TABLE 11.1

Parameters used to generate multivariate, random corn and
soybean prices and yields

Item	Mean	Standard deviation	Upper bound	Lower bound
Corn price, 1980	2.67	0.33	3.60	2.25
Soybean price, 1980	6.95	.84	10.04	5.90
Corn yield	0	19.00	44.9	-57.40
Soybean yield	0	6.10	14.8	-22.40
Summary of within-year correlations: 1/				
Corn and soybean prices		.50		
Corn and soybean yields		.55		
Corn yield and corn price		-.27		
Soybean yield and soybean price		-.27		
Summary of between-year correlations:				
Corn price$_t$ and corn price$_{t+1}$		.50		
Soybean price$_t$ and soybean price$_{t+1}$		.35		
Corn price$_t$ and soybean price$_{t+1}$		.30		
Soybean price$_t$ and corn price$_{t+1}$		.20		

1/ These parameters were used throughout a 40 x 40 matrix
that was used to generate 10 years of multivariate, random corn
and soybean prices and yields.

MSU Ag Model in a Monte Carlo fashion where world yields were the
random variables.[3] Although random values of prices and yields are
normalized to 1980 values, trends in prices and yield productivity were
used to adjust the random draws over the simulation period.

Deterministic simulation models usually assume land prices increase
at a specified rate over some planning period. Although this may be
an acceptable assumption when mean prices and yields also follow a
trend, it is not acceptable when stochastic prices and yields are used.
Prices and yields that result in either high or low returns directly affect
land price. Hauschew and Herr demonstrated that an Almon polynomial
lag model that uses the previous year's net earnings (above land cost)
and a capitalization rate could be used to explain U.S. farmland prices.
A second-degree, 6-year Almon lag regression model was estimated

based on the predictive properties and the behavior of the lagged relationships of the following variables: returns from growing corn and soybeans over the 1959–79 period from Illinois budget data, the lagged Federal Land Bank interest rates, and the consumer price index. The Almon lag model provided the feedback reflecting the probable impact of variant prices and yields on the price of land.

Despite efforts to introduce important market relationships between farm income and land prices, maintaining reasonable relationships between changes in input cost, inflation, land prices, and returns remains a problem. Since various trends and inflation rates must be compounded over time, it is particularly important to have reasonable relationships at the beginning of each planning period. In order to more fully model reality, a stochastic mode should allow for varying inflation rates over the planning period. For example, input costs may inflate at a greater level than output prices in one year and in the next output prices would inflate more. Although it is unrealistic to assume that all items will inflate at a constant rate over time (see Table 11.2), it is also difficult to develop the type of multivariate data that would be needed to generate random inflation rates. Therefore, it was considered essential to fix various inflation rates to reflect stable relationships over the time period (Table 11.3). This process was accomplished by use of a deterministic budget simulator that contained the Almon lag model and the flexibility to alter various inflation rates. The major objective was to achieve an inflation rate on returns at the same rate as that for the general economy and to peg the expected land price inflation rate at 3 percent above that for the general economy. Because of the random prices and yields, land prices do deviate around this 3 percent value for different simulations.[4]

Developing a Local Land Market

The demand side of the purchase land market is developed from the Almon lag model. It is assumed that the supply of land is inelastic.[5] Historical data on land transfers and the number of farms in the region were used to match farm numbers in the model to one parcel of available land per year. The question was: How many farms should be modeled when one parcel of land is available per year? The relevant land market is assumed to be a 10-mile radius circle. Thus, there are 201,050 acres in the market domain. There are about 567 farms in this circle and 19.5 parcels of land are sold per year. Of these 567 farms, only 66 have yearly gross sales of $100,000 or more. If it is assumed these farms are the primary competitors for purchased land, then there is roughly one parcel of land per 3.5 farms. Therefore, this study assumed

TABLE 11.2
Relationship among the Federal Land Bank mortgage interest
rate, the Consumer Price Index, and the Illinois Land Price
Index

Year	Federal Land Bank mortgage rate of interest (1)	Change in CPI (2)	Column 2 - Column 1 (3)	Change in Illinois Land Price Index (4)	Column 4 - Column 2 (5)
			Percent		
1979	9.25	10.0	-0.75	13.1	3.1
1978	8.50	6.5	2.00	10.5	4.0
1977	8.50	5.5	3.00	35.8	30.3
1976	8.66	5.7	2.96	24.4	18.7
1975	8.69	9.1	-.41	20.8	11.7
1974	8.14	10.9	-2.76	34.1	23.2
1973	7.48	6.2	1.28	11.2	5.0
1972	7.42	3.3	4.12	7.4	4.1
1971	7.86	4.3	3.56	.9	-3.8
1970	8.68	6.5	2.18	-1.8	-8.3
1969	7.82	5.4	2.42	4.8	-.6
1968	6.84	4.2	1.44	4.0	-.2
1967	6.02	2.9	1.82	6.4	3.5
1966	5.82	2.8	2.92	11.9	9.1
1965	5.60	1.7	2.80	7.7	6.0
1964	5.60	1.3	2.49	4.0	2.7
1963	5.60	1.2	4.30	5.6	4.4
1962	5.60	1.1	4.40	2.9	1.8
1961	5.64	1.0	4.54	-2.8	-3.8
1960	6.00	1.6	4.40	0	-1.6
Average, 1960-69			2.60		5.75
Average, 1970-79			1.50		8.80

Blanks indicate not applicable.
SOURCE: Compiled from Scott, "Forecasting the Prices of
Farmland for 1990."

that four contiguous farms compete for the same parcel of land each
year. A similar approach was used to justify using only one parcel of
available lease land for the four farms. Although there are more parcels
of land available for lease, there are also likely to be additional competing
farms.

TABLE 11.3
Summary of the most important assumptions concerning the
economic environment 1/

Variable	Annual percentage
Inflation rate of general economy	9.0
Inflation in--	
Cost of production	9.0
Returns to growing corn and soybeans	9.0
Corn and soybean prices	7.5
Trend corn yields	1.8
Trend soybean yields	.5
Interest rates--	
Federal Land Bank	11.0
Intermediate	12.0
Operating loan	13.0
Inflation in land prices	12.0

1/ Note that land inflates at 3 percent above the inflation
rate of the general economy. The 1970-79 average was 8.85
percent. Also note that Federal Land Bank interest rates are
set at 2 percent above the general inflation rate. The 1970-79
average was 1.5 percent.

The justification for using 4 farms to compete for the same parcel of purchase and lease land each year is tenuous. However, the intent of this research is to provide a reasonable accounting of land availability; it should be clear that an approach such as this is preferred to the typical approach of providing unlimited land availability. Frequency data by parcel size for farmland transfers from 1975 to 1980 were obtained from Illinois State Statistical Office of the Statistical Reporting Service. Frequency data for lease parcels were obtained (and adjusted) from an earlier study by Johnson (Table 11.4). A uniform random number generator determines the discrete size parcel available in any given year.

As most land is transferred at the beginning of the year, FLOSSIM begins each calendar year by determining which farms are eligible to compete for purchase and/or lease land. A number of criteria are used to check each farmer's financial ability to carry and service the debt

TABLE 11.4
Frequency data of available purchase and lease land, and
probability by land size

Acres	Purchase	Lease
	Percent	
40	0.21	0.20
80	.30	.35
120	.12	.15
160	.14	.20
200	.08	.05
240	.05	.05
360	.08	0
640	.02	0
Total probability	1.00	1.00

associated with land expansion and the matching increase in machinery. The parcel of land is then allocated to the farmer who is capable of placing the highest bid price per acre. Although FLOSSIM is capable of allocating land using a maximum bid price developed by the Lee-Rask land price model, this approach is not used because of its sensitivity. Instead, it is used to insure that the farmer can place a competitive bid (that is, a bid above the Almon lag price), and FLOSSIM then uses a modified version of the Lee-Rask model (excluding consideration of the salvage value of the land) to develop a net present value of the land. This net present value reflects each of the different farmer's expected returns and expected marginal tax brackets after purchase. The farmer with the highest net present value receives the land 75 percent of the time. A random number is used to allocate the land among those farmers who are financially eligible to bid at other times. This procedure reflects the roughly 25 percent of all land transfers occurring between family members that do not enter the competitive market. All land is sold at the price determined by the Almon lag model.

As share leasing dominates central Illinois, allocation of lease land proceeds from the viewpoint of the landlord. The most common arrangement involves an equal sharing of variable costs and returns. Thus, landlords are most interested in which farmer will be most likely to provide them with the highest return. In addition, it is assumed that a landlord will not lease land to a farmer who has just expanded through a new purchase.[6] Under these assumptions, FLOSSIM develops a proxy bid for those farmers who have not purchased land in the current year and for those who can afford a machinery complement to accompany the available parcel. The proxy bid reflects the expected landlord profit (given the share lease arrangement) adjusted by each farmer's exponentially averaged machinery complement.[7] If only the maximum proxy bid were used to allocate lease land, larger farmers would receive an inordinate proportion because of economies of size. As there is evidence of smaller farmers receiving lease land for a variety of reasons (friendship, family ties, or the landlord's inability to distinguish between prospective tenants due to imperfect information), a threshold criterion is created so that all farmers with a proxy bid within 5 percent of the maximum proxy bid have an equal chance to obtain the lease land. A random (0 through 1) number is finally used to determine which of the eligible farmers will receive the lease land.

Designing Farmer Behavior

Replicating farmer behavior remains a major limitation for farm-level models. However, it is primarily a limitation in the use of these models to predict actual behavior. It does not limit the important contribution of providing qualitative insights into otherwise hidden interactions. In turn, these insights can be used to develop inferences concerning the probable impact of specific institutional or economic changes. The primary objective of the research becomes the identification of the potential opportunity sets available to different size growth-oriented farmers under alternative environments. Consequently, farmer behavior within the model must be designed to reflect a utility function that reflects selection among specific multiple goals in alternative environments.[8]

Based upon the work of Fitzsimmons and Holmes, Smith and Captick, Patrick, and Harmon and others, the following hierarchy of farmer goals was developed: (1) remain in business, (2) increase net worth, (3) increase control over land with an emphasis on ownership, (4) maintain a minimum and reasonably stable level of family living, (5) maximize year-to-year income (given existing resources), (6) reduce federal tax commitments, and (7) rely on refinancing of land only rarely to cover negative cash flows. Because this study was concerned with growth, the major objective became that of achieving the maximum growth with a minimum risk of going out of business in an uncertain environment created by random prices and yields. For a single-farm model, optimization techniques such as MOTAD could have conceivably been used to reach these goals. However, introducing interaction between farms made simultaneously optimizing each farm technically unfeasible.

The most important behavioral-control mechanisms included the following behavioral assumptions:

1. Farmers will not borrow beyond the point where their internal leverage ratio is greater than 1.3—that is, 1.3 times their market value net worth.[9]

2. Farmers will not purchase new parcels of land if they expect the new purchase will cause the total farm cash flow to be larger than negative $50 per after-purchase tillable acre.

3. Family consumption is always above $10,000 (1979 dollars) and is a function of after-tax accrual income. It is bounded by 30 percent of the exponential average consumption on the high side and 20 percent of this average on the low side.

4. Farmers fully utilize the family labor supply even if this results in some labor being sold off-farm at the same wage on-farm labor is paid.

5. Farmers strive to store 70 percent of their crop with maximum and minimum adjustments around this value as a result of cash flows or expected income tax payments.

6. Although a number of criteria determine the current value level of machinery, the relationship between cash and credit reserves and the amount of reserves needed to cover 2 years of debt service constitute the primary decision rule. When reserves are larger than desired, an above-average amount of machinery will be purchased. If reserves are insufficient, machinery will be replaced at below-average values.

7. Farmers use double-declining balance depreciation techniques plus first-year accelerated depreciation. In addition, they use the investment tax credits and always consider income averaging.

These control mechanisms require that all land acquisition control variables be tied to a common denominator. Thus, ratios and tillable acres are commonly used. Although precisely similar behavior for different size farms or for the same farm as growth occurs is not insured, behavior is controlled so that it is quite similar.

Designing Controlled Experiments for Policy Analysis

Although FLOSSIM provides the flexibility to develop alternative initial endowments or behavioral responses for the 4 simulated farms, it is difficult to isolate important economic response relationships if many of these variables are allowed to differ. Therefore, rather than constructing representative farms that might reflect typical equity positions or behavior for different size farms, research conducted with FLOSSIM focused on developing controlled experiments. As the research has been most concerned with the impact of alternative tax policies on different-size farms, the 4 farms were designed so they would have the following characteristics: (1) equal increment acreage increases for each larger farm, (2) the same initial debt/asset ratio, (3) the same initial tenure position, (4) the same fully utilized family labor component, (5) similar managerial skills so that pecuniary economies in the input markets would be reflected in the cost data, and (6) similar behavioral responses. Initial-year deterministic prices and yields for base conditions are presented in Table 11.5. Changing the environment or farm characteristics and using the base results for comparison provides the type of qualitative information needed to develop deeper insights into important interactions attributable to farm size and the economic and institutional environment.

Developing Useful Measurement Variables

Recognizing the multiple goals of farmers in an environment of risk and uncertainty creates special problems in selecting appropriate measurement variables. Furthermore, a large number of scenario combinations could be evaluated because of the use of multiple farms and planning periods. The most useful output currently estimated by FLOSSIM is presented in Table 11.6. Monte Carlo means and standard deviations are presented. The first 5 measurement variables are for average present values over each planning period simulated; the remaining 6 values are calculated from ending values for each planning period. These results and a close examination of year-to-year variability were used to develop the behavioral control mechanisms ex post facto.

Frequency of refinancing land is an issue that merits further explanation. FLOSSIM uses 1-year loans to meet cash flow deficits up to the point where these loans exceed half the expected gross returns. At this point, land loans are refinanced to cover the residual negative cash flow. Therefore, the frequency of refinancing land indicates the number of times the farm has cash flow problems greater than $10 per tillable acre. Although this approach is typical, a preferred approach would include refinancing that also reduces the 1-year loans. This approach would reduce the frequency of refinancing land because negative cash flows encountered in following years could be covered by the 1-year loans.

Finally, each of the summary tables provides information on the local land market. For the base analysis, there are 250 opportunities to purchase land (that is, 10 years times 25 simulations). Only 189 of

TABLE 11.5
Summary of initial year for base conditions (debt/asset = 0.3 and land ownership = 33 percent)

Item	Farm 1 300 acres	Farm 2 600 acres	Farm 3 900 acres	Farm 4 1,200 acres
			Acres	
Owned acres	100	200	300	400
Leased acres	200	400	600	800
			Bushels	
Corn yields	141.2	140.7	140.7	141.2
Soybean yields	42.9	44.6	44.7	44.4
			1979 dollars	
Corn price	2.53	2.53	2.53	2.53
Soybean price	6.47	6.47	6.47	6.47
Market value balance sheet:				
Current assets--				
Savings	0	0	0	0
Cash on hand	9,318	10,410	11,136	11,795
Grain inventory	45,460	96,427	144,753	192,970
Total	54,778	106,847	155,888	204,765
Intermediate assets	72,712	137,432	210,542	274,672
Fixed assets (land)	350,070	700,140	1,050,210	1,400,280
Total assets	477,560	944,409	1,416,640	1,879,718

TABLE 11.5 (cont.)

Liabilities—				
Current	568	2,926	5,505	8,337
Intermediate	21,205	33,905	50,410	63,921
Long term	103,296	206,593	309,889	413,185
Contingent liabilities—				
Capital gains taxes	19,671	53,554	92,657	133,019
Total liabilities	144,740	296,978	458,461	618,463
Market value net worth	332,820	647,431	958,180	1,261,255
Income statement:				
Total cash revenue	58,738	125,934	189,038	252,045
Total cash expenses	37,339	81,792	123,400	165,641
Net cash farm income	21,399	44,142	65,638	86,404
Total depreciation	7,372	14,169	21,800	28,426
Crop inventory change	0	0	0	0
Net farm income	14,028	29,973	43,847	57,978
Total nonfarm income	6,579	4,411	2,135	0
Total net income	20,607	34,384	45,973	57,978
Taxable farm income	4,565	16,411	26,800	37,892
Net cash flow position	1,997	10,051	16,970	23,976

TABLE 11.6
Base analysis—Monte Carlo: Summary for 25 simulations with 10 years each 1/

Item	Farm 1 300 acres	Farm 2 600 acres	Farm 3 900 acres	Farm 4 1,200 acres
		1979 dollars		
Initial net worth	332,820	647,431	958,180	1,261,250
Annual accrual farm income:				
Mean values	18,301	28,070	31,197	29,575
Standard deviation	6,192	13,379	16,224	21,459
Annual taxable farm income:				
Mean values	8,937	15,768	17,285	14,156
Standard deviation	3,649	9,374	12,582	13,650
Annual tax and social security payments:				
Mean values	1,710	3,419	3,998	3,491
Standard deviation	1,019	3,248	2,880	2,497
Annual family living expenditures:				
Mean values	16,989	17,390	17,831	17,280
Standard deviation	721	1,607	1,750	2,188
		1989 dollars		
Ending net worth:				
Mean values	1,091,223	2,383,559	3,616,492	5,005,683
Standard deviation	148,707	296,658	504,100	714,594

TABLE 11.6 (cont.)

Annual real growth in net worth:		**Percent**		
Mean values	2.93	4.03	4.26	4.73
Standard deviation	1.26	1.17	1.27	1.34
Frequency of refinancing land over 10 years:		**Frequency**		
Mean values	.16	.20	.52	.68
Standard deviation	.62	.65	1.12	1.22
Ending debt/asset ratio:		**Ratio**		
Mean values	.20	.27	.30	.27
Standard deviation	.09	.08	.07	.04
Purchased acres:		**Acres**		
Mean values	19	128	253	338
Standard deviation	29	43	86	75
Leased acres:				
Mean values	322	269	262	208
Standard deviation	242	167	172	155
Number of bids on land	58	131	156	143
Percentage of possible bids	23	52	62	57
Total parcels purchased	9	37	66	77

1/ See table 6.8 for a summary of year 1.

these opportunities resulted in a land sale because of the financial barriers presented by acquisition of larger parcels of land. A great deal of interaction also occurs among the farms. In the early years, the larger farmers are able to purchase the available land. This changes their financial position by increasing their debt/asset ratio and reduces cash flow. Consequently, they are prevented from placing bids on available land for a few years. This provides the smaller farmers with an opportunity to obtain land.

Another measurement variable not included in Table 11.6 can also be used to measure the well-being of the different farms over time. Combining the family living expenses and the change in net worth provides the value of generated capital over the period. These values and other results are used below to demonstrate how FLOSSIM could test the sensitivity of specific assumptions and the possible impact of alternative income tax policies on farms of different size.

Testing Land Market Assumptions

Throughout the development of FLOSSIM, a major behavioral concern involved the allocation of purchased land among farms. A modified version of the Lee-Rask model was eventually incorporated to develop a bid price. FLOSSIM was further modified to allow each farmer financially able to bid on the current year's available parcel an equal opportunity to obtain that parcel. Results from this sensitivity analysis appear in Table 11.7 in sensitivity number 2. Comparing these results with the base run leads to the following conclusion: the financial constraints that prevent owners of different size farms from bidding for available land are more important than the land's capitalized earning capacity for different size farms. Although there is a redistribution of purchased acres (from larger to smaller farms), the values are surprisingly close. These results suggest that multiple-farm competition models should give greater attention to farmer behavior regarding willingness and ability to take financial risk than to developing more accurate allocative schemes based on potential earnings of different size farms.

Another important feature of FLOSSIM is its ability to consider expansion through lease land. As expected, this feature provides the smaller farms with greater opportunities for growth. Eliminating the ability to lease additional acres led to a decline of generated capital for all farms and a redistribution of purchased acres from the three smaller farms to the largest farm (see Table 11.7, sensitivity 3).

As a major feature of FLOSSIM involves the interaction among farms, experiments were designed to demonstrate the diverse outcomes of simulating individual farms separately (sensitivity 4 in Table 11.7). All

TABLE 11.7

Sensitivity of land market assumptions 1/

Sensitivity identification and item 2/	Farm 1 300 acres	Farm 2 600 acres	Farm 3 900 acres	Farm 4 1,200 acres
	1,000 1979 dollars 3/			
Generated capital:				
a	298	533	747	1,025
b	295	525	764	1,003
c	263	479	730	1,014
d	335	555	797	1,030
	Acres			
Purchased acres:				
a	19	128	253	338
b	42	134	250	323
c	14	125	229	362
d	54	166	270	376
	1979 dollars			
Annual accrual farm income:				
a	18,300	28,100	31,200	29,600
b	13,700	23,470	29,880	34,170
c	16,008	27,255	29,754	28,028
d	13,293	10,720	12,867	9,880
	Frequency			
Frequency of refinancing land over 10-year period:				
a	.16	.20	.52	.68
b	.72	.40	.52	.48
c	.28	.56	.68	.64
d	.76	1.68	1.40	1.76

1/ All values are means from 25 simulations of 10-year periods.

2/ a = Base run; all farms initial ownership of land = 1/3 and debt/asset = 0.3. b = Random land allocation is made to those financially able to place a bid. c = Expansion through lease land is not allowed. d = Simulating each farm separately, no interaction between farms.

3/ Generated capital is calculated by adding the net gain (in 1979 dollars) in the market value net worth to the present value of the stream of family living expenditures.

parameters were the same as in the base analysis. As expected, each farmer generated more capital and purchased more land. Without competition, the specified internal behavioral parameters are not capable of controlling growth as accurately and with the same plausibility.

Other sensitivity analyses were also performed to test the land market assumptions. The model was not overly sensitive to the land availability assumptions when 2 parcels of purchase land and 2 parcels of lease land were provided in separate experiments. Providing a larger number of available land parcels increased the smaller farmers' growth opportunities. The results of other sensitivity analyses with a greater number of smaller parcels of purchase land proved more sensitive. The financial constraints associated with larger parcels of land were relaxed when these parcels were broken into smaller sizes. This fact underscores the importance of accurately reflecting parcel size availability. Monte Carlo simulation is the only technique that will allow varying size parcels.

Using FLOSSIM for Policy Analysis

Most research conducted with FLOSSIM has focused on the impact of alternative tax policies on the growth of varied size farms.[10] Policies such as the use of cash accounting, investment tax credits, rapid depreciation techniques, the Reagan tax cuts, indexing taxes to inflation, and changes in the capital gains taxes have been analyzed. The emphasis throughout these analyses has been on conducting experiments using: (1) alternative initial farm structures (changes in debt/asset position and tenure), (2) alternative behavioral assumptions (changes in the internal credit rationing), and (3) alternative environmental assumptions (changes in mean commodity prices and inflation relationships). If consistent results were observed from these experiments, then hypotheses were developed deductively. This approach provides further insights into those assumptions that most influence the results and that may require further study.

Three alternative farm financial or tenure conditions were used to analyze the difference between double-declining balance plus accelerated first-year depreciation versus straight-line depreciation (Table 11.8). Although it has been shown that more rapid depreciation enhances growth opportunities for larger farmers more than for smaller farmers (Umberger and Whittlesey), this set of experiments provides many additional insights. As expected, farmers generally generate less capital if they are forced to use straight-line depreciation. However, when generated capital ratios are compared, in 12 of these 18 paired comparisons including different initial farm financial or tenure conditions as well as different depreciation techniques, the smaller farmers' generated capital is more severely affected when they are forced to use straight-line depreciation. Furthermore, the smallest 3 farmers purchased less land when forced to use straight-line depreciation. In 2 of these 3 farm conditions, some of this land was reallocated to the largest farmer. The most likely explanation is that these conditions represent aggressive growth where the larger farmers are in relatively low tax brackets because of interest write-offs associated with land expansion, and they do not derive as much benefit from rapid depreciation techniques as do smaller farmers. Testing this hypothesis involved a new set of experiments where the farmers were less growth oriented. (The maximum internal leverage ratio was decreased.) Under these conditions, the generated capital ratios suggest that the larger farmers would be relatively worse off if forced to use straight-line depreciation. The primary conclusion is that when farm competition for land is very active, forcing all farmers to use straight-line depreciation may actually enhance the larger farmers' competitive edge.

Although this set of experiments and results is conditional, the results would have been different if the farms had been simulated in isolation. Under such an approach, the reallocation of purchase land from smaller to larger farms would not have been discovered. However, it is highly likely that each farmer would have purchased fewer acres when forced to use straight-line depreciation.

Conclusions and Implications

FLOSSIM was designed to provide insights into the growth behavior and opportunities available to 4 competing and different size farms under alternative economic and institutional environments. As this research demonstrates, consideration of interactions between different farmers in a local land market is an essential component of the environment. Indeed, different and perhaps incorrect conclusions may be inadvertently reached if a single-farm growth model is used. Furthermore, the value of Monte Carlo simulation becomes apparent when attempting to account for the varying size parcels of available purchase and lease land.

FLOSSIM has been used only to develop a preliminary quantitative understanding of the primary factors that influence growth of different size farms. Understanding these factors will help develop hypotheses concerning the probable impact of policy changes on the ownership and control of agricultural resources. However, as the limited analyses presented here demonstrate, initial farm financial and tenure conditions will influence the outcome of different analyses. Therefore, the linkages

TABLE 11.8

Comparison of rapid depreciation techniques with straight-line depreciation for farms with different initial endowments

Farm identification and item	Initial endowments					
	Base 1/		Highly leveraged 2/		Greater ownership 3/	
	Depreciation technique					
	More rapid	Straight-line	More rapid	Straight-line	More rapid	Straight-line
	1,000 1979 dollars					
Generated capital:						
Farm 1 (300 acres)	298	282	288	280	412	409
Farm 2 (600 acres)	533	518	512	506	803	803
Farm 3 (900 acres)	747	759	717	717	1,213	1,181
Farm 4 (1,200 acres)	1,026	1,000	944	943	1,613	1,586
	Acres					
Purchased acres: 4/						
Farm 1 (300 acres)	19	13	5	3	45	32
Farm 2 (600 acres)	128	130	51	45	178	170
Farm 3 (900 acres)	253	226	112	117	310	285
Farm 4 (1,200 acres)	338	352	214	210	379	418

TABLE 11.8 (cont.)

Generated capital ratios: 4/

			Ratio				
Farm 1 vs. 2	.599 ( )	.544	.563 ( )	.553	.513 ( )	.509
Farm 1 vs. 3	.399 ( )	.372	.402 ( )	.391	.340 ( )	.346
Farm 1 vs. 4	.290 ( )	.282	.305 ( )	.297	.255 ( )	.258
Farm 2 vs. 3	.714 ( )	.682	.714 ( )	.706	.662 ( )	.680
Farm 2 vs. 4	.519 ( )	.518	.542 ( )	.537	.498 ( )	.506
Farm 3 vs. 4	.728 ( )	.759	.759 ( )	.760	.752 ( )	.745

1/ Debt/asset = 0.3; 33 percent owned.
2/ Debt/asset = 0.5; 33 percent owned.
3/ Debt/asset = 0.3; 67 percent owned.
4/ To make relative comparisons between different sized farms as a result of a policy change, the model developed capital ratios. These ratios compare the smaller farms with larger farms under a base analysis and an analysis with the policy change. A ( ) suggests that the smaller farm has relatively greater opportunities to purchase land or to generate capital under the more rapid depreciation technique. (Note that it is possible that all farms may generate absolutely more or less capital.)

between the micro-level and the current macro/aggregate structure of the farm sector must be more fully developed.

Finally, the issue of using representative versus "designed" farms needs to be discussed. This research demonstrates that designing farms with similar characteristics (other than size) provides an opportunity to isolate the impact of policy or environmental changes. Yet, representative characteristics may need to be used for different size farms to provide a more realistic forecast or for other types of analyses and research. This discussion revolves around the use of simulation techniques to explain how a complicated system of behavior is affected in contrast to using simulation for forecasting or predicting impacts. Nevertheless, at this point of development, given the heterogeneous farming sector and the estimation problems associated with creating representative farms, running controlled experiments that provide insight in how the system works is considered more valuable.

Acknowledgments

This research was conducted at Michigan State University with the aid of a cooperative research agreement with the National Economics Division, Economic Research Service, U.S. Department of Agriculture. The author is grateful for assistance form Dr. James Scheaffer and Dr. Stephen Harsh. Appreciation is also extended to Dr. Royce Hinton of the University of Illinois and the Illinois Statistical Office of the Statistical Reporting Service for their assistance in obtaining data.

Notes

1. The reader is referred to Skees for a thorough review of the model and all its assumptions.

2. The table look-up functions merely extrapolate between data points. Any desired observations beyond the available data points use the related outer data values.

3. The MSU Ag Model is an econometric model of the U.S. agricultural economy. It is partially driven by world prices. Accounting for variability in world yields, given the institutional and economic structure of the current markets, provides an estimate of price variability. This approach is preferred over use of only historical data since institutional arrangements change over time.

4. A simulation is one 10-year planning period that contains the set of random draws (that is, random prices, yields, and size parcels of purchase and lease land).

5. Since land is a fixed resource, there is likely to be little supply response to price changes.

6. Preventing a farmer who purchases land from obtaining new lease land in the same year is further justified by the fact that a high proportion of purchased land is land that was previously leased by the same farmer. The model does not allow for the relinquishing of lease land unless the farmer has a severe negative cash flow that cannot be covered by refinancing owned land.

7. Machinery complements are expressed as a fraction of the average, where 1.1 is the maximum and 0.9 is the minimum allowed.

8. Systems scientists such as Manetsch and Park define this approach as systems design. Systems design in this study proceeded by identifying a multiple set of goals, developing mechanisms using financial and economic data to aid in reaching these goals, and interacting with the model as a proxy farm manager to subjectively fine tune the behavioral mechanisms until the identified representative goals appeared to be satisfactorily achieved.

9. Market value net worth is adjusted by contingent capital gains taxes.

10. FLOSSIM has also been used to analyze the impact of more restrictive and less restrictive commodity policies (that is, different price distributions were used) on farms of different sizes. This work led to the conclusion that if farmers behaved as described by the model, less stable commodity prices probably provided larger farmers with relatively more growth opportunities.

References

Black, Roy, and Maurice Vitosh. *Impact of Corn: Soybean Mix on Returns to Machinery, Improvements, and Land.* Telplan Program 41. Michigan State Univ., Dept. of Agr. Econ., Feb. 1974.

Boehlje, M., and S. Griffin. "Financial Impacts of Government Support Price Programs," *American Journal of Agricultural Economics,* Vol. 61, No. 2, pp. 285–96.

Fitzsimmons, Cleo, and E. G. Holmes. *Factors Affecting Farm Goals.* Bull. 63. Indiana Agr. Expt. Sta., Purdue Univ., 1958.

Harmon, Wyatte L., Roy E. Hatch, Vernon D. Eidman, and P. L. Claypool. *An Evaluation of Factors Affecting the Hierarchy of Multiple Goals.* Tech. Bull. T-134. Agr. Expt. Sta., Oklahoma State Univ., Stillwater, June 1972.

Hauschew, Larry D., and William McD. Herr. "A New Look at the Relationship Between Farm Real Estate Prices and Expected Returns." Paper presented at the annual meeting of the American Agricultural Economics Association, Univ. of Illinois at Urbana-Champaign, July 1980.

Hoskin, Roger L. "An Economic Analysis of Alternative Saginaw Valley Crop Rotations—An Application of Stochastic Dominance Theory." Ph.D. dissertation, Michigan State Univ., East Lansing, Dept. of Agr. Econ., 1981.

Johnson, Bruce B. *The Farmland Rental Market—A Case Analysis of Selected Corn Belt Areas.* AER-235. U.S. Dept. of Agr., Econ. Res. Serv., Sept. 1972.

King, Robert P. "Operational Techniques for Applied Decision Analysis Under Uncertainty." Ph.D. dissertation, Michigan State Univ., East Lansing, Dept. of Agr. Econ., 1979.

Lee, W. F., and N. Rask. "Inflation and Crop Profitability: How Much Can Farmers Pay for Land?" *American Journal of Agricultural Economics*, Vol. 58, No. 4, 1976, pp. 984–89.

Patrick, G. F. "Effect of Alternative Goal Orientations on Farm Firm Growth and Survival," *North Central Journal of Agricultural Economics*, Vol. 3, No. 1, Jan. 1981, pp. 29–39.

Scott, John T., Jr. "Forecasting the Price of Farmland for 1990." Illinois Agricultural Economics Staff Paper No. 79 E-85. Univ. of Illinois, Urbana, Dept. of Agr. Econ., June 1979.

Skees, Jerry R. "A Multiple Farm Simulation Model of the Impact of Income Tax and Commodity Policies on the Opportunities for Growth of Varied Size Corn/Soybean Farms." Ph.D. dissertation, Michigan State Univ., East Lansing, Dept. of Agr. Econ., 1981.

Smith, Donnie, and Daniel F. Captick. "Establishing Priorities Among Multiple Management Goals," *Southern Journal of Agricultural Economics*, Vol. 8, 1976, pp. 37–43.

Umberger, D. E., and N. Whittlesey. *Effects of Selected Tax Provisions on Growth of Columbia Basin Farms: A Simulation Analysis.* Washington Agr. Expt. Sta. Bull. 787. Washington State Univ., Pullman, Dec. 1973.

Wilken, D. F., and R. P. Kesler. *50-55 Annual, Summary of Illinois Farm Business Records.* Coop. Ext. Serv., College of Agriculture, Univ. of Illinois, Urbana, 1975–80.

12. FAHM 1981:
A System Dynamics Model
of Financial Behavior of
Farming-Related Households

Lyle P. Schertz, Linda Calvin,
and Lois Schertz Willett

Abstract The experimental model, FAHM 1981, depicts financial behavior of farming-related households. Financial effects on individual households can vary significantly with alternative policies because of contrasting resource situations and different reactions to policies and their macro effects. Yet, policy analysts focus mainly on aggregate variables such as farm income. A focus on households points up the need for greater attention to markets for factors—land, labor, and capital inputs. Examining ways to link macrosimulation models to FAHM 1981 is an important step in eventually testing the usefulness of micromodels in policy analysis.

Keywords Farming-related household models, financial behavior

Our interest in micromodels was stimulated by our perception of a conceptual inconsistency. Benefits to households are cited to justify many agricultural policy proposals. Yet policy analysts focus mainly on macro variables such as farm income, production, government expenditures, and, on occasion, food prices. In fact, farm-related policies often have distributional effects among households. And, these distributional effects sometimes differ from those implicit to the original justification for the policy. Our perception led us to ask:

The authors are economists with the National Economics Division, Economic Research Service, U.S. Department of Agriculture, and an economist with Energy and Environmental Analysis, Inc.

- Does a focus on household behavior provide insights useful in comparing potential effects of alternative policies that affect farming?
- Are particular relationships between policy and behavior of households in greater need of research than are others?
- Are system dynamics methodology and the Dynamo computer language effective in modeling such households?

Our focus on households was stimulated by the substantial change in how resources used in farming are organized and managed. Farming for the most part is no longer organized around "one farm, one farmer, and one household." There is great heterogeneity among operator households. Also, ownership and "use" of land, labor, and capital may be increasingly separated (Schertz).

In turn, we are developing an experimental model to depict financial behavior of farming-related households and to simulate their related financial conditions over time.

By selectively changing the behavioral assumptions, economic conditional and household characteristics, the financial behavior of households with such varied relationships to farming as full owner operator, part-owner operator, tenant, landlord, and farm laborer can be examined. This modeling approach requires the explicit consideration of resources the modeled households own, use, rent out, and rent from others. In addition, an explicit consideration of nonfarm- as well as farm-related financial transactions of the modeled household must be made. The model can be further specified to represent what are perceived to be "typical" or "atypical" households with respect to characteristics and behavior.

System dynamics concepts were used to produce the behavioral model and we wrote the computer program in Dynamo (see Forrester, 1968, and Goodman for discussion of methods). The basic model was constructed so that it can be adapted to a variety of initial financial conditions, behavior of the household, economic conditions, and organizational arrangements of farm establishments. The model can be used to represent the "one farm, one farmer, one farm household" concept—an arrangement where one household provides land, labor, capital, and entrepreneurial skill for a related farm establishment (Upchurch). Alternatively, the modeled household might provide only one kind of resource such as labor, land, capital equipment, livestock, or managerial time to a farm establishment.

The model is specified so that any one household may shift from being one type of household to being another type of household over a period of time. Such shifts depend on economic conditions reflected in the model's data and reactive behavior.

Systems dynamics was selected because it emphasizes behavioral concepts, and because the related Dynamo language is relatively easy to learn and program. Linear programming can be utilized with Dynamo but was not used in this particular effort. Approaches contrasting with microfarm models using Fortran and linear programming were valued for several reasons.

With systems dynamics, difference equations can be easily used to describe the bounded system that is modeled. The explicit inclusion of nonlinear and linear feedback relations within the model causes cumulative adjustments by decisionmakers that often produce counterintuitive results.

Systems dynamics has been used extensively to model such social and economic systems involving the combined behavior of several decisionmakers as world population, world food economy, country economies, sectors such as agriculture or wheat within an economy, world commodity economies such as oil, and urban economies (Forrester, 1961; Dennis Meadows; Donella Meadows; Naill). The methodology discourages emphasis on point-specific projection results. It includes relationships considered to be important even though empirically estimated coefficients are not available. Expert judgment and sensitivity testing of alternative coefficients values are accepted when empirical estimates are unavailable.

The system dynamics approach has also been used to model micro-units. However, many of these approaches have focused on micro-units that experience monopolistic competition rather than purely competitive conditions; that is, managers of the modeled micro-units behave as if they expect their actions to have an effect on prices. In contrast, our approach assumes purely competitive conditions, in which households behave as if their actions do not affect prices received or prices paid or the quantity of inputs available to them.

Features of the FAHM 1981 Model

The assumed objective of the modeled household is increased wealth. The pursuit of this objective is subject to market and institutional constraints and behavior of the household. This behavior is determined by preferences, such as the desire to operate a farm, or the desire to own farmland and observations and experiences.

Assets—money balance, stocks and bonds, house, farmland owned, and rented farmland—denote the main parts of the model. Debts for house and for farmland are included. They are computed within the sections that focus on house and owned farmland.

The model is specified so that the household may add to or subtract from the specified assets owned each year. It may also increase or decrease land rented from others. In addition, the household uses money for consumption, borrows money for purchases of houses and farmland, may work off the farm, and may receive government payment. The household may or may not be an "operator" of farmland.

Behavioral characteristics of the decisionmakers of the household are included in the model. For example, the household may or may not desire to be an operator of farmland or to rent land from others.

The land-debt equity ratio desired by the household is influenced by observed and experienced changes in land prices. The preference for house ownership has priority claim on money that might also be used to purchase stocks and bonds or as downpayments for farmland. Level of income and age influence the value of house desired. It is assumed that a house at the desired price will be available for purchase.

The proportion of money made available for farmland downpayment as opposed to purchases of stocks and bonds is influenced by observed and experienced price changes and yields of stocks and bonds in relation to price changes and returns to owners of farmland. Age affects nonfarm wage rates, availability of family labor, land debt-equity ratio desired, and perception of expected cash flow conditions.

Availability of land parcels to buy and of land to rent are specified exogenously. Other variables specified exogenously include land and commodity prices, land rents, land taxes, wage rates, yields and costs of production. All of the land owned can be rented out at the current land rent.

Rules on credit availability for farmland purchases are consistent with those we understand the Federal Land Banks applied in 1980. Production receipts and production expenses are specified as "production cash flows." The production expenses used to compute these production cash flows vary on a per acre basis by size of farm. Land taxes, loan service payments, and cost of hired labor are not included in production cash flows. This approach facilitates cash flow calculations whether or not the relevant land is owned, operated, rented, or rented out. It also makes it possible to take into account differing availabilities of family labor.

Three sets of relationships demonstrate the dynamic relationships included in the model. Increased net worth leads to further accumulation of assets—money, house, stocks and bonds, and farmland (Fig. 12.1). If economic conditions are not adverse, the acquisition of assets leads to additional net worth and further asset accumulation. However, the acquisition of assets and accumulation of net worth raises debt, which in turn causes larger loan service payments and, in turn, leads to a

FIGURE 12.1
Relationships between net worth and acquisition of assets

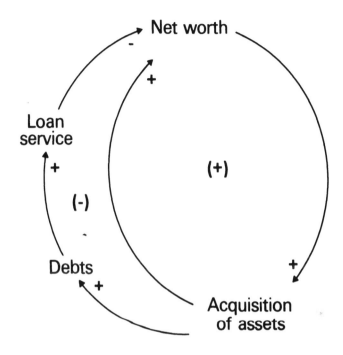

lower money balance which is part of net worth. Alternatively, reductions in net worth (and associated income streams) because of changes in asset prices may lead to the sale of assets and/or paying off of debts.

Money available for land downpayment is one of the determinants of whether available land can be purchased (Fig. 12.2). This amount reflects the amount of money available for both stocks and bonds and land downpayment and the preference for land relative to stocks and bonds. An initial preference is modified by observed and experienced returns of land relative to stocks and bonds. Learning response coefficients reflect the extent to which these observations and experiences influence the preference.

The desired land debt-equity ratio may influence the amount of land debt held by the household (Fig. 12.3). Changes of land price are observed, and if land is owned they are "experienced." These observations and experiences in turn affect the desired land debt-equity ratio. Increases in land price raise the land debt-equity ratio desired. Decreases have the opposite effect. However, the amount of land debt actually incurred is also influenced by the amount of land available and the

FIGURE 12.2
Key factors affecting land preference

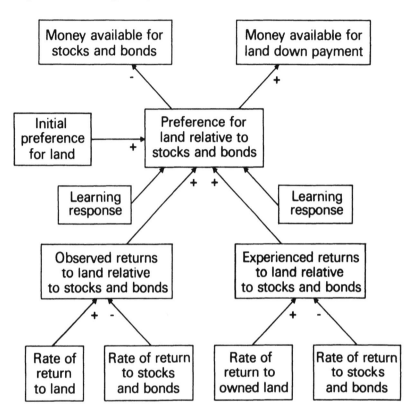

lender requirements relative to the land debt-equity ratio and expected cash flows.

The specific equations of the model are arranged around the concept of net worth (Fig. 12.4). Major parts of the model correspond to assets that might be held by the household. Debts incurred in the acquisition of some assets and the cash flows associated with these debts, as well as income streams associated with the assets, are included in the respective asset parts of the model.

Money Balance

Income from assets (including savings accounts), sale of assets, nonfarm wages, and government checks received increase the money balance. Changes in stock and bond values are shown as part of the income from assets instead of adjustments in the value of stocks and bonds.

FIGURE 12.3
Relationships between land price and desired land debt equity ratio

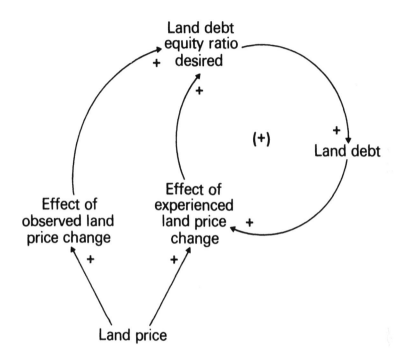

Income from farm assets (a part of income from assets) is determined by land rent received if the household owns land and does not operate it, land rent paid if rented land is operated, production cash flow for land operated, land taxes paid, and costs of hired farm labor.

Subtractions from money balance include consumption, house rent, downpayments for house and land purchases; loan service payments and stocks and bonds purchased decrease the money balance. Consumption expenditures do not include house rent paid, downpayments for house loans, or associated loan service payments. As now specified, consumption is influenced by additions to money balance except that sales of assets do not affect consumption.

Stocks and Bonds

Stocks and bonds are purchased with money not needed for a money reserve or for down payments for a house or for land. In contrast, stocks and bonds are sold if money is needed for consumption, loan service payments, or downpayments.

FIGURE 12.4
FAHM 1981: Major sections of model*

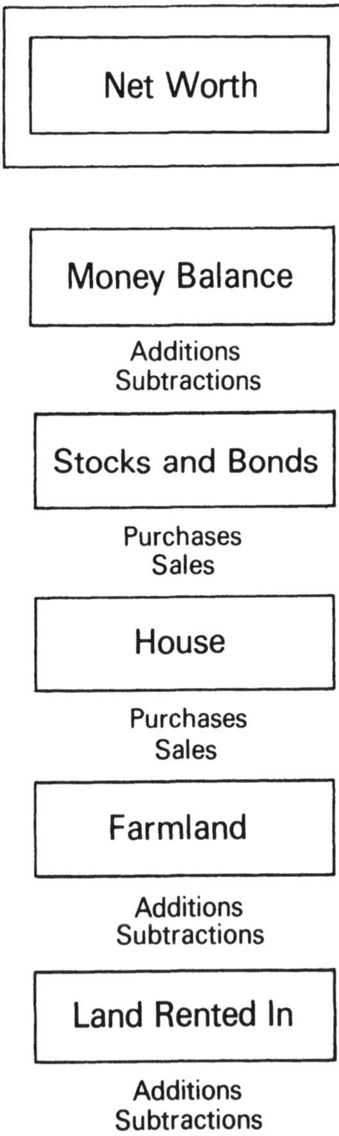

Net Worth

Money Balance

Additions
Subtractions

Stocks and Bonds

Purchases
Sales

House

Purchases
Sales

Farmland

Additions
Subtractions

Land Rented In

Additions
Subtractions

* All items in boxes, with the exception of Net Worth,
are technically "levels" and all additions/subtractions
and all purchases/sales are rates in systems dynamics
vocabulary.

House

A downpayment for a house has priority for the portion of money balance that exceeds the specified money reserve. The amount of this reserve increases with age of the head of the household.

The value of house purchased is dependent on age and income. A 15-percent downpayment is required, and a 25-year home loan for the unpaid balance is renegotiated annually at the then current rate of interest. Whenever the calculated house value objective exceeds the current value of the house owned (zero if a house is not owned) by $15,000 and money is available for a downpayment, a new house is purchased and the old house is sold.

Farmland

Availability of land parcels to purchase and the price per acre are specified exogenously. The price is also exogenous to the household. The household decides whether or not to buy the parcel. The purchase situation is modeled similar to when a parcel of land is listed with a real estate agent. Prospective buyers decide if they want to buy the listed land at the best price that can be negotiated by the seller. It is assumed that the exogenous price of available land is the lowest price that can be negotiated. Land is purchased when land is available, money is available to make a downpayment consistent with a calculated available loan which reflects expected cash flows, and the household desires to own land.

Available land to rent is similarly modeled. In contrast, the demand for renting land from those who own land is assumed to be perfectly elastic at the specified land rents.

The available land loan is the smaller of the following: the value of the available land, an amount consistent with the land debt-equity ratio rules of the credit institution, or an amount that can be serviced with the expected cash flow of the household. When the debt-equity ratio desired by households is larger than that required by the credit institution, the institution's maximum debt-equity ratio acts as the effective constraint. In contrast, when the debt-equity ratio desired by the household is less than that required by the credit institution, the household desired debt-equity ratio is controlling. The household's desired debt-equity ratio is influenced by observing changes in land prices, as well as by experiencing the price changes with land already owned and a rate of learning response. Money not held as a money reserve or used for a downpayment on a house is available for downpayment for land or for purchase of stocks and bonds. The proportion of this money that is available for land (preference for land relative to stocks and bonds) is influenced by an initial preference for land which may be specified from zero to one. The observed returns to land are compared with returns to stocks and bonds, and these comparisons are used to change initial preferences and are cumulative over time. Positive returns to owned land adds to the preference for owned land. Positive returns to owned stocks and bonds subtracts from the preference for land and vice versa.

Loans for land carry the current rate of interest with 35-year payments on unpaid principal. The interest rate for the loan varies from year to year. When another land parcel is purchased, a similar loan is negotiated with 35-year principal payment provision for all land owned.

Expected cash flows are calculated on the basis that the future will correspond to current conditions.

Land may be owned and rented out or owned and operated. However, land cannot be rented out if any land is operated. Additional land is rented in if: (1) land is available to rent; (2) either 200 or more acres are available, or 160 or more acres are already being operated; (3) the household has a money balance equal to or greater than $100 times the number of acres available to rent; (4) the household desires to rent and operate land; and (5) the expected effect on cash flow is positive if the available parcel is rented. The amount of land rented is reduced annually in 60-acre increments whenever the land rent per acre exceeds the production cash flow per acre.

Model Results

The FAHM 1981 model was run deterministically for the years 1961–80. Historic data were used for exogenous variables of the model that correspond to available data series. For example, 1961–80 U.S. average corn yields and prices received by farmers, Illinois land price, and Illinois land tax rates were utilized.

Financial conditions were simulated for 3 households:

Larger farm operator household
Nonfarm household
Smaller farm operator household

Each of the households owned farmland (Table 12.1). The nonfarm household, however, did not operate the land it owned. The larger farm operator household and the nonfarm household were assumed to be identical with respect to age in 1961, net worth, mix of assets and debts, potential wage rates for nonfarm work, response to changes in returns to land and to stocks and bonds, the rules applicable to credit availability, and the annual availability of land to each household for

TABLE 12.1

Beginning characteristics of simulated households, 1961

Variable	Unit	Type of household		
		Large farm operator	Nonfarm	Smaller farm operator
Age	Years	40	40	40
Net worth	Dol.	121,000	121,000	60,500
Assets	do.	171,000	171,000	85,500
Money balance	do.	30,000	30,000	15,000
Stocks and bonds	do.	40,000	40,000	20,000
House value	do.	40,000	40,000	20,000
Owned farmland:				
Value	do.	61,000	61,000	30,500
Area	Acres	200	200	100
Debts	Dol.	50,000	50,000	25,000
House debt	do.	30,000	30,000	15,000
Land debt	do.	20,000	20,000	10,000
Attitudes:				
Desire to own land	--	Yes	Yes	Yes
Desire to rent land	--	Yes	No	Yes
Desire to operate land	--	Yes	No	Yes
Nonfarm wage per week	Dol.	81	81	81

purchase. The smaller farm operator household differs from the larger farm operator household only in that its assets and debts were exactly half those of the larger. Consequently, in 1961 this household owned 100 acres of land instead of 200 acres, and its net worth was $60,500 instead of $1,210,000.

The financial growth of the 2 larger households was almost identical during the 1960s. But, their growth (net worth and acreages) paths differed in the 1970s (Figures 12.5 and 12.6), because the nonfarm household was able to buy additional 80-acre parcels of land in each of 3 different years in which the larger farm operator household was financially unable to do so. In comparison, the net worth and owned farmland of the smaller farm operator household increased. However, net worth and acreage size relative to the larger operator household decreased. Its 1980 net worth and owned farmland were approximately 45 percent of that of the larger operator household.

FIGURE 12.5
Net worth (1981 dollars), three simulated households, 1961-80

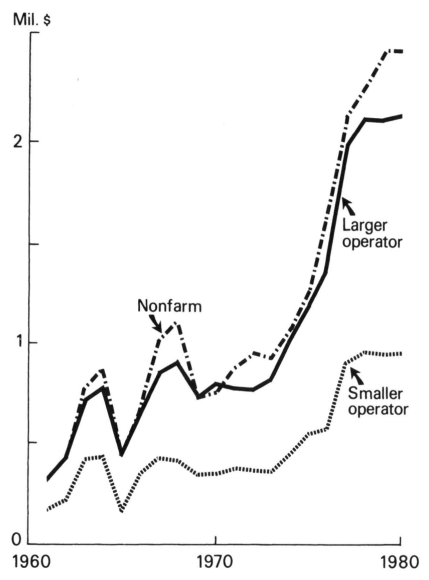

Mil. $

Nonfarm

Larger
operator

Smaller
operator

1960   1970   1980

Preliminary sensitivity and policy tests suggest that the growth in net worth is sensitive to rules applicable to credit availability and to the size of parcels available for annual purchase. However, the effects differ among the 3 households simulated. For the 3 simulated households, more restrictive credit rules such as 50-percent discount of model estimated cash flow rather than 25 percent and a maximum planned

FIGURE 12.6
Farmland ownership, three simulated households, 1961-80

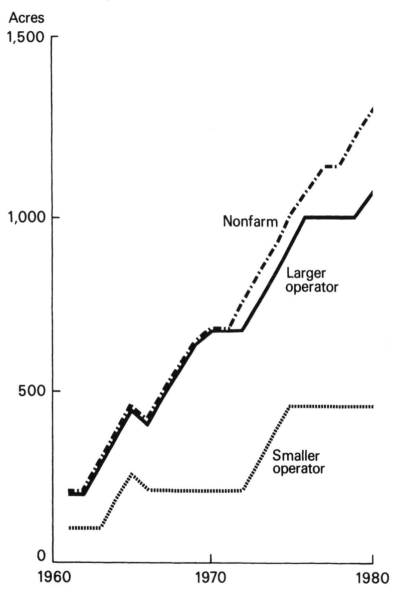

land debt-equity ratio of 0.40 instead of 0.65 exhibited only modest effects on net worth growth for situations in which available parcel size is 80 or 160 acres (Table 12.2). In contrast, restrictions on credit availability affect growth in net worth much more when parcel size availability is 320, 480, or 640 acres. For example, the 1980 net worth

TABLE 12.2

Net worth of households, 1980:   Larger farm operator, nonfarm, and smaller farm operator

Size of parcel available annually (acres)	Expected cash flow discount 1/		
	0.25	0.50	0.50 2/
	Million dollars		
Larger farm operator:			
80	2.1	2.0	2.0
160	2.5	2.4	2.4
320	3.0	2.1	1.7
480	4.3	2.1	1.4
640	4.3	1.7	.7
Nonfarm:			
80	2.4	2.4	2.4
160	3.4	3.4	2.1
320	5.7	4.8	2.1
480	6.0	5.1	2.4
640	6.6	4.6	1.8
Smaller farm operator:			
80	.9	.8	.8
160	1.0	.8	.7
320	.8	.7	.4
480	.4	.4	.4
640	.4	.4	.4

1/ Factor by which model estimates of expected cash flow were discounted before model calculated loan amount that could be serviced.

2/ In addition to 0.50 expected cash flow discount, maximum planned land debt equity ratio was set equal to 0.40 instead of 0.65 as used for other calculations.

of the larger farm was $4.3 million with a 25-percent discount on estimated cash flow, but only $2.1 million with a 50-percent discount given an availability of a 480-acre parcel annually. And, the 1980 net worth would have been $1.4 million with a combination of a 50-percent discount on estimated cash flow and a maximum planned land debt-equity ratio of 0.40 instead of 0.65.

The effects of the postulated restrictive credit rules are much less pronounced for the small farm operator household under most conditions simulated. The exception is when the parcel is 320 acres and the maximum planned land debt-equity ratio is 0.40 instead of 0.65. The more restrictive land debt-equity ratio makes it impossible to purchase 320 acres in 1973, even though the expected cash flow discounted at 40 percent would have made it possible to service the loan.

Summary, Conclusions, and Future Direction

In summary, our interest in modeling households related to farming was stimulated by the observation that farm-related policies often have distributional effects among households and that these distributional effects sometimes differ from those implicit in the original justification for the policy. System dynamics concepts were used to produce the behavioral model which was written in Dynamo. The basic model can be adapted to a variety of initial financial conditions, behavior of the household, economic conditions, and organizational arrangements of farm establishments.

The assumed objective of the modeled household is increased wealth subject to market and institutional constraints and behavior of the household. The format of the model is organized around assets—money balance, stocks and bonds, house, farmland owned, and rented farmland. Changes in prices of land and prices of stocks and bonds influence the desire to acquire land or stocks and bonds. These behavioral responses can have substantial effects on the growth in net worth.

Sensitivity and policy tests suggest that growth in net worth is sensitive also to rules applicable to credit availability and to the size of parcels available for annual purchase. The effects on different size households are not uniform especially when parcel size availability is large.

The preliminary conclusions of our experiment are as follows:

- Financial outcomes for individual households can vary significantly with the same policy because of contrasting resource situations and different behavioral reactions to policies and the macro effects of policies. Given the heterogeneity of farm households, we are optimistic that models of households may be able to provide useful insights about distributional effects of policy proposals.

- Macromodels capable of producing estimates of aggregate variables exogenous to micromodels such as inflation, interest rates, and land prices, as well as commodity prices, will be important to the effective use of household models.

- Research findings are limited for many behavioral relationships and decision criteria important within micromodèls. For example, determinants of household consumption expenditures remain obscure. What type of cash flows should be considered as influencing consumption? Payments on loan principal presumably do not immediately affect income calculated with commonly used procedures. Yet, principal payments affect money available for discretionary consumption, savings, and investment. Further, does net worth, or do changes in net worth, affect consumption? For example, is the consumption of two households, each with $30,000 of "income," likely to be different if the wealth of one is $30,000 and the wealth of another $1 million? Also, are the relationships between income and consumption of farm operator households likely to be different from farm labor households or those of landowner nonoperator households?

- A focus on households strongly suggests the need for greater attention to factor markets. For example, estimated cash flows for different types of households are tied closely to factor returns and prices. This circumstance points up the need to understand the relationships among policy alternatives and factor returns and prices. The sensitivity tests with FAHM 1981 that focused on the size of parcels available for purchase illustrate the importance of a more complete understanding of the characteristics of the land market. Surely the market is not accurately depicted when we postulate in FAHM 1981 that 80 acres or 640 acres are available for purchase annually.

- The system dynamics methods stimulated us to inquire about complex relationships on both the micro- and macro-level that may be important to understand the aggregate responses to policies. Dynamo computer language has worked well and the services of the Dartmouth College Computing Center have been very good. We have occasionally felt the need for greater flexibility in specifying some decision rules in the computer program than available in Dynamo. However, the Dynamo use of table functions is extremely useful in dealing with relationships among variables, especially threshold and nonlinear relationships.

Before drawing further conclusions from the experiment, further work on the following tasks needs to be completed:

- Develop relevant tax equations.

- Modify equations to accommodate "lumpiness" of capital equipment.

- Scrutinize model for consistency with systems dynamics method and for ways to simplify the model.

- Better define relationships and coefficients included in the model through literature surveys, discussions with knowledgeable observers and owners of resources, and examination of longitudinal census data and longitudinal farm record data.

- Explore the usefulness of combining linear programming routines with the model and techniques for running the model stochastically.

- Develop a classification of households meaningful in a policy context, including atypical kinds of households.

- Examine methods to link macromodels to FAHM 1981.

- Test model along with other micromodels in an actual policy analysis setting.

Acknowledgments

Many people have contributed in important ways to the preparation of this paper and to the research to which it relates. They include Nora Brooks, Sylvia Bryant, Terry McGuire, Dana Meadows, and Barry Richmond. Complete documentation of the model presented in this article is available from the authors.

References

Forrester, Jay W. *Industrial Dynamics.* Cambridge, Mass.: The Massachusetts Institute of Technology Press, 1961.

Forrester, Jay W. *Principles of Systems.* Cambridge, Mass.: Wright-Allen Press, Inc., 1968.

Goodman, Michael T. *Study Notes in Systems Dynamics.* Cambridge, Mass.: Wright-Allen Press, Inc., 1974.

Meadows, Dennis L., and others. *Dynamics of Growth in a Finite World.* Cambridge, Mass.: Wright-Allen Press, Inc., 1974.

Meadows, Donella H., and others. *The Limits to Growth.* New York: Universe Books, 1974.

Naill, Roger. *Managing the Energy Transition.* Cambridge, Mass.: Ballinger Publishing Co., 1977.

Schertz, Lyle P. "Relationships Among Households and Farm Establishments in the 1980's: Implications for Data," *American Journal of Agricultural Economics,* Vol. 64, No. 1, Feb. 1982.

Upchurch, M. L. *Steps Toward Better Data for the Food Industry,* Economic Research Report 2. U.S. Dept. of Commerce, Bureau of the Census, Feb. 1979.

Discussion of Part 6

Multiple-Use Farm Models

David W. Parvin, Jr.

FLIPCOM consists of 2 related experimental models, FLIPSIM and FLIPRIP. Both models are capable of analyzing the micro-level impacts of one or more macro-level policies either on a large number of farms or on one particular farm. Although the models are conceptually identical, they differ in their mathematical programming techniques. FLIPSIM uses a linear programming approach, while FLIPRIP uses a recursive interactive programming approach. As neither model is farm- or location-specific, each can be employed in a variety of settings.

Both FLIPCOM models extend far beyond traditional micro-level optimization models in that an optimal growth strategy for the farm can be determined in conjunction with any policy effects. Both models are capable of interfacing a vast number of macro policies with traditional farm-level structure data.

Baum and Richardson used 3 representative farms to stochastically simulate growth and adjustment behavior over a 10-year horizon. These 3 farms were very different in structure, resource mix, and alternative enterprise; furthermore, they represented different regions. A full complement of federal programs was assumed, and the model included target prices, a farmer-held reserve program, a price-support program with a 1-year Commodity Credit Corporation (CCC) nonrecourse loan, and an all-risk crop insurance program. Results of the simulation varied considerably among farms, and separation of the various effects was extremely complicated.

Regionally, both models appear useful for capturing acreage shifts over time and for evaluating the effects of alternative policies. At the farm level, the models should prove useful for planning purposes and for evaluating the alternative tax strategies of the firm over time (if the user is extremely talented).

Both models are complex and cannot be easily used without extensive training. The models also require detailed exogenous data which would not be obtainable for many farms. Consequently, the prospects for Extension's properly utilizing either model and interpreting results appear

---

The author is a professor of agricultural economics, Mississippi State University.

limited. Because both models are still in an experimental state, additional verification will be necessary if their full potential is to be realized.

Nonetheless, the authors are to be congratulated for accomplishing an outstanding piece of work in an area that has offered little or no professional reward.

Discussion of Part 6

Positive Farm Models:
A Comment on Aesthetic
Versus Functional Value

Katherine H. Reichelderfer

This paper uses the analogy of art as a vehicle for discussing some of the potential advantages, disadvantages, and trade-offs among various broad classes of positive farm models. It briefly explores the model as art and the modeler as artist. It focuses on my perspective as an art appreciator as that is basically how one must approach the review of models one has not used. Although my comments are fairly general, I draw upon the System Dynamics Model of Farm and Household Financial Behavior, FAHM 1981, to illustrate major points.

There are 2 aspects of art that are interrelated and between which the artist often must make trade-offs. These are: (1) aesthetics, as embodied in art by the great painters and sculptors and (2) function, as represented by the artistic disciplines of design and architecture. Each of these aspects can be related to positive farm modeling.

Aesthetic Aspects of Modeling

Aesthetics, in the modeling sense, might refer to the degrees to which a model captures reality and appeals to the user's perception of reality. Our general theory and many of our models fall into the realm of abstract art. The simpler, positive, econometric simulation models present the rudiments of a farm or agricultural market as based on economic theory and assumed relationships. Perhaps we can classify these as cubism because they necessarily must contain a large number of black boxes and must incorporate straight lines to represent more complex relationships. Larger, more detailed models also can remain abstract. The need to use "simplifying assumptions" to describe and link numerous interrelated economic units can so obscure the system described by the model that it is difficult for anyone other than the modeler/artist to

The author is an agricultural economist with the Natural Resource Economics Division, Economic Research Service, U.S. Department of Agriculture.

identify with and appreciate the black boxes. The results from these abstract models, however, tend to be precise and unambiguous.

At the other extreme is realism. We frequently speak of taking a "snapshot" of economic conditions, which implies static but real depiction. Actually, the state-of-the-art of modeling is such that true realism is not a feasible option. However, what I will refer to as "impressionistic modeling" is an alternative classification.

It is entirely possible for some of the black boxes of abstract modeling to be opened and explored and for some of our rigid assumptions to be relaxed. This may result in more realistic depiction of the variables affecting agricultural economic relationships. Certainly our knowledge of these relationships and the data available to test them are limiting factors. Nevertheless, I have little doubt a clearer impression of the agricultural economic environment can be painted with models. FAHM 1981 provides an illustration of this possibility.

FAHM 1981 explores a black box traditionally encasing all activities and preferences of the farmer including those that are distinct from onfarm profit (or utility) maximization. As Duncan and other conference participants point out in their papers, the relevant decisionmaking unit in reality is the farm household. Trade-offs occur in these households among such things as land acquisition, equipment purchase, leasing arrangements, house purchase, and leisure activities. Households learn, experience uncertainties related to economic conditions, and behave in reaction to their experiences. FAHM 1981 describes and uses these behavioral characteristics in an attempt to accurately simulate the effects of changes in economic conditions on the household. This appeals to me as an appreciator of impressionistic models. I understand, can identify with, and can relate to the depiction of farmers as people who have variable perceptions, learn from experience, and consider consumption items unrelated to the farm operation. It is apparent these considerations can affect farm-specific impacts of agricultural policy, and the paper by Schertz, Calvin, and Willett clearly illustrates this possibility.

A criticism of this kind of model is that its results are vague and qualitative, as opposed to precise and quantitative. This situation can, however, be seen as a relative advantage. Results are not solely defined by the formulator/artist. Rather, they are open to creative interpretation by the user/appreciator. Although neither person(s) may be able to use the model to predict events with any certainty, the impressionistic model is useful in indicating the direction and relative magnitude of impacts. This distinction begins to address the functional aspect of the art of modeling.

Functional Aspects of Modeling

The functional value of a model is determined by a number of things including: ease of operation, the degree to which users can understand results and their derivation, flexibility with respect to the range of issues that can be addressed, opportunity for adding to or adding on other models or new considerations, and the extent to which results can be validated.

It should be easy to trace through the logic and clearly understand models that fully and explicitly describe the system being evaluated. However, the amount of detail required for full explanation of relationships can be so great as to overwhelm the potential user. FAHM 1981, for example, is apparently easy to use. It also seems easy to understand, but it would take a relatively great investment of time on the part of the user to become familiar with and understand each of its 379 equations, and multiple equation relationships. Ideally, all users would make that investment because the analyst/user cannot adequately understand the results of a model unless the model itself is understood.

With respect to flexibility, FAHM 1981 (as well as FLIPCOM and FLOSSIM, the other two models presented in this session) has unique functional advantages. The capability to make selective changes is built directly into the model's design so that it can be used to estimate the effects of a variety of conditions on any type of farm-related household the user wants to examine. This characteristic enhances the potential value of the model in microanalytic policy evaluation.

All three of the models presented in this session seem amenable to integration with other microanalytic or broader models. FAHM 1981, in particular, describes a small set of agricultural economic activities and depends on a relatively large amount of exogenous information. Its linkage with a macromodel specifically designed to generate currently defined exogenous variables would be highly profitable and, apparently, fairly easy to accomplish with use of the Dynamo language. It is intriguing to me that, at least with respect to the issue of farmland acquisition, the three models presented in this session could benefit through some degree of integration with one another. FLIPCOM, in describing representative farms, assumes if farm operators are financially able to acquire land, they will do so. FAHM 1981 examines competition within farm households for land acquisition and other production and consumption items, and FLOSSIM describes competition among farm operators for fixed available land. A merger of the 3 is technically feasible and could realistically take into account all these interrelationships with respect to land. The integrated model would, unfortunately, probably be overwhelmingly complex.

The ability to statistically validate or otherwise confirm the results of models is a primary functional concern of agricultural economists/ users. The user must be able to feel confident that results are legitimate, particularly if those results are going to be used in the policymaking process. When a model is based on or includes intuitive relationships for which data are unavailable to test, it falls short in meeting this criterion. FAHM 1981 may be typical of the newly developed family of impressionistic models in that it cannot easily be statistically validated.

Schertz, Calvin, and Willett state that

> emphasis is given (in FAHM 1981) to the inclusion of relationships if they are considered to be important even though empirically estimated coefficients are not available. Expert judgment or sensitivity testing of alternative values of coefficients are accepted as substitutes for empirical estimates.

This judgment implies that validation of the model results also depends on expert subjective judgment. Many potential users will view this factor as a functional disadvantage. If economists move toward developing impressionistic models without simultaneously developing data sets or empirical estimation techniques, acceptance of this functional shortcoming may be a necessity. Alternatively, this functional consideration may lead to legitimate questions concerning the value of modeling exercises. If one must depend on intuition and expert judgment to develop and validate a model, then why is it necessary to model at all? Why not utilize the human capital that embodies that intuition and judgment to evaluate the impacts of various stimuli directly and more efficiently?

Trade-offs Between Aesthetics and Function

Aesthetics and function can strongly conflict in modeling, as in art. The complexity required to construct realistic models that capture a greater degree of the economic and behavioral factors affecting agricultural units can conflict with the perceived need to utilize simple, inexpensive, nonambiguous models for policy analysis. The greater flexibility offered by impressionistic models, while a functional advantage, must compete with the appeal of the "one question-one answer" class of models. In the real world, we rarely find simple answers to complex questions, and realistic models will reflect this ambiguity. While this is not an undesirable aesthetic trait, it can be viewed as a functional disadvantage.

The desire to model all relevant relationships is also constrained by our limited knowledge of farm household behavior, and, more critically, by a lack of data. During this period of budget constraints and survey respondent fatigue, ERS and others cannot expect to be able to collect data on every variable identified as important in the description of farms or farm households. Thus, to add these relationships, some expert judgment must be used to describe them. Expert judgment is and always has been important in constructing models and interpreting results. However, there is a substantive difference between incorporating intuitive judgment into models and modeling intuitive judgments directly. Formalization of what Weeks calls "the intuitive model we all carry around in our heads" is likely to be aesthetically pleasing to the modeler/artist and to those who agree with the model's intuitive relationships. But the purely or largely intuitive model has little functional value as a model. Identical objectives could be met by a verbal or qualitative description of one's thought process.

As we continue to explore the factors affecting farm operations and households, development of positive farms models that provide more "strategic detail" (as Day suggested) of the economic environment without losing too many of the functional advantages of the simple abstract model should be encouraged. Ideally, the most useful farm models for policy analysis will explain as much as possible about the real world without crossing the line between purely formalized intuition and objective, confirmable models which are fine-tuned by input from experts. This combination is a challenge not easily met, but well worth the effort.

Discussion of Part 6

Micromodeling: A Systems Approach at the National Level

John A. Miranowski

The papers presented in this session are on the frontier of advances in modeling the decision process of the farm firm. In addition to being stimulating, these papers also indicate the strengths and weaknesses of firm-level micromodeling. Generally, these models are detailed attempts to capture the complexities of farm decisionmaking, the interactions between farming units, and the influence of macroeconomic and institutional factors on farmers' decisions. Such models may provide helpful information to individual farmers and possibly to policymakers. Unfortunately, one is left with the feeling that the investigators are attempting to use a sledge hammer to crack a nut while hoping to save all the meat. Because of the complexity of these models, it is extremely difficult, if not impossible, to interpret the causal effects generating specific results. If one is to benefit from such a "systems approach," then understanding the linkages and causality is essential. Otherwise, we are dealing with "black boxes" which have to be taken on faith alone and which will not enhance the understanding of the decisionmaker or policymaker in any productive way.

Although my comments are directed to the paper by Professor Skees, many of the points are pertinent to the other papers as well. Skees's paper assesses the interactions between farms competing for limited land at the micro or local level. The dynamic growth of 4 firms is considered through a multiple-objective framework; when limited land is available for purchase each year, land is available in different size parcels, and alternative government policy schemes are specified. The conclusions are interesting. First, considering the interactions between farms in the local land market is an important dimension of modeling. Second, whether or not farmers can play in the bidding game for land is more important than the form of the bidding game. Third, the straight-line depreciation requirement transfers an advantage to larger farmers, but this advantage appears to be the result of the mix of debt and equity financing.

The author is Gilbert White Visiting Fellow, Resources for the Future, Washington, D.C.

The FLOSSIM model has some inherent problems that are common to this type of model. First, although a multiple-objective approach may be the appropriate framework, problems are encountered in modeling an individual's utility function. Until we develop better insights into the dimensions of satisfaction for an individual, attempts to model risk aversion and other objectives are going to be limited in their usefulness. Second, the limited information on some of the functional relationships (for example, yields) may introduce biases. Third, little discussion is given to the potential compatability problems encountered when algorithms are borrowed from other models. Finally, a model is unreasonably complex when the results can only be explained by devising indirect tests. For example, when straight-line depreciation was more advantageous to larger farmers, the original data should have helped to explain the results.

I also have a series of questions to ask with respect to FLOSSIM and to micromodels in general. First, is a micromodel such as FLOSSIM useful for policy evaluation? Probably not. The results are illustrative and not general. The analysis may need to be duplicated for every land market. Professor Skees indicated that "replicating farmer behavior is a major limitation in farm firm simulation models, but only in using models to predict" and that qualitative insights will permit inferences on the probable impacts of particular changes in the decision environment. That a simulation model is unable to replicate or predict behavior and can only aid our intuition in drawing qualitative conclusions is a very serious indictment.

Second, is the appropriate clientele clearly identified? Probably not. Although policy-oriented questions are being asked, one cannot generalize from the results to a region or the Nation. To be of use in policy formation and evaluation, a tremendous number of representative farms would have to be considered and an aggregation problem would still remain. If the purpose is human capital development (that is, to teach graduate students economic theory, production agriculture, and quantitative modeling), there appear to be more efficient methods of accomplishing this task. Basically, students should have learned the theory, institutions, and management environment before undertaking the model-building venture. The most appropriate audience may be farm firms or households planning longrun growth (or passing through the farm life cycle). These models are probably best delivered as a private good, adjusted for the behavioral characteristics of each individual farm, and marketed through the Extension Service or private farm planning agencies. For example, the Iowa State University Extension Service provides the background information for an integrated pest and crop management system, but leaves the onfarm delivery of information and recommended actions to the private sector. Similarly, Iowa State University charges for farm-specific estate plans produced by the estate planning model. Third, should ERS be allocating internal research funds to the micro-modeling effort or objective? Probably not. These firm models are specific to illustrative farms in selected state crop reporting districts. If the results of these models are useful in particular policy evaluations, then cooperative agreements should be used to acquire the information. The actual modeling is best done outside USDA in specific production environments.

Fourth, is the potential of these models being fully exploited? As James Richardson suggested, most of these models are designed for a single-purpose analysis. Both the Skees and Schertz models were developed to consider structural characteristics. With the exception of the FLIPRIP and FLIPSIM models developed by Baum and Richardson and by Richardson, respectively, these models usually lack the flexibility necessary to evaluate a broad range of policy issues.

In conclusion, micromodels do have a comparative advantage for some types of economic and policy analyses not easily handled by aggregate sector models. In addition, they are more useful in assessing the impacts of policies on individual farms than in formulating policy. For example, what is the impact of changes in tax policy or farm programs on a particular farm type or class in a region? If only qualitative assessments are forthcoming from simulation modeling as some of the authors have suggested, then we are using a sledge hammer to crack a nut when theory combined with intuition would be more appropriate. Finally, we will only obtain the correct "behavioral response" if we accurately model the farmer's objective function. Getting all the "meat" out of the nut will depend on what goes into the behavioral model. If our models do not replicate actual farmer behavior, then, given the desired goals and the environment, we cannot ascertain what is "appropriate" behavior.

PART 7

Optimizing Models

13. A Polyperiodic Firm Model of Swine Enterprises

Wesley N. Musser, Neil R. Martin, Jr., and Donald W. Reid

Abstract This paper discusses how polyperiodic mathematical programming models are used to analyze public policy impacts on farm firms. A firm model, which emphasizes swine enterprises in Georgia, illustrates the capabilities, limitations, and the assumptions incorporated into such models. This type of model can be used to assess the impact of changes in many important economic parameters such as income taxes, commodity prices, and credit policy. Adjustments to changes can be reflected in both production and investment responses.

Keywords Firm growth, linear programming, swine production

Polyperiodic firm programming models were developed to consider decisions inconsistent with standard static programming models. Static linear programming (LP) models of farms do not incorporate the effects of expected future events in later planning periods on decisions in the current planning period, nor do they incorporate the effects of current decisions on future decisions. Polyperiodic linear programming (PLP) models were developed to incorporate these simultaneous intertemporal aspects of farm firm decisions. Solution of a PLP model yields a program that optimizes a longrun objective over several time periods. These solutions have 2 important interpretations. In a decision context, the first period program can be considered the optimal course of action for

---

The authors are associate professor of agricultural economics at the University of Georgia, associate professor of agricultural economics at Auburn University, and assistant professor of agricultural economics at the University of Kentucky, respectively.

the current period given current expectations. Second, the solution can be interpreted as suggesting the dynamic growth equilibrium of the firm over the planning horizon. These 2 potential uses are not necessarily incompatible. However, PLP models are more seriously limited for the second use. This distinction is important in considering use of PLP and is stressed in this paper.

The potential of PLP models is relevant to modeling micropolicy impacts. Some policy issues are amenable to static, single-period models. Shortrun responses to policy changes obviously can be modeled with static models in the context of existing fixed resources. However, a PLP decision model would also be useful in examining investment and long-run financing responses. Although longrun equilibrium positions can be identified with a static model, analysis of the adjustment to this equilibrium requires a multiperiod model such as PLP.

Because PLP is not a new technique, the presumption is that most conference participants have some familiarity with PLP models. Nevertheless, a brief historical review of PLP models may provide perspective on their use in agricultural economics. Generally, Swanson is attributed with the first use of PLP in 1955. His model dealt with enterprise organization over a 5-year planning horizon. Most of the resources, except operating capital, were fixed over the planning horizon; the residual of net returns after consumption and fixed costs augmented operating capital in subsequent periods. In 1959, Loftsgard and Heady published a more detailed, but similar, model. In 1962, Irwin and Baker explicitly considered the use of external financing in farm decisionmaking. Their model considered only a single production period but did include quarterly treatment of operating capital. Martin and Plaxico completed a benchmark study of PLP in 1967 that allowed for growth in land and machinery capacity. Purchase and rental of assets along with equity and debt financing were components of this model; however, the production organization was fixed over time. Subsequently, Boehlje and White extended this firm growth framework to include both intertemporal investment and production decisions. Barry and Baker and Vandeputte and Baker made noteworthy advances in modeling finance decisions.

This paper focuses on a Georgia PLP model primarily concerned with expansion of swine enterprises. The model was initially completed in 1978 (Reid) and included most of the features of earlier models, plus several additional provisions of federal taxes so that impact analysis was possible. In reexamining the model for this conference, it became apparent that additional income tax provisions could be included and that other policy issues could be examined with the model. The details of the model have been published elsewhere (Reid, Musser, and Martin, 1980a). Therefore, this paper has the objective of discussing the assumptions incorporated in the components of the Georgia model in relation to general PLP methodology and the use of the model for impact analysis. From this perspective the conclusion considers the overall potential of PLP for impact analysis.

Components of a PLP Model

Discussion of the components of a PLP model can be organized around the standard definition of an LP problem:

$$
\begin{aligned}
\text{Maximize} \quad & c'x \\
\text{subject to} \quad & Ax \leqslant b \\
& x \geqslant 0
\end{aligned}
\tag{1}
$$

where c is the objective function, x is the vector of activity levels, A is the constraint matrix, and b is the vector of constraints. For a PLP model, each standard LP component can be partitioned into particular time periods with the additional activities and constraints defining resource transfers between time periods. However, a PLP model is not just a series of linked, independent single-period models. Obviously, the time value of money must be accommodated in the objective function. Furthermore, prices and input-output relations would be expected to be related among time periods because of such economic phenomena as inflation, price cycles, and technological trends. Assumptions concerning the effect of time is a major subject of this section. A fundamental consideration of time which affects all the components of the model is the length of planning horizon.

Length of Planning Horizon

The number of time periods to include in the model depends on the use of the model. Multiple time periods are included in a decision model to represent the influence of subsequent time periods on the current decision (Hicks). Boussard concluded that, for a firm which is to continue indefinitely, the planning horizon of a multiperiod model is long enough if additional periods will not influence the program for the initial period. A sensitivity analysis of the number of periods can help determine the appropriate number of periods to use. Past research can be utilized to select the initial length. For a growth model, the appropriate length may be the life of the firm, specified by the research context, or may be arbitrary, requiring some judgment. If the growth model is designed to represent firms in a particular area, the representative age of farmers may be helpful if the remaining life of the firm principle is to be used. A final consideration is the researcher's willingness to make long-term predictions of prices and other parameters.

The Georgia model has an 11-period planning horizon. The model was constructed for firm growth research over an arbitrary 10-year period. The additional year was added so that any special results from the termination year would not be reflected in the 10 years used for the results. One special problem that did emerge concerned the assumption about the status of the firm at the end of the planning period. Initially, it was assumed that the firm would continue operation after the 11 years; thus, final asset stocks were valued on a before-tax value. Consequently, solutions reflected a change in the farm organization to accumulate nontaxable inventories rather than after-tax wealth (Reid, Musser, and Martin, 1978). However, model solutions that assumed liquidation at the end of the planning period seemed more reasonable because after-tax asset values could be reflected in the objective function. Researchers who include income taxes in their models will need to consider this assumption.

Objective Function

Two different objective functions have been utilized in previous PLP models—discounted future cash inflows and final equity values. If perfect and complete capital markets are assumed, it can be demonstrated that the 2 criteria are equivalent (Reid, 1981). However, the latter criterion is easier to specify because it requires only one entry in the objective function. Furthermore, Boussard has provided a rigorous, but not easily understood, justification of terminal equity as an objective function. The Georgia model utilized terminal equity as the objective function. As Boehlje and White reported different solutions for the two different objective functions, some consideration of the justification for using the terminal equity criterion is warranted.

The use of discounted cash inflows as a decision criterion is based on neoclassical capital theory. Hirshleifer's treatment of this theory derives this criterion under the assumption that production and investment are only means for current and future consumption. With complete and perfect finance markets, production or investment decisions can be made independently of consumption because assets can be transformed into consumption at any date; the Fisher separation theorem formalizes this result. Maximization of current wealth, which can be interpreted as net present value of future cash inflows, is the appropriate production-investment criterion. It can be noted that the inclusion of consumption in PLP models would be unnecessary in this model of capital theory.

The problem with net present value as an investment criterion in agriculture arises from the lack of perfect finance markets for farmers.

As the authors have previously stressed (Reid, Musser, and Martin, 1980b), most farm firms do not have access to external equity financing so that total investment has a finite limit in any year. When the capital budget constraint is effective, the marginal returns on investment are likely to be greater than the market rate of return which is used both in the separation theorem and in the calculation of net present value. Modern finance texts, such as Brigham's, note that the implicit assumption involved in this criterion is that cash inflows from investments will be reinvested at the discount rate, which is obviously incorrect when capital rationing occurs. In finance theory, the correct capital budgeting criterion in this situation is to maximize the net present value of terminal equity. Maximizing terminal equity gives the same solution as the theoretical objective under any discount rate and is utilized in the Georgia model.

This conclusion has several important implications for firm growth models. First, use of net present value is appropriate if capital rationing does not occur. However, it is difficult, if not impossible, to determine whether or not capital rationing will actually occur external to the model. Under these circumstances, use of terminal equity as the objective function is a more general assumption. Second, the consideration of consumption is then necessary because the assumption of unlimited access to finance markets for consumption borrowing is not a valid assumption under capital rationing. Finally, derivation of the correct financial constraints for investment must take into account consumption withdrawals.

A Digression on Risk

With the recent attention to risk in the agricultural economics literature, it has become apparent that the assumption of certainty is a major limitation of PLP models. Compared with the more general theory of decisions under uncertainty, a PLP model has 2 important limitations: risk neutrality is assumed and the adverse impact of incorrect price and output expectations on financial constraints in subsequent periods are not incorporated. The Georgia model is subject to these limitations. This section evaluates some potential extensions of basic PLP methodology to account for risk.

One approach to accommodate risk is to adapt risk programming techniques to polyperiodic models. For example, Barry and Willmann and Kaiser and Boehlje have used risk programming to calculate the expected value-variance (E-V) frontier of net present value. The limitations of this approach can be illustrated with the objective function of such models:

$$\text{Maximize} \sum_{t=1}^{T} \sum_{j=1}^{N} \frac{1}{(1+r)^t} c_j x_j$$

$$- (\alpha/2) \left(\sum_{t=1}^{T} \sum_{j=1}^{N} \frac{1}{(1+r)^{2t}} x_j^2 v_j \right) + 2 \left(\sum_{t=1}^{T} \sum_{j=1}^{N} \sum_{k=1}^{N} \frac{1}{(1+r)^{2t}} x_j x_k v_{jk} \right) \tag{2}$$

where

c_j = expected value of returns for activity j,
x_j = level of activity j,
r = discount rate,
α = risk aversion coefficient,
v_j.   = variance of return for activity j, and
v_{jk} = covariance of returns from activities j and k.

Of course, an E-V formulation involves some fairly strategic assumptions concerning the decision process, which have been subject to considerable controversy. At most, an E-V formulation can be interpreted as an approximation of an expected utility objective (Levy and Markowitz). The polyperiodic E-V framework is also subject to some additional limitations. A complete development of these limitations is beyond the scope of this paper, but a summary may be helpful. The limitations arise from the problem of capital rationing discussed earlier. With capital rationing, not only is standard net present value analysis an inappropriate technique, but accommodating the intertemporal effects of stochastic returns becomes impossible. Because cash flows from production are stochastic, capital constraints in the second and subsequent periods are stochastic. The formulation in equation (2) does not reflect this impact on levels of feasible x_j. Without capital rationing, perhaps methods could be developed to incorporate the financial effects into the time series of returns to estimate v_j and v_{jk}. However, the appropriate market rate of return for stochastic returns cannot be determined when capital rationing occurs. Polyperiodic risk programming probably understates the variance of net cash flows from the firm because of this conceptual problem.

    Another procedure to accommodate risk in a polyperiodic framework has been to combine simulation with mathematical programming. Chien and Bradford combined a simulator with a PLP model. A growth path was then generated with a sequential use of the two models. The PLP model was solved to obtain an initial plan; then, a simulator was used to estimate the stochastic net income and financial outcome for the first period. The PLP model was then solved for the remaining life with the stochastic outcome of the first period reflected as the constraints for the second period. The two model components were then solved sequentially throughout the planning period. Mapp and others have also incorporated the enterprise organization from a static-risk programming model into a multiperiod simulator to test the longrun viability of particular organizations. Neither of these approaches completely incorporates risk into a multiperiod decision problem. Chien and Bradford do account for the financial effect of unfulfilled income expectations, but they retain the assumption of risk neutrality. Mapp and others do incorporate risk into the objective function and account for the effect of stochastic income on finance. However, polyperiodic production decisions are not modeled. Both approaches may have merit in modeling risk for impact analysis in growth models. However, the former does not improve PLP as a decision model, and the latter does not include an optimizing, multiperiod decision process.

Another promising approach is to constrain PLP models to reflect the risk responses of farmers. Such an approach is consistent with the general recommendations of Young and Sonka in their reviews of firm risk research. Leverage and liquidity constraints, which are often included in PLP models, constrain the financial risk. Production or marketing alternatives could also be restrained to reflect response to business risk. For example, the game theory approach of Boussard and Petit might be adapted to a PLP framework. Safety-first theory (Roy; Day, Aigner, and Smith) could serve as the formal theory for such an approach.

Initial Resource Constraints

The Georgia model was constructed to reflect a representative farm firm in the Georgia Piedmont. The basic farm situation was specified by use of standard methodology in farm management. The farm had 462 acres of land with 164 acres of cropland; 2,500 annual hours were available for labor or management of hired labor. A complement of machinery for production of corn and soybeans was owned at the beginning of the planning period. However, part of the machinery would have to be replaced if crop production was to continue beyond the fourth year. As the objectives of the research emphasized the assessment of impacts of different levels of leverage, initial debt was assumed to be zero. A more plausible assumption would have been to use a debt level estimated from Farm Credit System financial data or other sources. Initial cash was assumed to be $36,000; this amount was derived with parametric programming techniques as the minimum amount required for the firm to survive for 11 years with zero debt. Although few data exist on cash balances of farm firms, this amount is probably higher than the representative amount. Many farms of this size would have yearly off-farm income available, but it is assumed to be zero in this situation. With off-farm income and borrowing capacity, a lower initial cash level would have been necessary for survival.

Some comments concerning management ability are especially appropriate in the context of PLP models. As in most farm management research, assumptions concerning management are crucial. Penrose has argued that management becomes the ultimate limit to firm growth. Furthermore, standard neoclassical theory attributes decreasing returns to scale largely to increasing management problems (Ferguson and Gould). As in most PLP models, the treatment of management in the Georgia model was inadequate because above-average management, as reflected in input-output relationships, was assumed. This assumption, of course, translates into above-average cash inflows; Musser and White (1975) utilized such assumptions to model the dynamic effects of managerial ability on growth and survival. In addition, it was assumed that 1 hour of resident labor was required to manage 10 hours of hired labor, which placed an upper constraint on hired labor. Hindsight would suggest that additional managerial constraints would be desirable.

Activities and Related Constraints

Because of the relationship to other research activities at the University of Georgia, the expansion opportunities of the firm were limited to swine enterprises—feeder pigs, market hogs, and farrow-to-finish operations. Corn and soybean production activities were also included. The swine enterprises required investments in feeder pig and market hog facilities. As in all LP formulations, these facilities were assumed to be perfectly divisible. With the increasing availability of mixed integer programs and their use in machinery selection (Danok, McCarl, and White), these investments could be modeled as integer activities. Replacement gilts could be either raised or purchased. Standard procedures were utilized to transfer livestock, grain, machinery, and livestock equipment from period to period.

The financial component of the model included all of the important cash flows for the firm. Activities were included for long-term debt and short-term debt, which were given percentage limits of long-term asset values and the sum of input costs and short-term assets, respectively. For the solutions reported here, debt was limited to 30 percent of asset values. Long-term debt is refinanced each year. Although this assumption is an abstraction from the actual financial situation of farmers, it simplified the model specification. Provision for amortization of long-term loans, while allowing refinancing, would have added unnecessary complexity, given time and resource constraints, for the research. Activities were also included for investment of money into 1-year time deposits. Realistically, other off-farm investments also should have been included.

Finally, provision was made for a constant withdrawal for consumption and general farm overhead. Inclusion of a consumption function, which is a function of income or wealth, such as that of Vandeputte and Baker, would have been more consistent with economic theory and observed economic behavior. However, research on farm-family consumption functions is limited, especially for the South.

The federal tax component of the model was quite detailed. The basic federal income tax model of Vandeputte and Baker, which includes progressive ordinary rates and capital gains, was incorporated into the model. Straight-line depreciation was assumed. Tax rates were based on 4 exemptions, a standard deduction, and married filing jointly. The features of the investment tax credit were also modeled. The investment tax credit model is too complex to explain in this paper, but it has been reported elsewhere (Reid, Musser, and Martin, 1980a). State income tax was not included, but could be modeled in a manner similar to the federal income tax.

Trends in the Model

Multiperiod analysis requires a much larger forecasting effort than single-period analysis. In single-period analysis the current situation can be observed and represented in the model. Although this approach can be utilized for the first period, trends in these parameters must be included for future periods. As with all expectations, the trends will undoubtedly not be achieved; the longer the planning horizon, the larger the probability of errors in later years. However, farm managers are subject to the same errors in their decision process. If PLP is to be used as a decision model, forecasting errors are to be expected. However, PLP will probably overstate future cash flows in a firm growth context because production and investment choices are perfectly adjusted to prices and outputs realized (see Martin, Wise, and Musser for a static illustration of this effect).

With these caveats, discussion of the procedures used in deriving price trends follows. Standard farm management methodology was utilized to estimate prices and technical coefficients for the base year, 1975. The technical coefficients were assumed to be constant during the planning period. While this assumption was largely for computational ease, the recent slowdown in technological change supports this assumption. The price inflation rate for most purchased inputs and consumption was assumed to be 5 percent. This rate was based on the judgment that the economy would readjust to these rates after exogenous shocks of the 1971–74 period were accommodated. Interest rates of 7 percent for long-term credit, 8 percent for short-term credit, and 6 percent for time deposits (which represented existing credit conditions) utilized in the model support a low general inflation rate. If the real interest rate for agriculture is 4–5 percent, the market rates for interest were consistent with rates lower than 5 percent. The general inflation rate and interest rates were not completely consistent in the model which suggests needed adjustments (White and Musser). Soybean and corn price trends of 4 and 6 percent, respectively, reflected the long-range projections of the U.S. Department of Agriculture (USDA) under medium foreign trade export trends. Annual land value increases were 7.17 percent to reflect historical price increases in Georgia before the surge in land values in 1970 (Musser and White, 1976). The faster increase in land values than the general inflation rate is probably inconsistent without the assumption of technological change, given current concepts of agricultural land values (Melichar).

Because swine prices are so important in the model, a more detailed discussion of these prices is warranted. A projected market hog price series, which was based on historical price cycles, was specified with consultations with meat animal specialists in the Economic Research Service, USDA. This series is listed in Table 13.1. Feeder pig price forecasts were derived from these data and the projected price of corn with a regression equation estimated from historical data in Georgia. The regression equation is as follows:

$$FP = 1.6996MH - 3.4613C \quad R^2 = 0.99$$

$$(12.46) \quad (-1.73) \; D.W. = 2.09$$

(3)

where

FP = feeder pig price per hundredweight,
MH = market hog price per hundredweight,
C = corn price per bushel, and
numbers in parentheses = Student "t" ratios.

Table 13.1 also lists feeder pig price forecasts. Breeding stock prices were based on historical relationships with market hog prices.

In forecasting prices in the model, an effort was made to represent equilibrium price relationships. As Eidman has emphasized, multiperiod analysis with disequilibrium price relationships can lead to results with dubious normative or impact implications. Increased attention to specification of price interrelationships in future firm modeling research appears warranted.

Base Solution

Table 13.2 presents selected features of a base solution for the model. The farm specialized in a market hog operation with small acreages of

TABLE 13.1

Estimated prices of market hogs and feeder pigs, Georgia
Piedmont, 1975-85

Year	Market hog price	Feeder pig price
	Dollars per hundredweight	
1975	43.90	70.38
1976	41.00	59.85
1977	40.00	57.77
1978	35.00	48.86
1979	41.00	58.64
1980	40.00	56.49
1981	37.00	50.94
1982	35.00	47.05
1983	43.00	60.13
1984	42.00	57.95
1985	39.00	52.30

row crops. In the first year, 5,506 cwt of hogs were marketed. Rapid growth in production occurred until a stable production level of 127,776 cwt were marketed in 1980-84. Adjusted gross income for tax purposes was $54,480 the first year and reached a maximum of $446,977 in 1980. Equity at the end of the first year was $328,800 and at the end of 1984 was $2,073,386. Obviously, a strong growth bias was built into the model. Such a bias is common in PLP models partly because of the perfect knowledge assumptions.

Several major weaknesses relating to the model's specification of use of cash flows and financing arrangements contributed to the growth bias. The assumption of zero leverage in the beginning allowed the growth process to be initiated quickly. At the end of the first year, short-term debt was $2,237 and long-term debt was $117,451. If an initial debt level had been assumed, this capital used for expansion would have largely been unavailable. The 30 percent leverage constraint did restrain growth. Equity was $2,794,519 with a 70 percent constraint (Reid, Musser, and Martin, 1980a). The high initial cash position plus the absence of liquidity restraints provided initial capital for growth. The consumption level resulted in a too rapid accumulation of profits. For example, the consumption and overhead allowance in 1977 was $19,845, and net income after taxes (with tax definitions) was $87,142. Even if the whole allowance was for consumption, the average propensity to consume would have been only 22.77 percent, which is probably

TABLE 13.2
Base solution for Georgia swine model, 1975–84

Year	Market hogs	Corn	Soybeans	Depreci- ation	Adjusted gross income	Federal income tax	Equity
	Cwt	— — — Acres — — —		— — — — — — — — Dollars — — — — — — — —			
1975	5,506	89	0	10,640	54,480	16,500	328,800
1976	28,920	0	31	28,773	142,000	68,076	522,342
1977	52,955	0	31	54,375	180,200	93,508	824,909
1978	73,065	31	0	76,866	281,629	164,200	920,826
1979	94,451	0	31	101,984	305,429	180,860	1,326,153
1980	127,776	0	31	143,076	446,977	279,944	1,655,706
1981	127,776	0	31	143,076	222,602	122,881	1,704,992
1982	127,776	0	31	143,076	125,600	57,580	1,576,951
1983	127,776	0	31	143,076	105,600	45,180	1,923,762
1984	127,776	0	31	143,076	81,600	31,020	2,073,386

too low. Inclusion of a consumption function appears to be essential, even if the coefficients are arbitrary. Another feature which would have slowed growth would have been the inclusion of liquidity and risk management constraints in addition to the leverage constraints.

Finally, managerial ability is probably not adequately modeled. For a firm manager without experience in swine production to begin operation and produce 5,000 cwt the first year, 28,920 cwt the second year, and further large increases the next 3 years is probably not realistic. However, rapid output increases through specialization in an existing enterprise is also unlikely. But improved modeling of management decisions is much more difficult than recognizing the current observed deficiencies. Two approaches are possible. First, input-output relationships could reflect diminishing productivity in response to growth. Second, adjustment in production could be restrained with flexibility coefficients reflecting implicit institution risk constraints such as utilized in recursive programming (Day). Derivation of necessary coefficients at the firm level is difficult because of lack of empirical data. Observation of farm record data involving growth and adjustment situations, plus use of county or regional aggregate coefficients may provide some guidance in the specification of such relationships.

Impact Analysis

In the previous section, discussion emphasized the weaknesses of PLP models in general and the Georgia model in particular. For this conference, rather than correcting some of the later problems, a more fruitful approach seemed to be to concentrate on possible impacts of the model or similar models. Any policy that influences parameters included in the model or that could be included in the model, could be subject to impact analysis. This paper emphasizes income tax, price, and credit policy analysis.

Income Tax Policy Impacts

The authors had an interest in income tax impacts when the Georgia model was completed and results on the investment tax credit were published. This paper summarizes these results and presents some new results on accelerated depreciation, namely double declining balance. Other tax issues which could be analyzed are also discussed.

The investment tax credit provisions seem quite simple on the surface; a percentage of qualified investments can be credited to income tax liabilities. However, the provisions for carrying the credit backward and forward in years the credit is greater than current income tax liabilities are quite complex. Furthermore, the credit has a relatively small impact on net worth. In the particular model specification in which the investment tax credit was analyzed, equity was $1,165,809 with the credit and was $1,043,945 without the credit. The credit does have a greater relative effect on investment and production. For example, 70,769 cwt of hogs were marketed with the credit and 50,429 cwt without the credit (Reid, Musser, and Martin, 1980a). The investment tax credit, therefore, is an effective investment stimulus in agriculture.

Accelerated depreciation is another investment incentive which the model can evaluate. The theory of accelerated depreciation as presented in corporate finance can be illustrated with the net present value of the tax savings NPV_D from depreciation (D_t):

$$NPV_D = \sum_{t=1}^{T} \frac{I_t D_t}{(1+r)^t} \qquad (4)$$

where I_t = marginal income tax rate. In the corporate case, I_t is generally constant over time. Under this assumption, depreciation methods that yield larger D_t earlier in the planning period and smaller values of D_t later will increase the value of NPV_D and thereby provide an incentive for investment.

This reasoning has structural implications for the farm sector. Large farmers presumably have higher values of I_t so that accelerated depreciation gives them a relatively higher investment incentive. The problem with the universal application of this reasoning for farms subject to individual tax rates is that the use of accelerated depreciation can reduce marginal tax rates early in the planning period and increase rates later compared with straight-line methods. This differential in tax rates reduces the advantage of early depreciation allowances and can result in a NPV_D lower than one obtained using the straight-line method. Problems with using the net present value criterion in agriculture limit such analyses; earlier tax savings may have a higher value than implicitly assumed. Thus, accelerated depreciation in a PLP model may provide empirical evidence. Table 13.3 presents selected results from the Georgia model respecified with double declining balance depreciation schedules for the swine equipment and buildings. Equity is greater for all years in the planning period with accelerated depreciation and is about $200,000 larger in 1984. Hog production is also greater in the first part of the planning period for 1976–79, but is slightly smaller in the last 5 years. Income tax liabilities are less with accelerated depreciation for 1975–82, but after 1982 income tax liabilities are higher. The higher income tax liabilities reflect the different asset portfolios in two depreciation systems. A contributing factor is the reduction in depreciation allowance as the investment in hog facilities matures. In 1983–84, the accelerated de-

TABLE 13.3
Solution with accelerated depreciation for Georgia model,
1975-84

Year	Market hogs	Depreci- ation	Adjusted gross income	Federal income tax	Equity
	Cwt	- - - - - - - - - Dollars - - - - - - - -			
1975	5,506	21,280	43,840	11,348	333,955
1976	29,047	54,550	131,200	61,164	538,272
1977	54,456	100,225	125,600	57,580	888,632
1978	79,513	140,833	231,625	129,198	1,045,886
1979	108,751	187,958	573,621	368,595	1,383,999
1980	127,201	204,849	205,600	110,980	1,854,736
1981	127,201	173,990	165,600	83,580	1,946,969
1982	127,201	148,259	125,600	57,580	1,836,240
1983	127,201	126,789	185,600	96,772	2,155,316
1984	127,201	108,798	145,600	70,380	2,292,200

preciation is less than the straight-line depreciation. A tentative conclusion is that accelerated depreciation is beneficial in a rapid growth situation such as represented in the early part of this planning period. However, in a steady state environment, rapid depreciation may not favor investment as a result of higher tax liabilities for farmers as the investment matures. The issue of stimulus of accelerated depreciation on investment is obviously complex and requires more research.

Other tax issues can also be represented by a PLP model. The effect of cash accounting could be modeled by restriction of the income tax accounting constraints to an accrual basis. Effects of income tax rates, capital gains exclusions, and other tax provisions could be included. For example, the minimum tax provisions can be included with an additional activity and constraint for each year. The activity (MT_t) would have an entry for the tax rate in the cash flow rows, and the constraint would have the following form:

$$\sum_1 a_1 x_1 - MT_t \leqslant b_m \qquad (5)$$

where x_1 = activities that create minimum tax liabilities, a_1 = contribution of x_1 to base for minimum tax, and b_m = amount of base exempt from minimum tax. Additional methodological research may help develop methods to model additional tax provisions in an LP format. However, some of these tax provisions are probably not consistent with LP methodology, but are more accurately specified with mixed integer programming.

Credit Policy

The influence of monetary and other financial policies which affect farm credit can be analyzed with PLP models. Much more sophisticated methodologies than incorporated here have been developed and are reviewed in Baker and others. Even in this model, such issues as interest rate structure and credit availability can be evaluated. As discussed previously, changes in parameters should be consistent with market interrelationships. For example, changes in interest rates probably require changes in inflationary trends. Research with the Georgia model has been reported that analyzes different leverage constraints, which could reflect either external or internal capital rationing. This research gave the expected results; the higher the leverage ratio, the higher terminal equity and hog production. With zero debt, terminal equity was $513,918 with 6,310 cwt of hogs marketed. With a debt-to-asset ratio of 70 percent, terminal equity was $2,794,519 with 112,154 cwt of hogs sold; management was restrictive from 1976 to 1984 except in 1981 (Reid, Musser and Martin, 1980a).

Price Situations

The impact of farm commodity or foreign trade policy on farm firms can be analyzed in an LP framework. To analyze these impacts in a PLP model, the price interrelationships must be developed when a particular policy has direct influence on one price. The solution was prepared with a different hog/price cycle to demonstrate the effects with the Georgia model. To be consistent with base corn and other input price trends, base prices for market hogs in Table 13.1 were used but were sequenced differently. The price for market hogs in 1975 was $35.00; in 1976, was $41.00; in 1977, was $40.00; in 1978, was $37.00; and so on with $43.90, $41.00, and $40.00 being the final 3 prices. Other hog prices were recalculated using the same methods discussed earlier for the base model.

Table 13.4 reports selected results from the revised price situation. Both the equity and the level of hog production in 1984 are higher given the revised hog/price cycle. The level of production in 1976 and 1977 is lower, and the lower initial prices resulted in less profits for asset acquisition. Afterwards, the level of production increased. The higher level of production scheduled at the end of the planning horizon resulted from the elimination of crop production activities from the solution. Further growth past 128,510 cwt was impossible despite

TABLE 13.4

Solution for Georgia model with alternative hog-price cycle, 1975-84 1/

Year	Market hogs	Adjusted gross income	Equity
	Cwt	- - - - Dollars - - - -	
1975	5,506	27,400	312,291
1976	26,328	134,600	488,341
1977	48,997	166,000	774,844
1978	73,416	299,791	958,531
1979	78,938	185,800	1,060,276
1980	114,178	864,522	1,333,027
1981	128,510	634,092	1,599,121
1982	128,510	212,805	1,728,222
1983	128,510	287,926	2,157,153
1984	128,510	165,600	2,324,440

1/ 1978 market hog prices in table 7.1 became 1975 prices; 1979 prices became 1976 prices; 1975 prices became 1983 prices; 1976 prices became 1984 prices; and 1977 prices became 1985 prices.

favorable prices because the management constraint became effective. Given the basic profitability of market hog production and capital availability for growth, the sensitivity of the solution to alternative price relationships was not very large.

Conclusions

PLP models have both advantages and disadvantages in the determination of impacts of policy and institution changes on farm firms. Although most of these advantages and disadvantages have been previously reviewed in the firm growth literature, they have been discussed again in the context of the Georgia model presented here. Under the assumptions of the neoclassical theory of the firm of perfect knowledge and divisibility of investments, PLP models allow the impact of various changes on farm firms to be analyzed in a multiperiod context. This analysis is useful primarily in a decision context; for example, what response will the firm make in the current period when decisions with consequences for several future periods are involved? Regarding firm growth, PLP models will undoubtedly overstate the economic welfare of firms because expectations are never perfectly fulfilled. In situations where the above neoclassical assumptions are not fulfilled, alternative analytical techniques such as simulation or risk programming should be considered. For example, risk is a pervasive issue in analysis of the financial consequences of policy and PLP is probably an inappropriate tool for those problems. However, PLP does have the advantage of providing an efficient analysis of response to changes in economic parameters rather than a study of the impact of these changes on current business operations. This advantage is not to be dismissed lightly.

One should further consider some methodological issues when evaluating PLP. From the authors' perspectives, policy impact research is positive rather than normative analysis; the goal of such research is to predict impacts rather than to prescribe optimal decisions for the farmer. Clearly, many of the assumptions and aspects of the solutions vary from actual farm situations. However, such characteristics of PLP may increase its usefulness for analysis of policy impacts. This view is consistent with Friedman's famous methodological position that successful positive analysis must abstract from reality. Applying this position to farm management models, it is easy to become immersed in so many details that the model becomes descriptive rather than predictive. If responses are generated from a realistic model, the details can often be neither interpreted nor explained. Even a PLP model can become too detailed for easy interpretation. For example, it is very difficult to explain the switches from one crop to another that appear in some solutions of the model discussed here (see the base run in Table 13.2).

Some of the results reported here demonstrate the potential of PLP models for positive analysis. Despite the admittedly unrealistic growth patterns in the solutions, the qualitative impact of changes in parameters could be observed, especially in the tax impact analysis—probably the most innovative research reported here. Of course, other methodologies can trace the consequences of a parametric change, but most often the firm's organization is not given an opportunity to respond. Admittedly, the analyses of these impacts are primarily of qualitative value.

For researchers who elect to use PLP, some suggestions are apparent from experience with the Georgia model. Allowing increases in financial leverage and not including a consumption function definitely contributes to rapid expansion and probably needs modification. Management constraints were effective in this model, but should probably be strengthened if estimation procedures can be developed. Other risk management constraints in addition to leverage could also be included. Research on addition of limited integer activities may be worthwhile. With the rapid expansion in the solutions to this model, the divisibility of fixed assets was not a serious problem, but with slower expansion, small increments of investment response could be a problem. The integer activities also could be helpful in the modeling of some policy variables which are inconsistent with LP.

References

Baker, C. B., and others. *Economic Growth of the Agricultural Firm.* Technical Bulletin No. 86. College of Agriculture Research Center, Washington State Univ., Feb. 1977.

Barry, Peter J., and C. B. Baker. "Reservation Prices on Credit Use: A Measure of Response to Uncertainty," *American Journal of Agricultural Economics,* Vol. 53, No. 2, May 1971, pp. 222–28.

————, and David R. Willmann. "Risk Programming Analysis of Forward Contracting with Credit Constraints," *American Journal of Agricultural Economics,* Vol. 58, No. 1, Feb. 1976, pp. 62–70.

Boehlje, Michael D., and T. Kelley White. "A Production-Investment Decision Model of Farm Firm Growth," *American Journal of Agricultural Economics,* Vol. 51, No. 3, Aug. 1969, pp. 546–64.

Boussard, Jean-Marc. "Time Horizon, Objective Function, and Uncertainty in a Multiperiod Model of Firm Growth," *American Journal of Agricultural Economics,* Vol. 53, No. 3, Aug. 1971, pp. 467–77.

————, and M. Petit. "Representation of Farmers' Behavior under Uncertainty with a Focus Loss Constraint," *Journal of Farm Economics,* Vol. 49, No. 4, Nov. 1967, pp. 869–80.

Brigham, Eugene F. *Financial Management Theory and Practice.* 2nd edition. Hinsdale, Ill.: Dryden Press, 1979.

Chien, Y. I., and G. L. Bradford. "A Sequential Model of the Farm Firm Growth Process," *American Journal of Agricultural Economics,* Vol. 58, No. 3, Aug. 1976, pp. 456–65.

Danok, Abdulla B., Bruce A. McCarl, and T. Kelley White. "Machinery Selection Modeling: Incorporation of Weather Variability," *American Journal of Agricultural Economics,* Vol. 62, No. 4, Nov. 1980, pp. 700–08.

Day, R. H. *Recursive Programming and Production Response.* Amsterdam: North-Holland Publishing Co., 1963.

————, D. Aigner, and K. Smith. "Safety Margins and Profit Maximization in the Theory of the Firm," *Journal of Political Economy,* Vol. 79, No. 6, 1971, pp. 1293–1301.

Eidman, Vernon R. "Microeconomic Impacts of Inflation in the Food and Agriculture Sector: Discussion," *American Journal of Agricultural Economics,* Vol. 63, No. 5, Dec. 1981, pp. 962–64.

Ferguson, C. E., and J. P. Gould. *Microeconomic Theory.* 4th ed. Homewood, Ill.: Richard D. Irwin, 1975.

Friedman, Milton. *Essays in Positive Economics.* Chicago: Univ. of Chicago Press, 1953.

Hicks, J. R. *Value and Capital.* 2nd edition. London: Oxford Univ. Press, 1953.

Hirshleifer, Jack. *Investment, Interest, and Capital.* Englewood Cliffs, N.J.: Prentice-Hall, Inc., 1970.

Irwin, George D., and C. B. Baker. *Effects of Lender Decisions on Farm Financial Planning.* Research Bulletin 688. Illinois Agr. Expt. Sta., 1962.

Kaiser, Eddie, and Michael Boehlje."A Multiperiod Risk Programming Model for Farm Planning," *North Central Journal of Agricultural Economics,* Vol. 2, No. 1, Jan. 1980, pp. 47–54.

Levy, H., and H. Markowitz. "Approximating Expected Utility by a Function of Means and Variance," *American Economic Review,* Vol. 69, No. 3, June 1979, pp. 308–17.

Loftsgard, Laurel D., and Earl O. Heady. "Application of Dynamic Programming Models for Optimum Farm and Home Plans," *Journal of Farm Economics,* Vol. 41, No. 1, Feb. 1959, pp. 51–67.

Mapp, Harry P., Jr., and others. "An Analysis of Risk Management Strategies for Agricultural Producers," *American Journal of Agricultural Economics,* Vol. 61, No. 5, Dec. 1979, pp. 1071–77.

Martin, Neil R., Jr., James O. Wise, and Wesley N. Musser. "The Impact of Managerial Response in Enterprise Organization to Changes in Price Relationships on Profit Potential," *Journal of the American Society of Farm Managers and Rural Appraisers,* Vol. 40, No. 2, Oct. 1976, pp. 58–64.

Martin, J. Rod, and J. S. Plaxico. *Polyperiod Analysis of the Growth and Capital Accumulation of Farms in the Rolling Plains of Oklahoma and Texas.* TB-1381. U.S. Dept. of Agr., Sept. 1967.

Melichar, Emanuel. "Capital Gains versus Current Income in the Farming Sector," *American Journal of Agricultural Economics,* Vol. 61, No. 5, Dec. 1979, pp. 1086–92.

Musser, Wesley N., and Fred C. White. "The Impact of Management on Farm Expansion and Survival," *Southern Journal of Agricultural Economics,* Vol. 7, No. 1, 1975, pp. 63–69.

———. "Inflation and Farm Financing: Recent Trends and Projections," *Agricultural Finance Review,* Vol. 36, 1976, pp. 12–18.

Penrose, E. T. *The Theory of the Growth of the Firm.* Oxford: Oxford Univ. Press, 1959.

Reid, Donald W. "A Mathematical Programming Decision Model for Farm Machinery and Equipment Replacement." Ph.D. dissertation, Univ. of Kentucky, 1981.

———. "A Multiperiod Analysis of Intensive Farm Firm Growth in the Georgia Piedmont." Master's thesis, Univ. of Georgia, 1978.

———, Wesley N. Musser, and Neil R. Martin, Jr. "Consideration of Investment Tax Credit in a Multiperiod Mathematical Programming Model of Farm Growth," *American Journal of Agricultural Economics,* Vol. 62, No. 1, Feb. 1980a, pp. 152–57.

———. "Income Tax Aspects of Liquidation in Multiperiod Linear Growth Models," *Southern Journal of Agricultural Economics,* Vol. 10, No. 2, Dec. 1978, pp. 29–34.

————. *A Study of Farm Firm Growth in the Georgia Piedmont with Emphasis on Intensive Growth in Hog Production.* Research Bulletin 249. Georgia Agr. Expt. Sta., Jan. 1980b.

Roy, A. D. "Safety First and the Holding of Assets," *Econometrica,* Vol. 20, 1952, pp. 431–49.

Sonka, Steven T. "Risk Management and Risk Preferences in Agriculture: Discussion," *American Journal of Agricultural Economics,* Vol. 61, No. 5, Dec. 1979, pp. 1083–84.

Swanson, Earl R. "Integrating Crop and Livestock Activities in Farm Management Activity Analysis," *Journal of Farm Economics,* Vol. 37, No. 5, Dec. 1955, pp. 1249–58.

Vandeputte, J. M., and C. B. Baker. "Specifying the Allocation of Income Among Taxes, Consumption, and Savings in Linear Programming Models," *American Journal of Agricultural Economics,* Vol. 52, No. 4, Nov. 1970, pp. 521–27.

White, Fred C., and Wesley N. Musser. "Inflation Effects in Farm Financial Management," *American Journal of Agricultural Economics,* Vol. 67, No. 5, Dec. 1980, pp. 1060–64.

Young, Douglas L. "Risk Preferences of Agricultural Producers: Their Use in Extension and Research," *American Journal of Agricultural Economics,* Vol. 61, No. 5, Dec. 1979, pp. 1063–70.

14. A Recursive Goal Programming Model of Intergenerational Transfers by Farm Firms

Craig L. Dobbins and Harry P. Mapp, Jr.

Abstract Goal programming is an optimization technique that allows the simultaneous solution of multiple goals when modeling the choice criterion. The multiple goals can be included through the use of an objective function that ranks the goals by their importance or an objective function in which weights reflecting the trade-offs between goals are assigned to each of the different goals. These 2 objective functions were used to develop annual production plans for an Oklahoma farm. The technical weight structure chosen for the objective function substantively affects the results.

Keywords Farm planning, goal programming, multiple goals

This paper focuses on the development of a dynamic farm model that incorporates multiple goals or objectives to determine shortrun production decisions. This micromodel contains simulation and goal programming components and is used to simulate the impact of variations in goal structure and activity organization on the growth of the farm firm. The firm growth aspect of the model includes a longrun 15-year planning horizon and farm expansion and intergenerational transfer plan. This plan is invariant to change in shortrun goal structures. The paper is divided into 3 major sections. The first 2 trace the development of and the behavioral assumptions used in the model. In the final section, some empirical results are discussed under differing scenarios.

The authors are an associate professor of agricultural economics at Purdue University and a professor of agricultural economics at Oklahoma State University.

Development of Behavioral Assumptions

Dissatisfaction with the limiting assumptions of the static theory of the firm has spurred interest in incorporating dynamic and uncertain events in firm-level models. In early attempts to relax the assumptions of timelessness and perfect knowledge, decisionmakers were assumed to maximize the present value of net returns by discounting income for time and risk (Hicks). Subsequent efforts to incorporate decision-making under risk modified the behavioral assumptions of maximizing expected monetary values and maximizing expected utility (Dillon). Subsequent studies included eliciting utility functions based on choices between gambles with various probabilities of winning or losing certain cash sums (Anderson, Dillon, and Hardaker; Halter and Dean; Officer and Halter; Halter and Mason; Lin, Dean, and Moore; Baquet, Halter, and Conklin). These utility functions have been used to characterize decisionmakers as risk-averse, risk-neutral, or risk-preferring and to explain or predict decisions at the firm level under risky conditions.

Critics have argued that the direct elicitation method of obtaining utility functions is subject to interviewer bias and substantial measurement error (Binswanger). Others have argued that individual decision-makers are not completely rational, or do not have complete access to information, or cannot perform the calculations necessary to maximize a single-objective function, or reach a complete and transitive set of preferences with the single objective of maximizing utility (Simon, Lee). Simon argues that individual decisionmakers have in mind a "satisficing" level or "satisfactory" level of goal attainment rather than a maximum level. Ferguson has also suggested that decisionmakers have multiple goals or objectives and that a multidimensional utility function is required to indicate the interrelationships. Rather than assigning a utility value to each event, we can express the utility of an act as a priority-ordered vector having expected utility in each attitude dimension. Thus, a multidimensional, lexicographic utility function is based on the idea that the typical decisionmaker has multiple goals or objectives.

To illustrate the utility maximizing process of this utility function, assume managers have n goals, $Z_1, Z_2, \ldots Z_n$. *Each goal is ranked according to its importance,* Z_1 being the most important. Also, assume that there are 2 alternatives considered, A^0 and A^1, which provide the following levels of goal attainment:

$$A^0 = (Z_1^0, Z_2^0, \ldots , Z_n^0)$$

$$A^1 = (Z_1^1, Z_2^1, \ldots , Z_n^1)$$

The utility associated with each of the actions can be represented as $U(A^i)$ for $i = 0, 1$. The lexicographical nature of the utility function specifies $U(A^0) > U(A^1)$ if $Z_1^0 > Z_1^1$, irrespective of the relationship between Z_i for $i = 2 \ldots n$. If $Z_1^0 = Z_1^1$, then the choice between A^0 and A^1 is based on the relative values of goal attainment Z_2^0 and Z_2^1. If $Z_2^0 = Z_2^1$, the decisionmaker compares the level of goal attainment associated with the third goal, and so on.

This type of utility function has been extended to include the specification of goal aspiration levels or goal targets (Ferguson). The goal aspiration levels represent a satisfactory level of achievement for a particular goal. If Z_i^* represents a level of satisfactory goal attainment, it is assumed that the marginal utility of further achievement is less than or equal to zero. With this extension, the behavioral assumption that managers are "maximizers" has been replaced with the assumption that they are "satisfiers."

In modeling this type of objective function, assume that there are n goals which are ranked from 1 to n according to importance. A preferred alternative can be identified by the following problem:

$$\max Z_n$$
$$\text{subject to: } Z_i \geqslant Z_i^* \text{ for } i = 1, 2, \ldots, n-1$$

If a feasible solution does not exist, goal Z_n is deleted from consideration and a preferred alternative is selected by respecifying the problem as:

$$\max Z_{n-1}$$
$$\text{subject to: } Z_i \geqslant Z_i^* \text{ for } i = 1, 2, \ldots, n-2$$

If a solution is still not found, the next lowest ranked goal is deleted and replaced in the objective function with a successively higher goal until a solution is obtained (Halter and Dean; Hatch).

One of the major drawbacks of the lexicographic approach is that trade-offs between goals are not usually included. However, this restriction was relaxed in an early simulation study conducted by Patrick and Eisgruber. The decision criteria of their model incorporated goals related to living standard, farm ownership, leisure-children, and credit using or risk taking. A goal aspiration level was specified for each goal on the basis of previously achieved levels. One of 4 levels of satisfaction was assigned for each goal, depending on the level of expected goal achievement provided by a particular alternative. The level of satisfaction was then weighted by the importance of the goal and summed to give the overall index of satisfaction for the alternative. The alternative with the highest index was chosen for implementation.

Procedures for including multiple goals in linear programming models have also been suggested by assignment of weights to the goals (Candler and Boehlje). These weights are used as objective function values, and the model is optimized in the usual manner. If the model does not provide a satisfactory solution, the weights can be altered and the model again optimized. Specification of a satisfaction attainment level is not required in the linear programming model as the behavioral assumption is that managers are maximizers. This method of modeling goals also allows for simultaneous trade-offs among goals because all goal trade-offs are specified in terms of a common base.

Goal Programming

Another method of incorporating multiple goals into a model of the firm is through an extension of linear programming (LP) called goal programming (GP), (Charnes and Cooper). Others have extended the uses of GP to problems in accounting, production planning, financial decisions, academic planning, and medical care (Ijiri; Lee). Goal programming has been used in agriculture for the analysis of national policy issues (Neely and others; Vocke and others) and for firm-level modeling (Barnett; Harris and Mapp). The difference between GP and LP models is in the representation of the objective function. Although both the LP and GP objective functions are linear, the objective function in goal programming recognizes that decisionmakers have multiple and often conflicting goals.

Two methods of incorporating goal structures have been used in goal programming models. Although both these methods require the specification of a satisfying or target level for each of the goals, one approach assumes that the various goals can be ranked in priority order by the decisionmaker. This ranking is used to assign preemptive priority factors to the deviational variables in the model.[1] If 2 or more goals are identified as having the same priority, weights reflecting the trade-offs or rates of substitution allowed between them are assigned to the appropriate deviational variables. If all goals are assigned unique preemptive factors, the goal structure would represent a lexicographic utility function, and the usual LP solution procedure must be modified. The solution procedure of this type of GP model allows consideration of less important goals only after achievement of the higher priority goals has been sought. (See Ignizio or Lee for some examples.)

The second method of goal structuring assumes that all goals are of equal priority. Weights, which reflect the trade-offs or rates of substitution between goals, are then assigned to the deviational variables. This formulation of the goal programming model allows achievement levels of one goal to be substituted for any other. In this form, the GP model can be solved by use of standard LP solution algorithms.

The objective function of the goal programming model is to minimize the difference between desired levels of goal achievement and the actual level of achievement. Thus, goal programming uses satisficing criteria rather than the maximizing criteria of linear programming.[2]

The general goal programming model containing m goals with k priority levels may be represented by:

(1) Minimize $z = cd$

subject to:

(2) $Ax + Rd = g$

(3) $Ux \leq b$

(4) $X, d \geq 0$

where $c = P_{j1}w_1, P_{j2}w_2, \ldots P_{ji}w_i, i = 1, 2, \ldots 2m$ and $j = 1, 2, \ldots k$, is a row vector representing the products of the preemptive priority factors, P_{ji}, and the weights, w_i, assigned to deviational variables; $d = d_1^- d_2^- \ldots d_m^- d_1^+ d_2^+ \ldots d_m^+$ is a $2m$ component column vector of deviational variables, where deviations arising from overachievement are represented by d^+. Also, A is an m by n matrix of coefficients representing the interactions of production alternatives and goals; R is an m by $2m$ matrix composed of the identity matrix and its negative; g is an m-component column vector of goal satisficing levels; and U is a q by n matrix of production coefficients and b is a column vector of available resources. When all goals are assumed to be of the same priority (the substitution goal structure), $k = 1$ and assignment of the P_j's is not necessary. In this case, $c = w_1, w_2, \ldots w_{2m}$, and the model remains the same as in equations (1) through (4).

Model Development

Our firm-level model combines simulation and goal programming components when analyzing entry-exit coordination plans. A schematic diagram of the model is presented in Figure 14.1. The simulation component (developed by Roush) portrays the decisionmaking environment. It contains farm business accounting functions including those that transfer ownership of farm assets from the parents to the heir.

The simulation model represents the complex legal-economic environment in which the firm operates. The simulation model also allows the resources used in the goal programming model to be updated. These updated resource levels are then used to determine annual production plans through the recursive solution of the goal programming model.

The goal programming component represents the multiple goal objectives of the firm during the planning process. These goals can be classified as long-run, intermediate, or short-run. The long-run and

FIGURE 14.1

Schematic Diagram of Goal Programming/Simulation Model

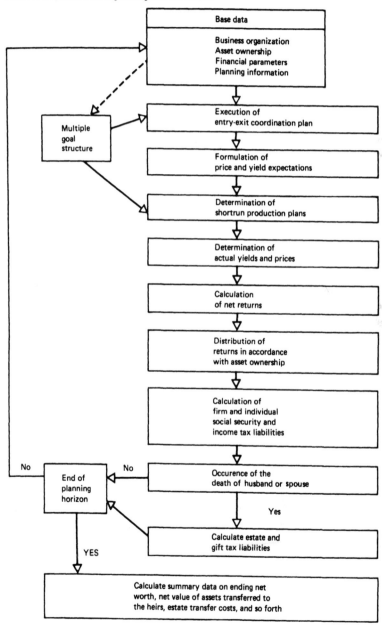

intermediate goals are used to formulate a coordination-growth plan. These goals are developed through interviews with a farm family, and the coordination-growth plan is specified exogenous to the model. The transactions specified by the coordination-growth plan represent the process by which the parent retires and the heir acquires the farm assets.

The short-run goals are used to develop annual production plans. A satisfying level and weight reflecting the importance of each goal relative to other (conflicting) goals is specified. This information is combined with expected prices and yields to determine the annual production plan. Resource availability is determined by the coordination-growth plan. Information on the production levels of alternative enterprises is then combined with a set of actual yields and prices to calculate actual net returns. These returns are distributed among individual members of the family in accordance with each person's resource contribution. Social security and income tax liabilities, and, in the event of a death or a gift, the gift, estate, and inheritance taxes are calculated.[3] If the end of the planning period has been reached, the simulation ends. If not, the adjustments in the following year of the coordination-growth plan are implemented. At the end of the planning period, ending net worth, net value of assets transferred, asset transfer costs, and other data are summarized.

Goals

Although goals are considered important in the farm planning process, little agreement exists on how goals should be described or on their relative importance. The importance (weights) of particular goals has been determined by use of 3 methods. The first method is to choose an arbitrary initial set of goals and weights, and then adjust the weights until the results either resemble the decisionmaker's observed behavior, or they are satisfactory to the decisionmaker. A second approach is to choose a set of goals and weights based on the decisionmaker's past activities by use of revealed preferences (see Eidman). The third method is to use survey techniques to elicit preferences from farmers directly.

Several studies have been conducted in an effort to determine the relative importance of goals to farmers (see Patrick, 1981). These studies indicate that although multiple goals are important to farmers, sets of goals are inconsistent over time or situations. The objective of most of these studies has been to simply identify goal importance or the relationship between farmer's goals and personal characteristics, without applying the results in models of firm decisions.

Six goals were selected for inclusion in the goal programming model: (1) level of year-end cash balances, (2) level of family consumption, (3) level of short-term borrowing, (4) amount of part-time labor hired, (5) amount of time available for leisure, and (6) value of current assets.[4] It is assumed that the operators are only concerned with deviations from the underfulfillment of a goal. If the level of achievement for a particular goal is exceeded, these positive deviations do not contribute to increased satisfaction through added marginal utility. The goal targets for the cash balance, family consumption, leisure time, and current assets are minimum levels of desired achievement. The targets for borrowing and part-time labor are maximum levels. Thus, the operators are concerned about the level of cash available for family consumption, year-end cash balances, the value of current assets, and the amount of leisure time only if they are less than the satisficing level. Similarly, they are concerned if short-term borrowing and seasonal hired labor become too high and are above the satisficing level. With this assumption, there is only one deviational variable of importance for the goals. The goal structure and target levels used in the goal programming model are summarized in Table 14.1.

Model solutions are found by use of an iterative approach. The first step of the process is to minimize the weighted value of the deviational variables associated with the first preemptive level, P_{1i}, or the most important goal(s). An additional constraint is then added to the model to prevent the weighted deviations from exceeding the minimum level established in the first step. The final step changes the objective function by including the next most important goal(s) at the next preemptive level. This process is continued until all preemptive levels have been considered by the solution process.

Alternative goal structures were simulated, each one representing a different general goal orientation. For example, a consumption-profit goal structure was represented by assigning consumption the highest priority. This was followed by cash balances, short-term borrowing, and current assets as second priority goals. The third priority goals included part-time labor and leisure. A consumption-labor goal structure was represented by assigning the highest priority to consumption followed by leisure and part-time labor goals. The third priority level was assigned to cash balance, short-term borrowing, and current assets goals. The weights assigned to represent the trade-offs between goals of equal priority are specified in Table 14.2.

When the substitution goal structure is used, changes in the objective function of the goal programming model are unnecessary. Because all goals are given the same priority level, the problem is solved directly. The substitution goal structures are specified in Table 14.2.

TABLE 14.1
Goal structures used in goal programming model

Goal	Production and transfer activities X_1 X_2 ......X_n	DG1 $P_{j1}W_1$	DG2 $P_{j2}W_2$	DG3 $P_{j3}W_3$	DG4 $P_{j4}W_4$	DG5 $P_{j5}W_5$	DG6 $P_{j6}W_6$	Satisficing levels 1/
Objective function 2/		$P_{j1}W_1$	$P_{j2}W_2$	$P_{j3}W_3$	$P_{j4}W_4$	$P_{j5}W_5$	$P_{j6}W_6$	
Cash balance	a_{11} a_{12} ......a_{1n} 1							84,600
Consumption	a_{21} a_{22} ......a_{2n}		1					24,828
Short-term borrowing				-1				0
Seasonal labor					-1			2,250
Leisure						1		160
Current assets	a_{61} a_{62} ..... a_{6n}						1	150,000

Blanks indicate not applicable.
1/ Satisficing levels for the goals represent target levels for each goal.
2/ The P_{ji} represents the priority level of the goals. When the ranked goal structure is used:
P_{j1} = P_{j3} = P_{j6} and P_{j4} = P_{j5}. When the substitution goal structure is used:
P_{j1} = P_{j2} = P_{j3} =P_{j4} = P_{j5} =P_{j6}.

TABLE 14.2
Goal weights used in ranked and substitution goal structures

Goal	Ranked goal structure		Substitution goal structure	
	Consumption-profit orientation	Consumption-labor orientation	Consumption-profit orientation	Consumption-labor orientation
Cash balance	0.25	0.25	0.25	0.05
Consumption 1/	--	--	.40	.50
Part-time labor	.05	.05	.05	.25
Borrowing	.10	.10	.10	.05
Leisure	.05	.05	.05	.10
Current assets	.15	.15	.15	.05

1/ As consumption was the only goal in the first priority level, a weight did not need to be assigned to it when the ranked goal structure was used.

Activities and Constraints

The constraints or resource restrictions in the goal programming model included crop and pasture land acreage, monthly labor, cash, and pasture animal unit months (AUMs). Resource availability was updated annually based on the coordination-growth plan and the results of the previous year's production plan. The production activities included wheat for grain, wheat graze-out, grain sorghum, alfalfa hay, bermuda pasture overseeded in the fall with small grain, bermuda hay, winter and spring stockers, and cow-calf production. Activities were also included to allow the sale and purchase of products and inputs, respectively.

The goal programming model spans a period from January to December of each year, *t*. Because some production enterprises were not completed during this 12-month period, additional activities representing the beginning and ending of wheat grain production, wheat graze-out, bermuda overseeding, and stocker grazing from November to March were also included. Each of the production activities in year *t* is initially fixed in the model at period *t*-1's level.

To evaluate production alternatives not yet completed, an estimate was made of expected income from completion of production. This return was calculated by subtracting cash expenses incurred during its completion from expected gross returns. If the labor supply was restrictive, labor expenses were also subtracted at a cost of $3 per hour.

Price and Yield Expectations

The returns for each of the enterprises are based on price and yield expectations and are updated each year. The equations used to represent these expectations were contained in the simulation model (Table 14.3). The expected prices, yields, and net returns were then passed to the goal programming model to update the objective function. The expected prices for wheat, grain sorghum, and alfalfa hay were based on the average observed prices received the previous year and U.S. aggregate ending stocks. The price expectations for feeder cattle were based on the price of feeder calves the previous year. The March stocker price expectations were based on the price of stockers the preceding November and on U.S. pasture conditions. The May stocker price expectations were based on the stocker price the preceding March. Yield expectations were based on the weighted average of past yields using an exponential smoothing function. The exponential smoothing function with a weight of 0.33 for the most recent observation results in a yield estimate similar to a 5-year moving average (Brown).

Evaluation and Adjustment

Adjustments in resource levels resulting from sales or purchases and changes in asset ownership resulting from gifts or estate settlement were assumed to occur on January 1 of each year. Debt repayments were also made at this time. This schedule simplified the calculation of ownership costs (depreciation, property insurance, property taxes) and interest on debts as all assets were assumed owned for a complete year. However, it was assumed that control of additional rented land was gained after wheat harvest in July.

After the annual production plan was determined, it was reevaluated from a set of stochastic prices and yields assumed to have actually occurred. Prices were generated from a set of equations representing commodity price cycles and random fluctuations. Crop yields were generated by use of the average crop yield for a 20-year period and random fluctuations. The random fluctuations in each commodity price were appropriately correlated with the procedure described by Clements and others. Crop yield deviations were also correlated. The price and crop yield deviations from their means were assumed to be independent (Dobbins).

TABLE 14.3
Price and yield forecasting equations

Wheat grain price (dol./bu.)

$$WP_t = 0.016 + 0.801WP_{t-1} - 0.00000028WS_t + 0.013t \qquad R^2 = 0.72$$

Grain sorghum price (dol./cwt)

$$GSP_t = -0.256 + 0.776GSP_{t-1} - 0.00735FGS_{t-1} + 0.0189t \qquad R^2 = 0.73$$

Alfalfa hay price (dol./ton)

$$AHP_t = 6.533 + 0.713AHP_{t-1} - 0.09127FGS_{t-1} + 0.329t \qquad R^2 = 0.79$$

Feeder calf price (dol./cwt)

$$FCP_t = 30.305 + 0.646FCP_{t-1} - 0.001CC_{t-1} + 0.0009CM_{t-1} \qquad R^2 = 0.79$$

March stocker price (dol./cwt)

$$MRSP_t = 14.262 + 0.89NVSP_{t-1} - 0.172PAST_t \qquad R^2 = 0.80$$

May stocker price (dol./cwt)

$$MYSP_t = 1.891 + 0.925MSP_t \qquad R^2 = 0.95$$

Crop yields per acre

$$EY_{ti} = 0.33Y_{ti} + 0.667EY_{t-1i}$$

NOTE: Variable abbreviations used were WP for wheat price, WS for January 1 U.S. wheat stocks, GSP for grain sorghum price, FGS for U.S. feed-grain stocks, AHP for alfalfa hay price, CM for total U.S. cattle marketings, CC for total U.S. calf crop, FCP for feeder calf price, PAST for U.S. pasture conditions, MRSP for March stocker price, MYSP for May stocker price, NVSP for November stocker price, EY for expected yield of crop i, Y for most recent yield, and t for the time period. Coefficients of the price expectation equations were estimated by use of ordinary least squares regression.

A revised data set containing new prices, yields, quantity of cropland and pasture land operated, hay inventory, cash on hand, short-term borrowing capacity, acres of alfalfa and bermuda meadow, hours of permanent labor available, level of production for enterprise completion activities, and the size of the beef herd was generated and the goal programming model was updated. The goal rankings, weights, and target values were assumed to remain constant over the planning period.

The representative farm used in this analysis was located in north central Oklahoma. The farm initially consisted of 2,400 acres, with 800 acres owned outright by the operator. The remainder of the cropland was rented on a cash basis. Initially, labor was provided by the operator and the heir in equal proportions. Part-time labor was hired as needed. The operator's labor contribution was reduced during the seventh year by 75 percent to reflect semiretirement. In year 10, the owner's labor contribution was reduced to zero to reflect full retirement, and a full-time employee was hired. The management of the farm was also shared. The operator initially provided 60 percent of the management. This percentage declined to 20 percent in year y and to zero in year 10.

The coordination-growth plan used in this analysis specified that the owners sold 160 acres of land to the heir the first year of the 15-year planning period and sold the farm machinery to the heir in the second year. Both these sales were made on an installment basis with small down payments. Land contract payments were spread over 30 years while the machinery contract payments were spread over 10 years. The son also received aid in the form of cash gifts from the parents during years 7, 8, and 9.

The size of the operation was expanded through the rental of an additional 320 acres during the second and fourth years of the planning period. Livestock production was expanded during years 10 through 15 by additional purchases of beef cows.[5]

Results of the Analysis

The effect of selecting alternative goal structures was evaluated by a comparison of the annual production plans, changes in net worth, and the financial position of the farm at the end of the planning horizon. The amount of wheat and grain sorghum planted, feeder cattle sold, and cattle placed on pasture in November for each of the goal structures are reported in Table 14.4.

The results of the ranked objective function indicate that the order in which the goals were ranked significantly affected the preferred production plan. Increasing the importance of the labor-related goals resulted in a cropping pattern with larger acreages of grain sorghum production. Plans under the consumption-labor ranking also reflected much more production stability, with smaller annual adjustments being made in wheat and grain sorghum acreage and stocker production. This increased stability resulted because the labor requirements for grain sorghum production were distributed more uniformly over the production period. Wheat's major labor requirements fell within the 4-month period of June through September. The more even labor distribution associated with grain sorghum production allowed a more complete utilization of the fixed labor supply and reduced the quantity of part-time labor. Requiring the deviational levels of the first two priority levels to remain at their minimum levels while attempting to minimize the deviations of the third priority level did not allow any feasible adjustments under the ranked consumption-profit goal structure.

Similar reasoning may be used to explain the variability exhibited under the ranked consumption-profit goal structure. However, in this case the production plan was primarily determined by the goals of family consumption, cash balances, short-term borrowing, and current assets. Because the cash balance goal was weighted most heavily in the second preemptive level, adjustments were made in an effort to achieve this goal.

Results from the substitution of objective function goal structures demonstrated a pattern similar to the ranked objective function goal structures, but the differences between the 2 goal orientations were less striking. Increased emphasis on using the seasonal labor goal combined with the reduced emphasis on the cash balances under the consumption-labor orientation did not result in the same production stability as with the ranked objective function. The change in goal weights did result in a larger quantity of grain sorghum production during years 4, 5, and 6, but annual production adjustments were still quite large. The increased variability in production activities under the substituting objective function resulted from the added flexibility of the model to substitute among goals.

Figures 14.2 and 14.3 show cash farm income associated with the production plans and firm net worth obtained under the consumption-profit goal structures. The cash income for both objective functions exhibited great variability and was quite similar over the planning period. It should be remembered that although the production plans were being formulated on the basis of expectations and although consideration was not given to the risk, these expectations would not necessarily be realized.

The net worth of the firm under these 2 objective functions was nearly identical (Fig. 14.3). The net worth initially fell, but then rose. The initial fall partly resulted from the increasing debt load of the heir as initial asset acquisitions occurred. Because these transactions increased only the debt level and not the value of the assets, the firm's net worth declined.

The cash farm income and firm net worth associated with the production plans obtained under the two objective functions by use of the consumption-labor orientation goals are presented in Figures 14.4 and 14.5, respectively. Again, the cash farm incomes associated with

TABLE 14.4

Acres of wheat and grain sorghum planted, feeder cattle sold, and stockers pastured for different objective functions

	Ranked objective function 1/							
	Consumption-profit				Consumption-labor			
Year	Wheat planted	Grain sorghum planted	Feeder cattle sold	Stockers pastured 3/	Wheat planted	Grain sorghum planted	Feeder cattle sold	Stockers pastured 3/
	- - Acres - -		- - Number - -		- - Acres - -		- - Number - -	
1	1,172	113	0	564	1,131	113	0	538
2	1,313	0	0	556	950	40	0	335
3	1,313	0	0	442	818	363	0	300
4	1,525	6	90	0	995	546	90	0
5	1,503	0	0	90	918	511	0	90
6	346	3	90	0	341	563	90	0
7	349	1,160	90	0	343	1,140	90	0
8	307	1,158	90	0	301	1,138	90	0
9	378	1,173	53	37	310	1,155	90	0
10	637	1,078	0	212	304	1,156	86	4
11	342	810	113	0	346	1,162	113	0
12	644	1,114	0	171	406	1,120	136	0
13	1,431	812	0	686	485	1,070	160	0
14	538	25	183	0	547	1,001	183	0
15	894	918	1	205	638	939	197	9

TABLE 14.4 (cont.)

	Substitution objective function 2/							
	Consumption-profit				Consumption-labor			
Year	Wheat planted	Grain sorghum planted	Feeder cattle sold	Stockers pastured 3/	Wheat planted	Grain sorghum planted	Feeder cattle sold	Stockers pastured 3/
	- - - Acres - - -		- - - Number - - -		- - - Acres - - -		- - - Number - - -	
1	1,172	113	0	565	1,172	113	0	565
2	1,313	0	0	556	1,313	0	0	556
3	1,313	0	0	441	1,313	0	0	441
4	1,531	0	90	0	568	51	90	0
5	1,531	0	0	90	398	963	35	55
6	351	0	90	0	349	1,125	90	0
7	354	1,180	90	0	352	1,174	90	0
8	321	1,202	90	0	319	1,196	90	0
9	331	1,210	0	0	477	1,204	0	90
10	459	1,175	0	90	646	1,021	0	207
11	352	1,021	113	0	352	835	113	0
12	902	1,129	0	346	605	1,129	0	136
13	1,403	533	0	687	1,431	851	0	687
14	537	25	183	0	538	25	183	0
15	895	917	0	206	619	917	206	0

1/ Consumption-profit ranking assumes that family consumption is the goal of highest priority; cash balance, short-term borrowing, and current assets are of second priority; leisure time and part-time labor are the third priority. Consumption-labor ranking assumes that goals in the second and third levels are interchanged. 2/ Consumption-profit goal weights were assigned as: cash balance = 0.25, family consumption = 0.40, short-term borrowing = 0.10, part-time labor = 0.05, current assets = 0.15, leisure time = 0.05. Consumption-labor weights were assigned as: cash balance = 0.05, family consumption = 0.50, short-term borrowings = 0.05, part-time labor = 0.25, current assets = 0.05, leisure time = 0.10. The target level of seasonally employed labor under weight set 2 was reduced to 1,125. 3/ Stockers were placed on pasture in November.

326

FIGURE 14.2
Cash Farm Income Under the Consumption-Profit Goal Orientation for Ranked and Substitute Goal Structures

1,000 dollars

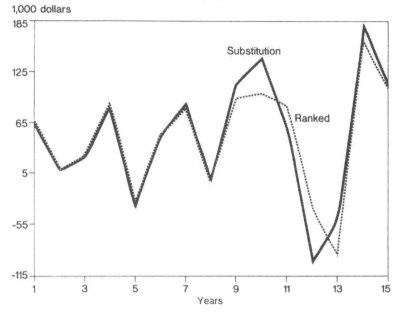

FIGURE 14.3
Farm Net Worth Under the Consumption-Profit Goal Orientation for Ranked and Substitute Goal Structures

1,000 dollars

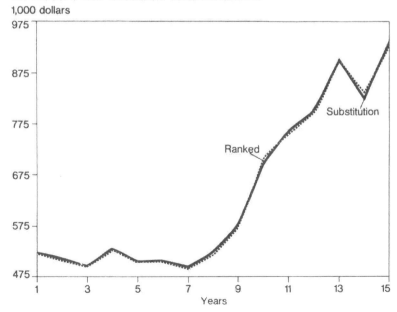

FIGURE 14.4
Cash Farm Income Under the Consumption-Labor Orientation
for Ranked and Substitute Goal Structures

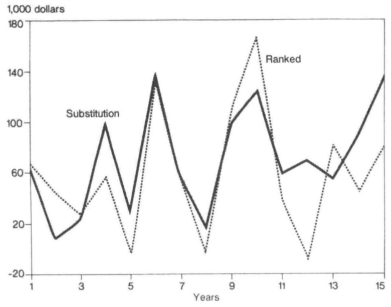

FIGURE 14.5
Farm Net Worth Under the Consumption-Labor Goal Orientation
for Ranked and Substitute Goal Structures

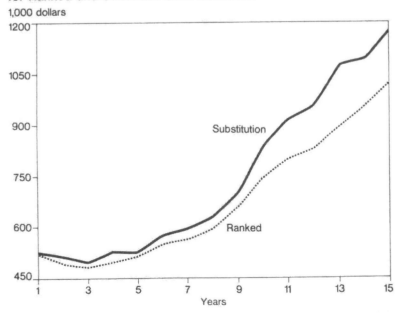

these plans showed much variability. Farm net worth exhibited similar trends.

Summary and Conclusions

The methodological procedures and modeling techniques used have the ability to handle farm-level problems involving multiple goals. The goals could involve income-leisure, cash balances-borrowing or consumption profit choices or focus on other policy alternatives. The type of objective function selected for use in the goal programming model will have an important effect on the results. Use of a ranked objective function results in the first unachievable goal having an extremely important impact on the final solution. Thus, selection of appropriate goal target levels and goal priorities must be given careful consideration. However, when the substitution objective function is used, the importance of the goals is reflected through the selection of goal weights. The goal weights selected could lead to the solutions being dominated by one or more goals, but this is less likely to occur than when the ranked goal structure is used. Nevertheless, the use of the substitution goal structure requires the careful specification of the goal target levels and goal weights.

Goal programming could be applied to many of the same problems to which other optimization routines, such as linear programming, have been applied. However, if GP is to become a useful tool in research related to farm-level policy analysis, new efforts are needed to determine the goals of farmers, and trade-offs among these goals, and the relationship of these goals to government programs. The stability of farmer goals is also important. Because of the uncertainties associated with the farmer's goals, their appropriate goal structures, goal weights, and goal target levels, agricultural economists concerned with policy analyses need to give careful thought to defining a systematic procedure for evaluating these items. Even with these difficulties, goal programming can still be used to provide useful qualitative information for policy analysis. The GP model can be used to analyze the credibility of a goal structure. It can be used to determine economic conditions required to achieve a set of goals, and provide information on the degree of goal attainment possible with a given set of resources.

Finally, the goal programming approach is not a methodological solution for all managerial decision problems. Its use requires defining, quantifying, and ordering objectives. It provides a solution, given constraints and a prioritized goal structure. But, research questions concerning the identification, definition, and ranking of goals still remain, and need to be addressed to enhance its usefulness.

Notes

1. The relationship between preemptive priority factors, P_i, is defined by Lee as follows: If $P_i > P_{i+1}$, then multiplication of priority factor P_{i+1} by n will never make the product greater than or equal to P_i, regardless of how large n may be.

2. However, if the satisfying levels of the goals are equal to zero, when all goals are of equal priority and the objective function is maximized, the GP model would be equivalent to the multiple goal LP formulation proposed by Candler and Boehlje.

3. The rules used in these calculations were those enforced prior to the Economic Recovery Tax Act of 1981.

4. These goals were selected on the basis of previous research. Harmon and others identified 8 goals as important to farmers: control more acreage by renting or buying, avoid being forced out of business, maintain or improve family's standard of living, avoid years of low profits or loss, increase time off from farming, increase net worth, reduce borrowing needs, and make the most profit each year. Smith and Capstick identified 10 basic goals: provide a college or vocational education for children; reduce borrowing needs; increase net worth; increase time devoted to family, personal, church, and community needs; increase efficiency of production; operate the farm to realize the highest long-run profit possible; improve family living standard; increase farm size by expanding acreages; avoid being forced out of business; and organize the farm to stabilize income. Patrick and Eisgruber conclude that goals can be grouped into 4 major areas: living standard; farm ownership; leisure and children, referring to the desire for leisure time and a family; and credit-using, risk-taking behavior that is characterized by the willingness to sacrifice in the farm operation in order to achieve other goals.

5. The coordination-growth plan discussed here is based on a case farm situation used in a detailed estate planning evaluation performed by Roush. He determined growth and coordination plans by interviewing family members. Other plans could be evaluated with this approach.

References

Anderson, Jock R., John L. Dillon, and J. Brian Hardaker. *Agricultural Decision Analysis*, Ames: Iowa State Univ. Press, 1977.

Arthur, Jeffrey L., and A. Ravindran. "PAGP, A Partitioning Algorithm for (Linear) Goal Programming Problems," *ACM Transactions on Mathematical Software*, Vol. 6, No. 3, Sept. 1980, pp. 378–86.

Baquet, A. E., A. N. Halter, and Frank S. Conklin. "The Value of Frost Forecasting: A Bayesian Appraisal," *American Journal of Agricultural Economics*, Vol. 50, No. 3, Aug. 1976, pp. 511–20.

Barnett, Douglas. "A Study of Farmers' Goals and Constraints: Their Effects on the Cultivation of Crops in the Sine Saloum, Senegal." Unpublished M.S. thesis, Purdue Univ., Aug. 1979.

Binswanger, Han. P. "Attitudes Toward Risk: Experimental Measurement in Rural India," *American Journal of Agricultural Economics*, Vol. 62, No. 3, Aug. 1980, pp. 395–407.

Brown, Robert Goodell. *Smoothing Forecasting and Prediction of Discrete Time Series*. Englewood Cliffs, N.J.: Prentice-Hall, Inc., 1963.

Candler, Wilfred, and Michael Boehlje. "Use of Linear Programming in Capital Budgeting with Multiple Goals," *American Journal of Agricultural Economics*, Vol. 59, 1971, pp. 325–30.

Charnes, Abraham, and William W. Cooper. *An Introduction to Linear Programming*. New York: John Wiley and Sons, Inc., 1956.

Clements, Alvin M., Harry P. Mapp, Jr., and Vernon R. Eidman. *A Procedure for Correlating Events in Farm Firm Simulation Models*. Tech. Bull. T-131. Oklahoma Agr. Expt. Sta., Aug. 1971.

Dillon, John L. "An Expository Review of Bernoullian Decision Theory in Agriculture: Is Utility Futility?," *Review of Marketing and Agricultural Economics*, Vol. 39, No. 1, Mar. 1971, pp. 3–80.

Dobbins, Craig L. "The Intergenerational Transfer of the Farm Firm: A Recursive Goal Programming Analysis," Ph.D. thesis, Oklahoma State Univ., 1978.

Ferguson, C. E. "The Theory of Multidimensional Utility Analysis in Relation to Multiple-Goal Business Behavior: A Synthesis," *Southern Economic Journal*, Vol. 32, Oct. 1965, pp. 169–75.

Halter, A. N., and Robert Mason. "Utility Measurement for Those Who Need to Know," *Western Journal of Agricultural Economics*, Vol. 3, No. 2, Dec. 1978, pp. 99–109.

Halter, Albert N., and Gerald W. Dean. *Decisions Under Uncertainty*. Cincinnati: South-Western Publishing Co., 1971.

Harmon, Wyatte L., Roy E. Hatch, Vernon R. Eidman, and P. L. Claypool. *An Evaluation of Factors Affecting the Hierarchy of Multiple Goals*. Tech. Bull. T-134. Oklahoma Agr. Expt. Sta., June 1972.

Harris, Thomas R., and Harry P. Mapp, Jr. "A Multiobjective Analysis of Energy Tradeoffs: Costs of Crop Residue Collection." Professional Paper No. P-900. Dept. of Agricultural Economics, Oklahoma State Univ., 1981.

Hatch, Roy E. "Growth Potential and Survival Capability of Southern Plains Dryland Farms: A Simulation Analysis Incorporating Multiple-Goal Decision Making," Ph.D. thesis, Oklahoma State Univ., 1973.

Hicks, J. R. *Value and Capital*. Oxford: The Clarendon Press, 1946.

Ignizio, James P. *Goal Programming and Extension*. Lexington, Mass.: Lexington Books, 1976.

Ijiri, Yuji. *Management Goals and Accounting for Control*. Amsterdam: North-Holland Publishing Co., 1963.

Lee, Sang M. *Goal Programming for Decision Analysis*. Philadelphia: Auerbach Publishers, 1972.

————. "Goal Programming for Decision Analysis of Multiple Objectives," *Sloan Management Review*, Vol. 14, No. 2, Winter 1972–73, pp. 11–24.

Lin, William, C. W. Dean, and C. V. Moore. "An Empirical Test of Utility vs. Profit Maximization in Agricultural Production," *American Journal of Agricultural Economics*, Vol. 56, 1974, pp. 497–508.

Neely, Walter P., Ronald M. North, and James C. Fortson. "An Operational Approach to Multiple Objective Decision Making for Public Water Resources Projects Using Integer Goal Programming," *American Journal of Agricultural Economics*, Vol. 59, Feb. 1977, pp. 198–203.

Officer, R. R., and A. N. Halter. "Utility Analysis in a Practical Setting," *American Journal of Agricultural Economics*, Vol. 50, May 1968, pp. 257–77.

Patrick, George F. "Effects of Alternative Goal Orientations on Farm Growth and Survival," *North Central Journal of Agricultural Economics*, Vol. 3, No. 1, Jan. 1981, pp. 29–39.

————, and Brian F. Blake. "Measurement and Modeling of Farmer's Goals: An Evaluation and Suggestions," *Southern Journal of Agricultural Economics*, Vol. 2, No. 1, July 1980, pp. 199–203.

————, and Ludwig Eisgruber. "The Impact of Managerial Ability and Capital Structure on Growth of the Farm Firm," *American Journal of Agricultural Economics*, Vol. 50, Aug. 1968, pp. 491–506.

Roush, Clint E. "Economic Evaluation of Asset Ownership Transfer Methods and Family Farm Business Arrangements After the Tax Reform Act of 1976," Ph.D. thesis, Oklahoma State Univ., 1978.

Simon, H. A. *Models of Man*. New York: John Wiley and Sons, Inc., 1957.

Smith, Donnie A., and Daniel F. Capstick. *Evaluation of Factors Affecting the Ranking of Management Goals by Farm Operators*. Arkansas Agr. Expt. Sta. Report Series 232. Nov. 1976.

Vocke, Gray, Earl O. Heady, and Orhan Saygidegar. "Application of a Multiple Goal Linear Programming Model to Measure Trade-Offs Between Cost and Environmental Variables in Agriculture." Contributed paper, American Agricultural Economics Association meeting, San Diego, Calif., July 31–Aug. 3, 1977.

15. Generalized Risk-Efficient Monte Carlo Programming: A Technique for Farm Planning Under Uncertainty

Robert P. King and George E. Oamek

Abstract Generalized risk-efficient Monte Carlo programming is a technique for identifying preferred choices under uncertainty. It combines random search methods, stochastic simulation, and evaluation of alternatives by the criterion of stochastic dominance with respect to a function. This approach is applied to the problem of choosing among flexible forward-pricing strategies and a range of crop insurance options for a typical dryland wheat farm in eastern Colorado. The results show that decisions concerning these 2 important aspects of a risk management strategy are relatively independent for the example considered.

Keywords Flexible decision strategies, risk programming, stochastic dominance

In agricultural production processes, resources must often be committed long before returns are realized. During the production process, environmental, economic, and institutional factors beyond the control of the individual decisionmaker may have a significant and unpredictable effect on production levels, costs, and product prices. Thus, farmers operate in a highly uncertain environment that affects farm-level decisions. Agricultural economists have developed and refined a number of risk programming models that can be used to analyze decisions under uncertainty. In this paper, we describe a recently developed method

The authors are an associate professor in the Department of Economics at the University of Minnesota, and a graduate research assistant in the Department of Economics at Iowa State University, respectively.

for identifying preferred choices under uncertainty and apply that method to a farm planning problem.

Uncertainty affects decision processes in 2 important ways. First, it affects the rules decisionmakers use to order choices. When decisionmakers are confronted with probabilistic rather than known outcomes, the simple and familiar rule that "more is better" cannot be applied, since individual attitudes toward risk taking affect the weights placed on each possible outcome. Decision rules for ordering uncertain choices must, then, integrate information on the probability distributions of outcomes associated with alternative choices and information on the decisionmaker's risk preferences. Expected utility maximization is, from a theoretical standpoint, the most widely accepted decision rule for ordering choices under uncertainty, but operational problems make this rule difficult to use in a practical context. Therefore, a number of alternative criteria, such as game-theoretic and safety-first rules and several approximations of the expected utility criterion, such as mean-variance and MOTAD decision rules, have also been developed.

A second effect of uncertainty on the decision process is that the very character of the decision to be made may change. In situations involving a sequence of actions, optimal choices under uncertainty are often flexible strategies, which make the pattern of future actions contingent upon future events that the decisionmaker can observe but cannot control (Massé; Murphy; Rae; Day, 1975). Thus, the choice becomes one of identifying rules for responding to conditions as they evolve rather than one of determining a fixed sequence of actions. For example, most farmers do not follow a fixed sales pattern when marketing grain. Instead, they base their decisions on general guidelines, or rules of thumb, that incorporate information about past and projected market conditions.

These impacts of uncertainty are not important in all decision situations. Ideally, though, it should be possible to consider both in a risk programming model. The model should allow for the representation of preferences and expectations in a manner that conforms with the decisionmaker's own beliefs and with the computational requirements of the choice rule. The model should also be flexible enough to allow the evaluation of both flexible and fixed strategies. Unless these requirements are met, farm planning or strategy recommendations derived from the results of risk programming models may be unrealistic and unreliable.

In the following sections we review the most commonly used risk programming models. Next, we describe the structure of a new method for identifying preferred choices under uncertainty. We use that method to identify preferred risk management strategies for a typical eastern

Colorado dryland wheat farm. Finally, we evaluate the potential of this approach for analyses of farm firm adjustments to changes in agricultural policy.

Risk Programming Models

Mathematical programming models are commonly used in the analysis of complex decision problems with the assumption of perfect knowledge. They are analytically elegant, computationally efficient, and easily adapted to a wide range of decision situations. In general, risk programming models are simply special formulations of standard mathematical programming models. The most widely used in agricultural economics applications include quadratic programming (Freund, Markowitz), MOTAD (Hazell), safety-first (Roumasset, Benito), and game-theoretic (McInerney) models. Quadratic programming has gained the widest acceptance on conceptual grounds. The other models are designed to be solved by standard linear programming algorithms, however, which are more widely available and less expensive to run.

Given the impacts of uncertainty on the decision process, 3 important criteria for the evaluation of any risk programming model are: suitability of the choice criterion implied by the model, flexibility in the representation of complex stochastic processes, and ability to evaluate flexible as well as fixed strategies. With regard to the first of these criteria, if consistency with the expected utility hypothesis is a desired property for a risk programming model, each technique identified at the outset of this section has potentially serious shortcomings. Quadratic programming uses a mean-variance criterion for the ordering of alternative choices. It is fully consistent with expected utility maximization when all random factors are normally distributed or when the decisionmaker has a quadratic utility function. Given the often skewed nature of agricultural price distributions caused by commodity program provisions and of crop yield distributions (Day, 1965), however, the assumption of normality may be suspect. Because quadratic utility functions imply increasing absolute risk aversion, their use is also questionable. Therefore, quadratic programming may not identify the expected utility-maximizing strategy. The linear risk programming models are even more weakly linked to the expected utility hypothesis and so may also fail to identify a truly optimal strategy.

The limitations of the most widely used risk programming models are more serious with regard to the representation of complex stochastic processes. Stochastic processes involving random factors that interact in a nonadditive fashion (for example, prices and yields) are difficult to represent in all of these models, although Paris (1981) has made progress in resolving this problem in quadratic programming models. A related limitation of these techniques is that they tend to restrict a problem's stochastic factors to the coefficients in the objective function. In a typical farm planning problem, then, the random factors are often limited to net returns for each of the possible enterprises. In many such problems, though, resource requirements and availability levels represent an equally important source of uncertainty. These constraints may indirectly affect net returns by making some otherwise attractive strategies unfeasible. Although Paris (1979) has demonstrated how this type of uncertainty can be incorporated into quadratic programming models under certain conditions, other researchers typically either ignore this problem or handle it in a rather unsatisfactory way with chance constrained programming methods (Charnes and Cooper).

With regard to the third evaluative criterion identified above, none of the widely used risk programming modeling techniques is well suited to analyze flexible strategies, which allow for the use of information as it becomes available. Stochastic programming (Cocks, Rae) provides a conceptual framework for the identification of optimal flexible strategies, but if the problem under consideration is even moderately complex, the size of the input-output matrix quickly expands to an unmanageable level. Therefore, the usefulness of this approach in an applied decision analysis is limited.

Linear decision models, as developed by Holt, Modiglioni, Muth, and Simon, represent a more workable alternative. Under this approach, which has been further refined by Zellner, Chow, and McRae, dynamic programming techniques are used to determine analytical solutions for a special class of optimal control problems. Having assumed quadratic cost (utility) functions and linear relations between state and control variables, linear decision rules are derived that can be used to determine optimal levels for control variables on the basis of forecasts of key factors. The parameters of these rules remain invariant for as long as the design of the system studied is unchanged. These models have the same analytical limitations imposed on preferences and probability distributions by quadratic programming models. Because there is usually a high cost of performing the initial analysis needed to determine optimal decision rules for a particular process in a complex model, the applicability of this approach is limited to situations where similar decisions are made repeatedly.

The risk-efficient Monte Carlo programming (REMP) model developed by Anderson is an attractive alternative to the more widely used mathematical programming models described in this section for several reasons. The REMP model employs Monte Carlo programming techniques (Donaldson and Webster) to construct a large number of feasible strategies. The distribution of total net returns associated with each strategy is determined analytically under the assumption that the distribution of net returns for each activity and the distribution of total net returns for each strategy are members of the beta distribution family. The criterion of second-degree stochastic dominance is used to evaluate strategies sequentially. As such, the REMP model allows for considerable flexibility in the representation of probability distributions, as the beta distribution can assume a variety of forms. Also, the criterion of second-degree stochastic dominance is fully consistent with expected utility maximization and holds for all decisionmakers who are risk averse at all income levels. The REMP model does, however, have some of the same serious limitations as the previously discussed techniques. First, second-degree stochastic dominance is not a very discriminating evaluative criterion. In a study by Anderson, 20 of only 48 randomly generated strategies were in the efficient set. In addition, the REMP model limits the random factors in a problem to the coefficients of a linear function that determines the value of the performance measure. Finally, the REMP model is not designed to evaluate flexible decision strategies. Although decision problems under uncertainty are often inherently dynamic in nature, the REMP model and the other widely used risk programming models are designed to solve essentially static problems.

Each of the risk programming models discussed here is well suited for analyzing some decision problems, but none is generally applicable to a wide range of practical problems. Furthermore, many rather common types of decision problems cannot be adequately analyzed with any of these procedures. Consequently, a more general approach that permits greater flexibility with respect to problem formulation, the determination of probability distributions, and the representation of decisionmaker preferences is needed.

Generalized Risk-Efficient Monte Carlo Programming

The procedure for identifying preferred choices described in this section can be viewed as an extension of the REMP model and is called generalized risk-efficient Monte Carlo programming (GREMP). As in the REMP model, Monte Carlo programming techniques are used to generate the strategies considered, and a stochastic efficiency criterion is used to evaluate alternative strategies. However, many of the problems of the REMP model are eliminated with this more general approach.

King gives a detailed description of the GREMP model and a listing of the computer program used to implement it. The basic structure of the model is outlined in the flow chart in Figure 15.1. During the

FIGURE 15.1
General flow chart of program GREMP

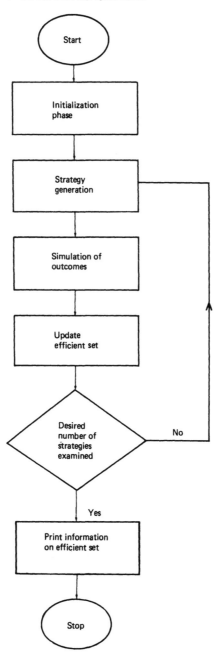

initialization phase, problem-specific information is provided as input to the model through a user-supplied subroutine. This includes information needed to determine the feasible set of strategies for the problem being considered, such as ranges of permissible values for each choice variable and values of parameters required to define constraints on combinations of choice variables. Also, because the GREMP model involves a random search, the number of strategies to be generated and evaluated must be specified. In addition, information on the process being managed must be supplied. This includes parameter values and initial conditions that may vary from one decision situation to another and sample vectors drawn from the multivariate distribution of the stochastic factors that have a significant impact on the outcome distributions associated with alternative strategies. Finally, interval measurements of the decisionmaker's risk preferences based on the procedure developed by King and Robison (1981a,b) must be specified.

When the initialization phase is completed, the process of strategy generation and evaluation begins. A strategy is defined by the values of a set of choice variables that may directly specify control actions or may be the parameters of flexible decision rules that determine forthcoming control actions as information becomes available. Specified acreage levels for particular crops are examples of the first type of choice variable, since they directly indicate the action to be taken. The variable X in the following statement illustrates the second type of choice variable: "Irrigate to field capacity when the level of available soil moisture falls below X." In this example, X is the parameter for a decision rule that uses current information to determine a control action.

Choice variables can be either discrete or continuous in the GREMP model. The value of each choice variable is typically assigned a randomly selected level drawn from a uniform distribution defined over that variable's permissible range of values. Linear, nonlinear, or logical constraints on the levels of the choice variables are repeatedly checked to ensure that none is violated. When a constraint is violated, the strategy is adjusted until it is feasible.

When the number of feasible strategies is relatively small, the GREMP model can be modified to permit a complete enumeration and evaluation of all possible combinations of choice variables. The GREMP model can also generate a set of strategies in accordance with a predetermined, rather than random, search design.

The performance of each strategy is simulated through each of the sample states of nature specified in the initialization phase. As was noted earlier, each state of nature is defined by a sample vector drawn from the multivariate distribution of stochastic factors in the problem.[1] For example, if the price and yield for each crop enterprise and the days available for fieldwork are the random factors for a typical farm planning problem, a sample vector would contain one value for each of these variables. The set of outcome levels resulting from these repeated simulations—one for each sample state of nature—can be used to construct an approximate cumulative distribution function for the outcome distribution of the strategy considered (Barnett).

A user-supplied subroutine performs the simulations required during this phase of the procedure. This subroutine can be as simple or as complex as the problem requires. Because outcome distributions are determined by simulation, complex stochastic processes involving linear and nonlinear interactions between choice variables and random factors can be modeled. The effect of flexible strategies can easily be represented if they are built into the simulation subroutine. For example, if days available for fieldwork are random in a farm planning problem, it may be neither possible nor desirable to plant the entire intended acreage of corn when spring weather greatly delays necessary fieldwork. A producer's response to such a situation can be modeled with a flexible rule of the form: "If all of the intended corn acreage has not been planted by May 18, stop planting corn, and start planting soybeans" (King and Robison, 1980). This rule uses current information on corn acreage planted, as affected by random weather, to alter a previously defined crop plan. The date, May 18, can be viewed as a choice variable in this example, as the selection of a cutoff date for corn plantings is likely to affect the performance of an overall farm plan.

Strategies are evaluated as they are generated using the criterion of stochastic dominance with respect to a function (Meyer). This stochastic efficiency criterion orders uncertain choices for classes of decisionmakers whose absolute risk-aversion functions (Arrow, Pratt) lie within specified lower and upper bounds.[2] Unlike other commonly used efficiency criteria, such as first- and second-degree stochastic dominance, which hold only for quite broadly and inflexibly defined classes of decisionmakers, stochastic dominance with respect to a function does not impose restrictions on the shape of the relevant risk-aversion interval. Therefore, it can discriminate more effectively among choices than can other efficiency criteria.

An important advantage of stochastic dominance with respect to a function is that it typically identifies more than a single efficient strategy, yet it does not yield an unmanageably large efficient set. In normative applications, the GREMP model allows the individual decisionmaker to choose from several good alternatives rather than be presented with a single "best" solution. In positive applications of the model, the presence of multiple strategies in the efficient set can help policy analysts by identifying the probable range of farmers' responses to a change in economic conditions or government policy.

In summary, the GREMP approach has several distinct advantages over other more commonly used risk programming techniques. It is well suited for problems in which stochastic factors interact with choice variables in a complex manner. It employs an evaluative criterion that is fully consistent with the expected utility hypothesis, which can be quite discriminating, and yet permits explicit consideration of the imprecision with which decisionmaker preferences are measured. The greatest strength of the GREMP modeling approach, however, is that it is flexible enough to let the analysis conform to the problem at hand. In past studies, the GREMP model has been used to identify optimal post-harvest marketing strategies for pinto beans (King and Lybecker, 1981a) to identify combined strategies for crop production and forward contracting on a Michigan corn-soybean farm (King) and to evaluate alternative weed management strategies for corn in eastern Colorado (King and Lybecker, 1981b).

Nevertheless, the GREMP model is more difficult to use than quadratic or linear risk programming models. Several problem-specific subroutines must be supplied by the user, including a simulation model. The GREMP model may also be costly for some problems. With regard to the output the model provides, there is no guarantee that the efficient set of strategies identified by the model will contain a true optimum, since all feasible strategies are usually not evaluated.

An Application

An example will help illustrate how the GREMP model can be applied to a practical decision problem. The problem considered is to identify a preferred risk management strategy for a typical Colorado dryland wheat farmer using the GREMP modeling approach. Forward pricing through hedging on the futures market and the purchase of federal all-risk crop insurance are the two risk management tools considered. Anticipated provisions for the Agricultural Stabilization and Conservation Service (ASCS) commodity program have been programmed into the model.[3]

We selected this problem because it demonstrates some of the special features of the GREMP model. The effects of both the federal crop insurance program and the ASCS commodity program would be difficult to model analytically, given their asymmetric impact on the distribution of net returns realized by a producer. These effects are easily represented in a simulation submodel in the GREMP model. Hedging decisions are sequential in nature. Choices for the current period depend on all available information. Nevertheless, most studies have considered only fixed hedging strategies (for example, Klinefelter, Sonka, and Baker; Leuthold and Peterson). Within the GREMP model flexible hedging strategies can be evaluated using an approach developed by King and Lybecker (1981a). Finally, the criterion of stochastic dominance with respect to a function provides an ordering of strategies with potentially non-normal distributions, which is consistent with the expected utility hypothesis.

This example also considers a methodological issue of general importance in micromodeling—that of the degree to which individual components of an overall management strategy can be analyzed separately. Clearly, large problems become more manageable if they can be divided into smaller, independent subproblems. However, significant interactions may be ignored if the subproblems are solved independently. In the analysis which follows, then, the extent to which hedging and crop insurance decisions are affected by each other and by the structure of the ASCS commodity program will be explored. If these decisions are independent, the structure of models designed to identify preferred strategies for individual producers or to predict the responses of groups of producers to possible policy changes can be greatly simplified.

The Problem and the Problem Setting

A typical dryland wheat operation in northern Weld County, Colo., serves as the empirical focus for this analysis. Of 2,500 tillable acres, all of which are owned, 1,250 are planted in wheat. The remaining 1,250 acres are kept in summer fallow. Variable cash preharvest production costs (fuel and repair costs for machinery, purchased inputs, and interest on operating expenses) are set at $30 per acre. The farmer is assumed to combine his own wheat. A fixed cost of $8 per acre and a marginal harvest and hauling cost of 20 cents per bushel are charged for harvest and hauling operations. Cash fixed costs (land payments, machinery payments, and property taxes) are set at $50,000. The producer is assumed not to have access to on-farm storage, and the entire crop is marketed at harvest.

The risk management strategies considered in this analysis determine efficient patterns of futures contract transactions during a preharvest marketing period that extends from the beginning of September to harvest in early July as well as the level of crop insurance protection purchased.[4] The hedging strategy limits transactions to the July, Kansas City contract and is defined by a simple feedback control rule that determines actions sequentially on the basis of currently available information. That rule is of the following form:

$$\text{NSOLD}_t = \text{MAX}(0, \text{NDES}_t - \text{NCON}_t) \tag{1}$$

where $NSOLD_t$ is the number of contracts sold in month t, $NDES_t$ is the number of contracts one would desire to have sold by month t, and $NCON_t$ is the number of contracts already sold. $NSOLD_t$ is a positive integer. The restriction that it cannot be less than zero ensures that the producer will not take a speculative position or lift his hedge early. $NDES_t$ is determined by the following expression:

$$\text{NDES} = [V_1(P_t - \text{EHP}_t - V_2) + V_3(\text{PF}_{1t} - P_t) + V_4(\text{PF}_{2t} - P_t)$$
$$+ V_5(\text{PF}_{3t} - P_t) + V_6 \text{DP}_t] \text{ECROP}/5000$$
$$\text{s.t.}$$
$$(2)$$

$$\text{NDES} \leqslant 1.2 * \text{ECROP}/15000$$

where EHP_t is the expected harvest price; P_t is the current July futures contract price; PF_{1t}, PF_{2t}, and PF_{3t} are expected July futures prices 1, 2, and 3 months ahead; DP_t is the current percentage rate of price change; and $ECROP$ is the expected size of the crop. EHP_t, PF_{1t}, PF_{2t}, and PF_{3t} are updated monthly with a simple linear expectations model of the form:

$$\text{PF}_t = b_{0t} + b_{1t}P_t + b_{2t}P_{t-1} \tag{3}$$

where PF_t is the price to be forecast, and P_t and P_{t-1} are the July futures prices for the current and previous month. Parameters were estimated for each price forecast equation for each month with data for 1970–81. Again, $NDES$ is an integer indicating the number of contracts to be sold. To prevent speculation, its value is constrained to levels less than or equal to the possible number of contracts in 120 percent of the expected crop.[5]

Equations (1) and (2) define a flexible rule that is consulted at the beginning of each month to determine whether or not additional contracts should be sold. If $NSOLD_t$ is positive, $NSOLD_t$ contracts are sold at price P_t. Therefore, both the timing and level of forward pricing decisions are determined by the rule. The hedge is automatically lifted in early July before the contract delivery date. Within this rule, the parameters V_1 through V_6 in equation (2) can be viewed as choice variables that will determine the strategy's behavior.

Producers can choose not to purchase federal crop insurance or they can select one of 9 coverage levels defined by 3 possible price guarantees—$2.50, $3.50, and $4.50 per bushel—and 3 possible yield guarantees—12.5, 15.0, and 17.5 bushels per acre. Premiums range from $2.60 to $11.95 per acre, and if coverage is purchased, it must be purchased for all planted acres.[6] The crop insurance decision can be described, then, by 3 choice variables: V_7, which is equal to zero if no insurance is purchased and to 1 if coverage is purchased; V_8, which is set equal to 1, 2, or 3 to indicate the price guarantee selected; and V_9, which is set equal to 1, 2, or 3 to indicate the yield guarantee selected. When V_7 is equal to zero, the levels of V_8 and V_9 have no meaning.

Producers would consider the distribution of outcomes possible under each alternative and their own risk preferences when selecting a risk management strategy, as defined by the values of the choice variables V_1 through V_9. In this study, the performance criterion used to evaluate alternative strategies is net cash flow—that is, receipts from crop sales, futures transactions, crop insurance indemnities, and deficiency payments minus cash fixed and variable production costs, cash costs for futures transactions, and insurance premiums. In early September, when the decision concerning crop insurance must be made and when a hedging strategy is selected, the pattern of futures contract price levels for subsequent months, the cash harvest price, crop yield, and the national average price for the 5 months after harvest cannot be known with certainty. Consequently, the producer can view net cash flow as a random variable, whose probability distribution is determined by a complex interaction of these stochastic factors, given the selected risk management strategy. The problem for the GREMP model is to determine a strategy defined by the values of parameters V_1 through V_9 that will be preferred by a decisionmaker whose absolute risk-aversion function lies everywhere within specified bounds.

The Simulation of Outcome Distributions

The impacts of a flexible hedging strategy, federal crop insurance coverage, and ASCS commodity program deficiency payments on the probability distribution of net cash flow would be difficult, if not impossible, to model analytically. Under the GREMP approach, however, a user-supplied simulation model is used to determine the outcome distribution associated with each strategy considered.

The simulation submodel developed for this analysis has two major components—a forward pricing component and a crop production and sale component. The forward pricing component implements the hedging strategy defined in equations (1) and (2). Each month it forecasts the expected harvest price and July futures prices for the next 3 months and determines whether or not additional contracts should be sold. This component also models margin account transactions, adding to the account when margin calls are made and withdrawing from it when possible. Interest charges are assessed monthly on the margin account. Finally, this submodel component lifts the hedge in July, repurchasing at the current market price the contracts previously sold. A net gain or loss on futures transactions is determined by the value of contracts sold, the cost of repurchasing them, the interest fees on the margin account, and broker's fees.

The crop production component determines crop sale receipts, assesses fixed and variable production costs, determines insurance costs, and calculates indemnities when the yield per acre falls below the yield guarantee. In addition, it determines the size of deficiency payments when the national average wheat price for the 5 months after harvest falls below the target price.

Each strategy is simulated through 30 sample states of nature. Each state is defined by a randomly selected vector of values from the multivariate distribution of July futures prices for September through July, the cash harvest price, the national average price for the 5 months after harvest, and the crop yield. Futures price levels were generated by use of a stochastic simulation model that recursively predicted prices 1 month ahead with the appropriate expectations model discussed earlier. A normal random variate generated from the distribution of the expectation model's disturbance term was added to each price. These randomly generated prices were then used to generate subsequent prices. The cash harvest price was determined in a similar manner from the May and June futures prices, and the national average postharvest price was based on the cash harvest price. This approach ensured that all random prices in the model were correlated. Yields were assumed to be uncorrelated with prices. Their distribution was based on detrended yield levels for 1951–80.

The model was run for conditions in the fall of 1981. July futures prices for wheat at the beginning of August and September were $4.92 and $4.54 per bushel, respectively. The loan rate and target price had not yet been determined in early September. Levels from the then-current House version of the farm bill, $3.55 and $4.20, respectively, were used in modeling the ASCS program.

Results

The GREMP model was used to identify efficient strategies for 2 classes of decisionmakers. The first class is composed of individuals whose level of absolute risk aversion lies within the interval (0, 0.00001) at all outcome levels. This class includes risk-neutral decisionmakers who maximize expected net cash flow and those who are slightly risk averse. The second class of decisionmakers includes farmers whose levels of absolute risk aversion lie within the interval (0.00001, 0.00002) at all income levels. Given the range of possible outcomes in this problem—from a loss of $85,000 to a gain of $160,000—these decisionmakers can be termed moderately risk averse.

To simplify the analysis in this example, we considered only 3 of the 10 possible insurance options: no coverage, coverage at the minimum price and yield guarantee levels, and coverage at the maximum price and yield guarantee levels. To facilitate analysis of the interdependence between hedging and crop insurance decisions, we specified the strategy generation procedure so that the same 250 hedging strategies would be considered with each insurance coverage option. Thus, 750 strategies were considered for each decisionmaker class.

Table 15.1 gives information on the strategies in the efficient set of each class of decisionmakers. These results show, first, the ability of the stochastic dominance with respect to a function criterion to discriminate among alternative strategies. The 2 efficient sets contain 3 and 4 strategies, respectively, from a total of 750 strategies considered.[7] The results show that risk preferences do have a significant impact on the choice of a risk management strategy, especially on the decision concerning crop insurance coverage. Although decisionmakers in the first class unanimously prefer not to insure, each of the coverage levels appears in the efficient set for the second class of decisionmakers.

The results for the second class of decisionmakers indicate that insurance decisions have little effect on preferred hedging strategies, as the same set of parameters is efficient regardless of the level of insurance coverage selected. Further analysis indicates that the insurance choices for each class of decisionmakers remain unchanged when hedging opportunities and the deficiency payment component of the ASCS commodity program are not considered. In addition, hedging strategies are largely unaffected within each class of decisionmakers when deficiency payments are eliminated from the analysis. Although these results are by no means conclusive, they indicate that the degree of interdependence among these aspects of a risk-management strategy is relatively weak for a Colorado wheat producer in Weld County. The degree of interdependence may be quite different in other regions and for other crops.

Potential Uses of the GREMP Model in Policy Research

Most recent applications of the GREMP model emphasize its value as a tool for normative analysis (King and Lybecker, 1981a,b). The model also has considerable potential, however, as a tool for positive analyses of producer responses to policy changes. A model similar to that described in this paper will be used in a forthcoming study designed to determine the responsiveness of Colorado dryland wheat farmers to

TABLE 15.1

Effecient risk management strategies for two classes of decisionmakers

Item	Class 1 (0 ≤ r(y) ≤ 0.00001) efficient strategy			Class 2 (0.00001 ≤ r(y) ≤ 0.00002) efficient strategy			
	1	2	3	1	2	3	4
Hedging strategy parameters:							
V_1	-0.80	-0.40	-0.60	-0.60	-0.60	-0.80	-0.60
V_2	.10	.05	.30	.30	.30	.10	.30
V_3	-1.40	-1.20	-1.40	-1.40	-1.40	-1.40	-1.40
V_4	-.20	-.40	-.60	-.60	-.60	-.20	-.60
V_5	-2.20	-1.60	-.80	-.80	-.80	-2.20	-.80
V_6	-3.60	-2.80	-1.60	-1.60	-1.60	-3.60	-1.60
Insurance coverage	none	none	none	none	lowest	highest	highest
Net cash flow:			Dollars				
Mean	42,558	42,796	41,940	41,940	39,804	34,680	34,062
Standard deviation	51,918	52,716	50,614	50,614	48,603	43,646	42,760
Minimum	-79,717	-80,177	-78,729	-78,729	-58,952	-25,575	-33,250
Maximum	157,844	157,844	160,578	160,578	157,392	143,201	145,935

changes in the federal crop insurance program. In this study, risk preference measurements will be made for each of a panel of farmers, and relevant information will be collected on the characteristics of each operation. The GREMP model will then be run under alternative insurance program provisions for each farmer to determine how this cross section of the state's farmers will respond to possible changes in the program. Although relatively costly, this analysis should provide valuable insights concerning the way farmers adjust to policy changes. It will allow us to predict both the aggregate impacts and the distributional effects of alternative program provisions.

Techniques for modeling long-term farm firm adjustment and growth have not been emphasized in this paper because the GREMP modeling technique has not yet been applied to this problem.[8] It should be noted, however, that the example discussed here is, itself, a multiperiod problem. Just as the sequential decisions that comprise a hedging strategy can be modeled with flexible decision rules, this same approach can be used to model farm firm adjustments through time. For example, key financial ratios could be used to drive flexible rules for farm expansion and machinery investment. The parameters of these rules would be choice variables selected in accordance with the decisionmaker's risk preferences. The resulting long-term adjustment strategies could help to explain in greater detail the mechanisms by which farms respond to changes in policies and economic conditions.

Notes

1. See King for a discussion of procedures for generating sample vectors from multivariate distributions with normal or non-normal marginal distributions.

2. The absolute risk aversion function, $r(y)$, is defined by the expression, $r(y) = -u''(y)/u'(y)$, where $u'(y)$ and $u''(y)$ are the first and second derivatives of a von Neumann–Morgenstern utility function. Although such a utility function is unique only to a positive linear transformation, the absolute risk aversion function represents preferences uniquely.

3. At the time this paper was written, the 1981 farm bill had not yet been passed. We assumed that the Disaster Assistance Program, which provided protection against low yields, would be eliminated from the ASCS wheat program. Provisions for deficiency payments and the farmer-owned reserve were assumed to remain in effect.

4. We assume that the producer complies with all requirements for participation in the ASCS commodity program and so benefits from its provisions.

5. The expected total harvest is 33,750 bushels in this example. Therefore, a maximum of eight 5,000-bushel contracts can be sold.

6. The farm being analyzed is assumed to be in zone 2 of Weld County. These are actual price and yield guarantees and premium levels for that area.

7. Given the range of possible outcomes in this problem, the intervals specified are relatively broad.

8. Abkin and Ingvaldson have used the GREMP model to analyze energy supply investment decisions under uncertainty.

Acknowledgments

This research was funded by the Colorado Agricultural Experiment Station.

References

Abkin, M. H., and G. R. Ingvaldson. "An Empirical Model for Energy Supply Investment and Disinvestment Decision Making under Uncertainty," in *Theoretical and Practical Models for Investment Supply Industry* (ed. L. J. Robison and M. H. Abkin). Agricultural Economics Report 390. Michigan State Univ., Mar. 1981, pp. 127–54.

Anderson, J. R. "Programming for Efficient Planning Against Non-normal Risk," *Australian Journal of Agricultural Economics*, Vol. 19, 1975, pp. 94–107.

Arrow, K. J. *Essays in the Theory of Risk Bearing.* Chicago: Markham Publishing Co., 1971.

Barnett, V. "Probability Plotting Methods and Order Statistics," *Applied Statistics*, Vol. 24, 1975, pp. 95–108.

Benito, C. A. "Peasants' Response to Modernization Projects in Minifundia Economies," *American Journal of Agricultural Economics*, Vol. 58, 1976, pp. 143–51.

Charnes, A., and W. W. Cooper. "Chance Constrained Programming," *Management Science*, Vol. 6, 1959, pp. 73–79.

Chow, G. C. "Effect of Uncertainty on Optimal Control Problems," *International Economic Review*, Vol. 14, 1973, pp. 632–45.

Cocks, K. C. "Discrete Stochastic Programming," *Management Science*, Vol. 15, 1968, pp. 72–79.

Day, R. H. "Adaptive Processes in Economic Theory," in *Adaptive Economic Models* (ed. R. H. Day and T. Groves). New York: Academic Press, 1975, pp. 1–38.

————. "Probability Distributions of Field Crop Yields," *Journal of Farm Economics*, Vol. 47, 1975, pp. 713–41.

Donaldson, G. F., and J.P.G. Webster. "An Operating Procedure for Simulation Farm Planning—Monte Carlo Method." Univ. of London, Wye College, Dept. of Agr. Econ., 1968.

Freund, R. J. "The Introduction of Risk into a Programming Model," *Econometrica*, Vol. 24, 1956, pp. 253–63.

Hazell, P.B.R. "A Linear Alternative to Quadratic and Semivariance Programming for Farm Planning Under Uncertainty," *American Journal of Agricultural Economics*, Vol. 53, 1971, pp. 53–62.

Holt, C. C., F. Modigliani, J. F. Muth, and H. A. Simon. *Planning Production, Inventories and Work Force.* Englewood Cliffs, N.J.: Prentice-Hall, Inc., 1960.

King, R. P. "Operational Techniques for Applied Decision Analysis Under Uncertainty." Unpublished Ph.D. dissertation, Michigan State Univ., 1979.

————, and D. W. Lybecker. "Adaptive Decision Models." Paper presented to annual meeting of NCR113, North Central Regional Committee on Farm and Financial Management, St. Louis, Mo., Oct. 1981a.

————. "Generalized Risk Efficient Monte Carlo Programming: An Application to Weed Management." Paper presented at annual meeting of GPC-10, Great Plains Regional Committee on Farm Management and Production Economics, Fort Collins, Colo., May 1981b.

————, and L. J. Robison. *Implementation of the Interval Approach to the Measurement of Decision Maker Preferences.* Agr. Expt. Sta. Research Report 418. Michigan State Univ., 1981a.

————. "Implementing Stochastic Dominance with Respect to a Function," *Risk Analysis in Agriculture: Research and Educational Developments.* AE-4492. Univ. of Illinois, Urbana-Champaign, Dept. of Agr. Econ., June 1980, pp. 65–81.

————. "An Interval Approach to Measuring Decision Maker Preferences," *American Journal of Agricultural Economics,* Vol. 63, 1981b, pp. 510–20.

Klinefelter, D. A., S. T. Sonka, and C. B. Baker. "Selection and Screening of Marketing Options for Risk Evaluation," *Risk Analysis in Agriculture: Research and Educational Developments.* EA-4492. Univ. of Illinois, Urbana-Champaign, Dept. of Agr. Econ., June 1980, pp. 131–47.

Leuthold, R. M., and P. E. Peterson. "Using the Hog Futures Market Effectively While Hedging," *Journal of American Society of Farm Managers and Rural Appraisers,* Vol. 44, 1980, pp. 6–12.

MacRae, E. C. "An Adaptive Learning Rule for Multiperiod Decision Problems," *Econometrica,* Vol. 5-6, 1975, pp. 893–906.

Markowitz, H. *Portfolio Selection: Efficient Diversifications of Investments.* New York: Wiley and Sons, 1959.

Massé, P. *Optimal Investment Decisions.* Englewood Cliffs, N.J.: Prentice-Hall, Inc., 1962.

McInerney, J. P. "Linear Programming and Game Theory Models," *Journal of Agricultural Economics,* Vol. 20, 1969, pp. 269–78.

Meyer, J. "Choice Among Distributions," *Journal of Economic Theory,* Vol. 14, 1979, pp. 326–36.

Murphy, R. E. *Adaptive Processes in Economic Systems.* New York: Academic Press, 1965.

Paris, Q. "Expectations in Quadratic Programming Models." Invited paper, Western Agricultural Economic Association annual meeting, Lincoln, Nebr., July 1981.

————. "Revenue and Cost Uncertainty, Generalized Mean-Variance, and the Linear Complementarity Problem," *American Journal of Agricultural Economics,* Vol. 61, 1979, pp. 268–75.

Pratt, J. W. "Risk Aversion in the Small and in the Large," *Econometrica,* Vol. 32, 1964, pp. 122–36.

Rae, A. N. "Stochastic Programming, Utility, and Sequential Decision Problems in Farm Management," *American Journal of Agricultural Economics*, Vol. 53, 1971, pp. 448–60.

Roumasset, J. *Rice and Risk*. Amsterdam: North-Holland Publishing Co., 1976.

Zellner, A. *An Introduction to Bayesian Inference in Econometrics*. New York: John Wiley and Sons, Inc., 1971.

Discussion of Part 7

A Comparison of Optimizing Modeling Methodologies

Robert House

In reviewing these 3 papers, I find little to quarrel with, either regarding technical detail or exposition. I view the session as an opportunity to showcase several types of models and I feel that the papers were presented well. What interests me most in considering these papers is how they are related or unrelated to one another and how they are relevant or irrelevant for particular uses and users. I have briefly classified and compared these and related approaches to help organize my thoughts and discussion.

Classification of Programming Techniques

In considering classification criteria, I run into time and dimensionality problems, so I am, therefore, limiting my observations to only several classification criteria: (1) preferences are known or unknown; (2) values in the model are deterministic or stochastic; and (3) optimization is of single or multiple objectives. Time dimensions and a variety of other potential criteria are beyond the scope of my discussion.[1] The selection of these 3 criteria is not meant to be general. They reflect my interests and what I feel are important issues at this time.

The figure is a rough substitute for a 3-dimensional classification of model techniques. In this quick classification, I refer only to the stylized version or generic types of the authors' and others' models; the subtle and unique features will have to be ignored. In the remainder of this section I will discuss the 3 classification criteria and comment on how some of the less familiar techniques fit into the figure.

The number of target variables refers to how many objectives are under consideration. Many models have used the single objective of maximizing income. Other models include additional objectives such as maximizing leisure time. A farm decisionmaker may in reality be

The author is an agricultural economist with the National Economics Division, Economic Research Service, U.S. Dept. of Agriculture.

A Classification of Several Programming Techniques

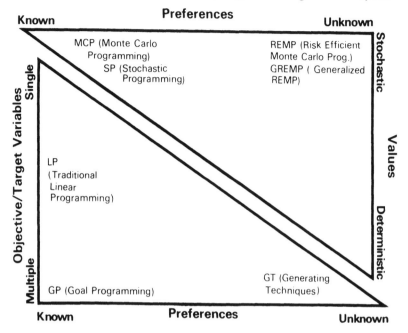

concerned with maximizing income, leisure, crop yields, minimizing soil loss, equipment failure, and so forth.

In most traditional models, all values such as yields, prices, or input availability have been deterministic or known with certainty. Some modeling techniques recognize that several or all of these values are stochastic variables. In such models the values of such variables are selected from a probability distribution. This usually involves simulating the effects of many different values of the stochastic variables in order to generate a distribution of expected values for the objective variable(s).

The state of knowledge about preferences determines the extent to which one can specify a model objective function which will yield a unique solution. Most traditional modeling techniques have assumed that preferences are known sufficiently to uniquely determine a solution. Under certain circumstances, this is a difficult condition to satisfy. For example, when there are many objectives to be maximized, it may be impossible for the decisionmaker to prespecify relative weights of the objectives without first knowing the rates of technical substitution among objectives. Stochastic variables create another situation where it may be impossible to prespecify preferences enough to guarantee a unique model solution. In this case a model may generate a probability distribution of results. An interesting methodology which combines multiple objective decisionmaking, uncertainty, and varying levels of preference specification is presented in Goicoechea.

Goal programming raises the issue of multiple objectives. It also assumes known preferences. Risk-efficient Monte Carlo programming (REMP) and King's generalization, GREMP, involve the generation of a set of strategies and corresponding probability distributions of an objective variable such as an income measure. A partial ordering or screening mechanism is then used to filter out inefficient strategies, leaving only a set of efficient strategies. GREMP assumes partial knowledge of preferences. REMP can assume some or little knowledge of preferences.

Multiple Objective Analytical Techniques

Some multiple objective analytical strategies are analogous to REMP or GREMP in that they generate a feasible set of multiple objective satisfaction states and then employ a partial ordering screen to filter out all but the noninferior subset of strategies. I find this similarity very interesting because each approach starts with the realization of an incomplete knowledge of preferences and then provides the decision-maker with a set of efficient or noninferior strategies from which to choose.

Multiple objective analysis (MOA) did not receive much attention at this conference aside from Mapp and Dobbins' discussion of goal programming. I will, therefore, describe generating techniques (GT), another MOA approach that can be useful in farm level modeling.[2] Generating techniques were developed from ideas presented in 1951 in Kuhn and Tucker's paper on nonlinear programming. My description will be brief; further study of the appropriate MOA references may be useful.

If an ordinary, mathematical programming optimization problem may be described as a scalar optimization problem (for example, maximizing net income), then the vector optimization problem is:

$$MAX \; z(x) = G[z_1(x), \; z_2(x), \; \ldots, \; z_k(x)]$$
$$\text{subject to } x\varepsilon X, \; x \geq 0$$

where the vector $z(x)$ contains the scores or values of the k objectives. Obviously, we cannot "optimize" the vector; this requires some sort of preference structure. But, we can use a partial ordering to get the noninferior set of solution values. To do this with the GT approach, the mathematical programming problem is first specified with one objective as the scalar objective function (as in a standard programming problem). The remaining objectives are included in this model as nonrestricted equations. These equations are sequentially constrained and a standard parametric procedure is run on the right-hand side constraint value. From these k-1 parametric solutions, the noninferior set of solutions traces out the multiple-objective frontier in k-dimensional objective space. This frontier displays absolute objective levels and marginal rates of technical substitution between pairs of objectives. With this noninferior solution set information, decisionmakers can be expected to exercise their preferences and select a strategy more easily. With only a few objectives to deal with, the trade-off functions can be presented graphically—a point that cannot be overemphasized.

Comparison of Programming Techniques

Let me now turn to some comparison of the approaches I hope may prove useful when one considers what type of micromodel to use in different circumstances. The table presents a summary comparison of several programming techniques according to 4 criteria developed in Willis and Perlack: computation expense, quantification of trade-offs, quantity of information, and validity of interaction between decisionmaker and analyst. Computational expense is a straightforward concept. Quantification of trade-offs refers to how well the analysis demonstrates the trade-offs among multiple objectives or goals. Quantity of information is an important criterion to consider because both too little or too much information can be a disadvantage. Validity of interaction is a criterion which summarizes how well the information and analysis provided by the analyst help the decisionmaker to make "good" decisions.

The techniques compared are several of the more important programming approaches available for farm modeling. Goal programming refers to minimizing the weighted sum of deviations from a set of goals. Generating techniques were described above. The Monte Carlo programming (MCP) technique randomly selects strategies and evaluates them with respect to a scalar (though perhaps quite complicated) objective function. With respect to stochastic programming (SP), I mean problems with stochastic constraints or technical coefficients that are tractable with MCP techniques. REMP and GREMP are as described in King and Oamek. I am also assuming that other objective variables are included in the REMP models in an accounting sense and would be provided with each efficient strategy.

The computational expense comparisons are meant to indicate roughly relative relationships. GP, which assumes prespecified fixed objective weights, would require the least computation. Many parametrics (GT)

A comparison of several programming techniques

Item	Goal programming	Generating techniques	Monte Carlo programming (MCP) stochastic programming	Risk-efficient MCP(REMP) generalized REMP
Computational expense	$	$$	$$	$$$
Quantification of tradeoffs	None (without sensitivity analysis which can be difficult with GP—technical problems exist).	Good; noninferior subset of solutions; rate of substitution between goals may be developed and presented.	None (without sensitivity analysis).	Good; dominant subset with respect to risk; rate of substitution between goals is presented.
Quantity of information	Little.	Most; can be too much, giving rise to "information overload."	Little.	Most with respect to risk; can be too much; "information overload" possible.
Validity of decisionmakers (DM)/analyst interaction	Very good if DM commands situation (that is, all choice variables) and is very familiar with the system, and if the system tradeoffs are stable over time. Less so for complicated systems or for situations where a group must make the decisions.	Can show tradeoffs and assist with formation of weights and preferences by the DM.	Key is the specification of preferences regarding risk—if a single-valued measure (for example, net return variance) is adequate then this is acceptable.	Most useful when measure of risk less complicated than the full distribution of outcomes may: 1. Be unable to adequately represent preferences 2. Yield demonstrably inferior solutions (but the stochastic dominance rules may still be fairly rough).

or random-strategy solutions (MCP, SP) may entail moderate computation expense for a single analysis. With the REMP procedures, there is a substantial compounding of computations, since for each random strategy, a univariate probability distribution of outcomes must also be generated.

Little or no quantification of trade-offs is provided by standard goal programming. Furthermore, extensive sensitivity analysis may be required to ensure that the GP solution is actually on the efficiency frontier (Cohon and Marks, p. 213). In the case of GT, the maximum possible trade-off information is made available: the noninferior subset of solutions. A decision may be made based on more or less complete knowledge of objective levels and trade-offs. When MCP or SP are applied to optimize a single variable, trade-off information is not generated. As with GP, some type of sensitivity analysis would be required to provide data on trade-offs among objectives. In the case of REMP and GREMP, objective trade-offs corresponding to the stochastically risk-efficient (noninferior) solution set may be presented. A strategy from the efficient set may be selected according to the decisionmaker's preferences regarding performance on the other objectives. It should be noted that this solution set is efficient only with respect to risk.[3]

Quantity of decision information presented is minimal with GP, MCP, and SP because only one solution is developed in most situations. However, a complete set of decision information is developed in a GT analysis. There is substantial information developed with the REMP techniques, at least with respect to the risk variable. This can be a disadvantage if there is so much efficient solution information that the selection of a single strategy is difficult. When this situation occurs, some type of further screening technique will be required.

The final criterion, validity of decisionmaker/analyst interaction, depends on a variety of factors, such as complexity of the problem, completeness of the decisionmaker's preference structure, quantifiability of the objective variables, and others. GP is generally quite appropriate if the decisionmaker has fairly complete knowledge and control of a stable situation. If the system is changing over time, if the situation is unfamiliar, or if the decision must be a consensus among several persons or groups then an approach such as GT is superior. For the randomized models, if any decision strategy can be expected to generate a reliable, single valued measure of net return (or whatever), then MCP or SP may be the appropriate technique. If, however, there is a quantifiable distribution of returns associated with any strategy, the more sophisticated REMP techniques may be justified. More specifically, we would want to be sure that analysis based on the full distribution of outcomes would result in superior decisions. Beyond this, there is perhaps an inherent superiority, or at least greater realism, in the GT and REMP approaches which admit that a decisionmaker's preference function can only be partially elicited and quantified. This incomplete preference function can be used to screen out the inferior strategies, and thus display the trade-offs more clearly. The decisionmaker can then select the final strategy, perhaps by using an unidentified and more sophisticated preference function.

Notes

1. See Tauxe and others for a useful discussion of multiple objective programming.
2. For a useful recent discussion of 2 MOA techniques, see Willis and Perlack. Much of their discussion is an update and abstraction from an older, more complete paper by Cohon and Marks. A more general review of multiple objective planning techniques is presented in Thampapillai.
3. An extension not discussed here would be to generate a noninferior solution set with respect to all objectives, including risk.

References

Cohon, Jared L., and David H. Marks. "A Review and Evaluation of Multiobjective Programming Techniques," *Water Resources Research*, Vol. 11, No. 2, Apr. 1975.

Goicoechea, A., L. Duckstein, and M. M. Fogel. "Multiple Objectives Under Uncertainty: An Illustration of Protrade," *Water Resources Research*, Vol. 5, No. 2, Apr. 1979.

Kuhn, H. W., and A. W. Tucker. "Nonlinear Programming," in *Proceedings of the Second Berkeley Symposium on Mathematical Statistics and Probability* (ed. J. Neyman). Berkeley, Calif.: Univ. of California Press, 1951, pp. 481–92.

Tauxe, G. W., R. R. Inman, and D. M. Mades. "Multiple Objective Dynamic Programming: A Classical Problem Redressed," *Water Resources Research*, Vol. 15, No. 6, Dec. 1979.

Thampapillai, D. J., "Methods of Multiple Objective Planning: A Review," *World Agricultural Economics and Rural Sociology Abstracts*, Vol. 20, No. 12, Dec. 1978.

Willis, C. E., and R. D. Perlack. "A Comparison of Generating Techniques and Goal Programming for Public Investment, Multiple Objective Decision Making," *American Journal of Agricultural Economics*, Vol. 62, No. 1, Feb. 1980.

Discussion of Part 7

Optimizing Models:
Some Observations and Opinions
on the State of Progress

Garnett L. Bradford

The comments of this paper are, by design, primarily philosophical and general; and thus, they do not deal solely with the 3 papers of this session. I cite the papers by King and Oamek and by Dobbins and Mapp in several instances to illustrate specific opinions, premises, or conclusions.

Model Efficacy

> He had been eight years upon a project for extracting sunbeams out of cucumbers, which were to be put into vials hermetically sealed, and let out to warm the air in raw inclement summers.
>
> *Jonathan Swift, 1714*

The most common criticism leveled at any sort of research, especially at economic research in recent years, is its relevance. But, often what is relevant, or pertinent, information for some users of research results is not for others. Hence, I believe that criteria for evaluating models should focus on a combination of other features, particularly what some observers call model efficacy (see the figure). I do not intend to deny the critical importance of conducting relevant research. Rather the criteria used in this paradigm emphasize that a commonly accepted measure of relevance is not definitive.

Model efficacy is defined here as a relative measure of model performance combined with its theoretical detail and degree of simplicity. Theoretical detail is defined as the dual of simplicity. For each pre-determined level of theoretical detail (and simplicity), an optimization model is more efficient as its performance is improved.

Logical validity is the set of theoretical conclusions following de-ductively from a set of premises and is a necessary condition for the

The author is a professor of agricultural economics, University of Kentucky, Lexington.

Criteria for Evaluating Decision Making Models

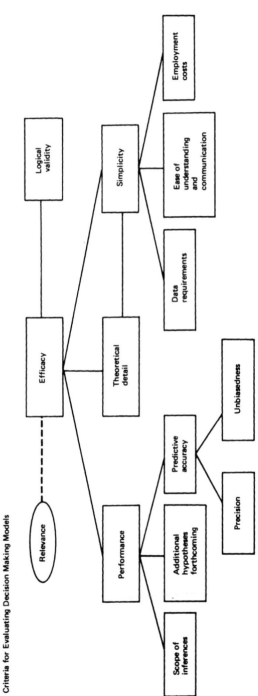

acceptance or use of any model. Researchers spend considerable time meeting this requirement by devoting much of their manuscripts to delineating and discussing a model's logic. I don't believe we should significantly delimit this time and effort, but we should recognize it for what it is—only a necessary beginning.

Theoretical detail (sophistication) is perhaps the most emphasized of all evaluation criteria. It is closely related to, and frequently confused with, performance. For example, King and Oamek's review of risk programming models employs "three important criteria": (1) suitability of the choice criterion, (2) flexibility in representation of complex stochastic processes, and (3) ability to evaluate adaptive as well as fixed strategies. These criteria, in my opinion, deal almost exclusively with theoretical detail. Notwithstanding their language, "each of the risk programming models is well suited for the analysis of some decision problems but none can be said to be generally applicable in a wide range of practical contexts. . . ." I believe they are in effect advocating more theoretical detail in the belief that the resulting performance of the model will be enhanced.

Economic and statistical theory is often set forth as some sort of ideal toward which models should strive. But theoretical sophistication in no way guarantees better performance! The study by Taylor and Chavas, cited by King and Oamek, employs Monte Carlo experiments and argues cogently for additional use of this methodology. A simpler "open-loop feedback strategy" consistently gave estimates (predictions) which were at least as accurate as the more sophisticated models. King and Lybecker's paper on adaptive decision models (cited by King and Oamek in this conference) illustrates a case where a fixed-control strategy of marketing pinto beans, that is, a fixed, predetermined percentage of the year's production sold in the months following harvest, would result in higher farm profits than would a linear (more sophisticated) control strategy. Dobbins and Mapp suggest the theoretical superiority of multiple goal programming models compared with a model with a single-objective function, but they present little evidence regarding performance relative to other simpler models. What level of performance (particularly predictive accuracy) would have ensued if they had employed a standard one-period LP model that incorporated the 5 subordinate goals as constraints within their recursive simulation structure?

Simplicity is often inversely related to theoretical detail, although the trade-offs are seldom easy to identify in a continuous, direct, linear relationship. Most of the optimization research reported or presented at this conference does recognize the value of simplicity. King and Oamek argue for simplicity at several points, for example: "Once set up, the GREMP model may be more costly to run. . . . In order to simplify the analysis . . . , only 3 of the 10 possible insurance options were considered." Still, when choosing among models most researchers rarely weight simplicity as heavily as theoretical detail. Many researchers apparently believe that added theoretical detail somehow improves model performance. Obviously, this is a debatable maxim.

Perhaps researchers should strive for more explicit determination of an optimal ratio between simplicity and theoretical detail. This ratio might, by necessity, remain subjective, but would depend on the type of research goals in each project (and thereby, project funding). Simplicity is the essence of the standard argument for tailoring a model to the individual research project. In contrast, we often hear the argument for a generalized model, applicable to a wide assortment of policy issues. King and Oamek contend: "A more general approach is needed—one which permits greater flexibility with respect to problem formulation, the determination of probability distributions, and the representation of decisionmaker preferences." Dobbins and Mapp, though not explicitly arguing for either a generalized or a special-purpose approach, seem to lean toward special-purpose models.

Performance of Decision Models in Policy Situations

If researchers are to develop usable optimization models for purposes of policy applications, then performance must be evaluated, improved, and reevaluated. First, the scope of inferences that are possible must be extended well beyond their current status, whereas most models apply to only illustrative, case-study farm situations. This deficiency suggests a need to carefully apply selected, different models to a cross-section of farm situations. Certainly, not all farm situations need to be modeled. Rather, the logic of stratified sampling theory must be applied to fairly small (and perhaps roughly defined) samples of farm firms, as researchers identify population attributes.

Finally, micromodelers must come to grips with the realization that decisionmaking models, with the usual ties to normative economics, cannot be independent of the need for predictive accuracy and positive economics. As Friedman noted in 1953: "Any policy conclusion necessarily rests on a prediction about the consequences of doing one thing rather than another." The papers of this session and several other papers presented at this conference underscore this realization. Theoretical tools of operations research and those of statistical inference (and econometric models) must be blended to develop efficient models for meaningful policy analyses.

Some General Impressions

The literature in farm business and farm financial management exhibits an impressive number of models describing a varied assortment of decisionmaking processes. However, in terms of quality, or what could be called efficacy, these models indicate that our efforts remain in an elementary stage of analysis. The following outlined comments are intended to summarize my impressions regarding the general state of progress in this area:

1. Most of the models adequately handle only single-firm situations. They are purposefully decisionmaking models with an individual farm firm serving as the decisionmaking unit.
2. Few models deal with the knotty problems of capital investment and replacement in a very comprehensive manner. By design, most concentrate on handling short-run decisions primarily in a partial equilibrium. Versions that do account for several production periods (that is, "one-pass" MLP models or recursive multiperiod models) frequently attempt to accommodate capital decisions, but their emphasis is still usually short-run.
3. Because most models are computer-oriented, they are synthetic versions of researchers' perceptions and imaginations. Experimental findings and actual conditions often comprise only a small part of a model's genesis. Furthermore, such models are often not rooted in any specific body of economic theory but are constructed from some type of reasoned experience or empirical ideal.
4. Although many models are carefully developed with what seems to be considerable parallel to empirical states, their thorough and comprehensive validation is usually lacking. We continue to pay homage to the "real world" almost as if it exists light years away in some other galaxy.
5. After models are developed and partially validated, they are rarely applied beyond a few farm firms for a limited timespan. Accordingly, we should not be surprised when extension economists develop and use their own much simpler and heuristic versions of the same decision alternatives.
6. Thus, consistent with the above thoughts, we do not (and should not) seriously rely on our models to provide predictions regarding the aggregate effects of changes in government policies and regulations. We may employ the model to provide some qualitative

analysis of what might happen, but usually we blend common-sense experience with static economic theory.

References

Friedman, Milton. *Essays in Positive Economics.* Chicago and London: University of Chicago Press, 1953.

King, Robert P., and Donald W. Lybecker. "Adaptive Decision Models." Colorado State Univ., unpublished manuscript, Oct. 1981.

Taylor, C. R., and J. P. Chavas. "Estimation and Control of an Uncertain Production Process," *American Journal of Agricultural Economics*, Vol. 62, Nov. 1980, pp. 675–80.

Discussion of Part 7

Optimizing Farm Models: Some Concerns

Verner G. Hurt

The 3 papers presented in this session represent different approaches to calculating or finding a set of activities for a farm firm. Behavioral characteristics or optimizing criteria for the decisionmakers were different even though based on a combination of utility theory, the satisficing assumption,[1] or statistical decision theory. Algorithms used to calculate the so-called optimal set or sets of activities were simulation, traditional linear programming, modified linear programming, and quadratic programming. Each author advanced substantial, logical, convincing arguments on the merits of his procedures for modeling the decisionmaking processes of a farmer. An interesting exercise might be to apply their methods to the farm firm.

There are other similarities between the models, particularly in data requirements. Technical coefficients, prices, and yields must be provided for each of the models. Prices for the King and Oamek model were estimated utilizing "a simple linear expectations model" plus "a normal random variable generated from the distribution of the expectations model's disturbance term." Their estimated distribution of yields was based on detrended levels for a historical period. They did not state whether historical yields were for an individual farm or for farms in the aggregate.

Price expectations were estimated from a set of linear equations by Dobbins and Mapp. An exponential smoothing function with heavy weighting of recent years was used to estimate expected yields.

Other similarities and differences among the 3 models could be enumerated and are important in any evaluation of their usefulness. However, one should consider several broader issues in determining the potential application of these models in solving problems. Presumably, it is my function as a discussant to raise these issues, but I'm not to resolve them.

The author is head of the Department of Agricultural Economics at Mississippi State University.

The first issue concerns the intended users of the models. The authors, other than Musser and others, suggest that models are useful tools for research and for policy analysis, but identify neither the hypotheses tested in the example they present nor the kinds of hypotheses that might be tested when used by other researchers. Neither did they communicate to me how the models would be applied to policy issues, except that King and Oamek assert that the forthcoming study of Colorado dryland wheat farmers "should provide valuable insights concerning the way farmers adjust to policy changes." In my judgment, there is still a wide gap between the researchers' perception of the usefulness of models in resolving policy issues and their actual use by the individuals who draft and pass legislation or otherwise establish policy. If you look at the question normatively, then we could probably agree that the results should be useful to policymakers. But in my judgment, there is an exceedingly low probability that they will be used. A tremendous communication gap exists between the economist modelers and the policymakers.

Even so, if we presume that these models are intended for use by researchers in providing knowledge that can be useful to those who establish farm policies, then we must ask whether or not their results are representative of effects on all farmers, a specified class of farmers, or even any individual farmer.

In the papers presented, the models were applied to a specific farm resource situation with the decision by the farmer and the results determined by ex ante prices, yields, technical relationships, and behavioral parameters not necessarily representative of those applicable to any given individual farm situation. At best, the results would appear to be a plan. No verification ex post facto of the model was presented, although such future application was proposed by King and Oamek. Given these and other limitations identified by the authors, how can policymakers utilize the results generated?

Another question is: "How are the models to be used by farmers in managing their operations?" Obviously, they can be used to generate alternate sets of activities for the decisionmaker to consider. But, how many farmers can satisfy the insatiable data needs of the model, afford the expense of the calculations, or have the expertise to utilize it? Can these data needs, computing facilities, and technical expertise be provided, and at what cost? Are the plans generated sufficiently better than those developed by less expensive, less sophisticated processes to be justifiable? How is the plan to be revised to incorporate information (that is, updated within the production period(s)) to provide for adaptive strategies?

Researchers and research administrators should also consider which of the following strategies is preferable: to continue devoting resources to the development and refinement of essentially static planning models or to shift these resources to developing models that can be used by farmers in the day-to-day management of their businesses.

If models are developed whose results are used by many farmers in their decisionmaking and management processes (that is, they become conditioned to basing their decisions on the model's results), then won't these same models, appropriately modified, be superior research and policy analysis tools? Certainly, following the latter strategy will result in the establishment of a known client base.

Presentation of the above comments by a not-so-recent practitioner of the fine art of model building may be interpreted as undue criticism of past and current practitioners, in general, and of the authors of the papers in this section, in particular. I would like to assure these authors this is not my intention. Nor did I intend to suggest that model building is not a highly professional, extremely valuable endeavor. As professionals, we must continue our efforts in developing complex research tools. However, once these tools are developed, we must also accept the responsibility for verifying their results and for specifying who the intended users are and what kinds of problems the models can solve.

Notes

1. For a discussion, see Herbert A. Simon, "A Behavioral Model of Rational Choice," *Quantity Journal of Economics*, Vol. 69, 1955, pp. 99–118. Also see Herbert A. Simon, *Models of Man: Social and Rational* (New York: John Wiley & Sons, Inc., 1957).

PART 8

Use of Farm Models in Extension Activities

16. Specification and Use of Farm Growth and Planning Models in Farm Management Extension Activities

Ted R. Nelson

Abstract Since the advent of computers 25 years ago, it has been possible to use formal models in assisting farmers with periodic management decisions. The experiences of the past 15 years had laid a foundation for development of practical computerized decision models. The recent availability of desktop computers provides the opportunity to employ such models on the farm and to more appropriately relate national policy decisions to decisions made at the individual firm level. However, the effective use of optimizing models in providing farm management advice requires input-output and other data relevant to the particular farm situation and reliable estimates of variables exogenous to the particular farm situation.

Keywords Farm management, microeconomic models, small computers

Farm management extension programs advise farmers in making wise management decisions about their resources. Growth and planning models have a definite place in these types of extension programs. Both optimization and simulation type models have been used. Some models that have been tried include the whole-farm programming approach encompassing crops, pasture, and livestock. Some models have also focused on machinery selection and planting and harvest dates. Sim-

The author is an extension economist in farm management and a professor of agricultural economics at Oklahoma State University.

ulation systems such as the Kansas Transitional Planning model have been available for remote processing on telephone terminals. The use of these models has met with some success. However, problems have been encountered with all of them, and their adoption and use by farm management extension people have been limited.

Two of the major problems encountered with the models used to date are: (1) the difficulty of incorporating individual farm data (which were in most cases not in a sufficiently accurate form) and (2) a slow response time between the farm manager's submitting the problem and obtaining the results. We are still some distance from solving these problems. But, better tools are now available for overcoming them. It will take time to develop appropriate data gathering, sorting, and editing software to transform data from the form in which farmers are prepared to make it available into the form required for input to models. Model research and development will need to focus on availability of data as well as on relationships included in the data and model output.

Farmer Needs

Farmers are first and foremost concerned with records that will satisfy the immediate operating demands imposed by the environment. Income tax, payroll, accounts payable and receivable, and related financial statements required by the tax official, the banker, and the estate planner are mentioned by nearly every farmer who calls to ask whether an on-farm computer can be helpful. A minority are equally concerned about production records for commercial livestock and crops and genetic records for purebred herds. Mailing operations are high on the list for the limited few who have specialized enterprises such as certified seed and purebred livestock.

Simulation results are the second most popular service desired by farmers. Services requested are those requiring models that use sketchy data and are at best elementary. Crop and livestock budgets projecting net returns from "top of the head" estimates of cost items, livestock growth models evaluating various rations, crop growth models evaluating production practices, and models evaluating least-cost combinations of inputs are well accepted. Comprehensive planning programs are attractive to only a few of the most sophisticated farmers who have academic training in a closely related area and to a few who are temporarily infatuated by reading accounts that gloss over the data requirements of an effective model.

Microcomputers are relevant to each of these 3 needs. They are attractive to farmers because once the problem is in mind, they want the solution immediately. When basic data are organized and needed programs are on line, micros provide the response in a few minutes if at home or in an hour or two if at a service bureau in the local town. Cost of running successive problems is also low. Results can be acquired within a matter of minutes or a few hours rather than days, and rerun solutions can be obtained with small operating costs after the gremlins in the data have been located.

The appearance of desktop computers and the existence of comprehensive models that work will not bring these models into general use by farmers. Training opportunities are needed. Learning to use the hardware is time consuming. Although it is not terribly difficult, most laypeople will not bother to learn the rudiments without personal instruction. Classes in high schools and community colleges in the winter or at night will have to fill this need; extension services have neither the personnel nor the equipment to do it.

An introduction to the purposes, objectives, and data requirements of economic models is needed for their successful use. The agricultural research and extension complex has a special responsibility in designing approaches to record keeping, data collection, and farm planning models. Further, their work in introducing farmers to the purposes and objectives of models and the related data requirements can complement personal instruction activities.

Model Construction, Data Requirements, and Computer Operations

The primary management objective of farmers is to optimize use of their resources. Consequently, models that indicate how farmers can use their resources to increase their income and wealth are especially attractive in farm management extension activities. Effective application of such models requires data particular to the farmer. However, input-output data for similar farming situations can be substituted on occasion. Goals can be specific to the particular farmer. Alternatively, models can be solved for alternative goals providing the farmer with perspectives that may facilitate deciding among alternative goals and required resource and behavioral commitments given expected environmental constraints. Suppose that an optimizing model produces a plan requiring a different set of machine sizes or types than is presently on the farm. The expansion path from the current farm organization to the optimal needs mapping and cash-flow feasibility for the disinvestment and reinvestment of this capital would then need to be developed. This growth problem illustrates a set of implementation problems often encountered with planning models.

The application of micromodels to this type of problem requires considerable information about variables that the individual farmers will, in most cases, not be able to significantly influence. Input and factor availabilities, nominal prices, and nominal commodity prices are examples. Effective use of micromodeling in supporting wise farm management decisions is heavily dependent on the optimizing features of the models and also on reliability of anticipated values of variables exogenous to the individual farm operator.

The first hurdle is to create an efficient set of input-output data adequate from a computational standpoint and worth the cost and effort of accumulating it. With present technology, motivation of farmers to use computers is of far more cause for concern than the costs of data storage, intermediate transformation, formatting, scaling, and so forth.

Decisions about the amount of data directly related to an individual farm situation must reflect a balance between: (1) The natural desire of farmers to be able to solve problems with as little recordkeeping and keyboard entry as possible, (2) Demands of the model for detailed individual farm data to obtain useful results, and (3) Availability of reliable data for other farm situations similar to the one being studied. Disk storage and distribution and telephone-accessible data bases enable the downloading of county or state data to the local processor, facilitating the use of data developed from experiments and farm records systems as well as centrally produced estimates of exogenous variables. Multiple use of data files (compatible records) are inserted in this process. Farmers will not tolerate repetitive entry of information. Results from farm records, depreciation schedules, budgets, simulations, operations programs, and projected cash flows must be coordinated.

On-farm or local collection of input-output coefficients that will be used by farmers, spouses, and their employees requires that model programs be structured for easy operation. Concise and complete documentation of models is difficult and developed only with careful thought and lots of time for both design and actual programming.

Data entry and revision is moving into the age of keyboard entry where data preparation is to be shared by the layperson at the keyboard and the professional who prepares the programs to be used by that layperson. The style used in Microsoft's VISICALC illustrates the use of forms for the data input displayed on a computer user's screen (Table 16.1) The operator using the arrows on the keyboard moves the cursor on the screen to the value to be changed and changes the particular element. Data from this type of screen can then be converted to machine, labor, and fuel requirements and stored on a disk file for later inclusion in the appropriate reports. Such data can also be retrieved for later use in an analytical planning model.

TABLE 16.1
Input screen with illustrative numbers

Operation							Times over the field					
	Jan.	Feb.	Mar.	Apr.	May	June	July	Aug.	Sept.	Oct.	Nov.	Dec.
Plow												
Disk												
Harrow					1							
Planter					1							
Cultivate						1	2					
Irrigation inches							3.5	10.6	5.3			
Harvest										.25	0.75	

Different programs for data files for various-sized farms may be required to obtain useful results at the farm level. The coordination of data from these programs will require a good deal of attention. Conceptual and operational problems include the following.

1. Classification of cohort groups by size and type of farms:
   a. Use of land classifications to separate farms into manageable groups while preserving yield and input validity, and
   b. Definition of risk acceptance levels and number of farms in each of these groups;
2. Data transfer from collection points to storage and retrieval locations:
   a. Compatible disk formats on various brands of machines,
   b. Compatible transmissions protocols for telephone transmission, and
   c. Recognizably similar analyses at farm and policy levels.

Impacts of Policies and Institutions on Farmers

The use of models that address policy and institutional changes poses some challenging conditions. Farmers want to be able to assess the effects of proposed changes on their particular level of operation. The requirements for application, therefore, change rapidly during formulation of new policy.

Models useful in guiding farm management decisions should ideally be structured so that the effects of proposed farm income and supply-management programs can be assessed within a short time. Farmers could then base their position of being "fer em or agin' em" on a better estimate of the expected effect of the program change on their own peculiar situation. Devising models capable of handling a majority of the production rules that will be actually implemented will be a real challenge. Most of us find it difficult to be sure what those rules are several months after legislation has passed and USDA has wrestled with translating the law into operational rules for field application.

There would be a great advantage to having models that handle these parameters as well as provide reliable data for such variables. However, an objective of flexibility and ease of rapid adjustment of models to accommodate considerations of changing parameters would be helpful.

Conclusions

The complexity of developing and then effectively integrating input-output and other data for individual situations as well as providing reliable estimates of variables exogenous to a particular farm situation seems overwhelming. Yet much more complex systems requiring more coordination have been developed when sufficient resources have been committed. Whether the agricultural research and extension complex can meet this challenge is an open question.

References

Bogle, T. Roy. "Computerized Whole Farm Planning—A Budgeting Process–Transition Planning for Up to Five Years." Kansas State Univ., Dept. of Economics, Cooperative Extension Service, 1979.

Brant, William L., and Ted R. Nelson. "User's Guide to LP-FARM." A.E. 7205. Oklahoma State Univ., Dept. of Agricultural Economics, 1972.

Hardin, Mike L., and Ted R. Nelson. "COSTFINDER User's Manual." Oklahoma State Univ., Cooperative Extension Service, 1976.

Hardin, Mike L. "A Simulation Model for Analyzing Farm Capital Investment Alternatives." Ph.D. dissertation, Oklahoma State Univ., 1978.

Kletke, Darrel D. "User's Manual: Oklahoma State University Budget Generator." Research Report P-661. Oklahoma State Univ., 1972.

———. "Operation of the Enterprise Budget Generator." Research Report P-790. Oklahoma State Univ., 1979.

Nelson, Ted R. "Capital Optima on Pump Irrigated Farms." M.S. thesis, Univ. of Nebraska, 1959.

———. "User's Instructions: Farm Management Microcomputer Programs." AGEC 420. Oklahoma State Univ., Dept. of Agricultural Economics, 1980.

———. "Now That You Have Decided to Buy a Microcomputer—What Next?" AECO 4.204.0. Oklahoma State Univ., Dept. of Agricultural Economics, 1981.

Walker, Rodney L., and Darrel D. Kletke. "User's Manual for the Oklahoma State University Crop Budget Generator." Progress Report P-656. Nov. 1971, revised 1972.

Appendix—Outline of a Data and Program System for Planning and Growth

Storage File Codes:

IV = Inventory record basically used for income tax reports, but used as the basis for building the resource situation file

RS = Resource situation file built from the inventory file and used as input for planning and decisionmaking

FR = Farm record file composed of income and expense transactions identified by enterprise as entered weekly, but subject to modification during transfer to the activity budget file

AB = Activity budget file containing information from source files used in planning and decisionmaking programs

Other Codes:

KB = Keyboard or original entry

PO = Printout (not stored)

Components of Data and Program Systems:

	Source	File
1. Basic farm data		
a. Land area, slope, soil type, productivity index	IV	RS
b. Machinery on farm, cost, size, speed, and so forth	IV	RS
c. Annual inputs for production activities (enterprises)	FR	AB
d. Monthly distribution of income and expenses	FR	AB
e. Monthly distribution of labor requirements by activity	LB	AB
2. Programs		
a. Ledger, income statement, inventory update, payroll	KB	FR
b. Budget builder	FR	AB
c. Depreciation-tax return	FR	PO
d. Optimizing farm planner-simulator	IV	PO
e. Financial feasibility and annual cash flow	FP,AB	PO
f. Monthly cash flow and credit planner	AB	PO

17. Specification and Use of Farm Growth and Planning Models in Policy Extension Activities

Otto C. Doering, III

Abstract The farm firm has often not been a critical component of policy issues covered by policy extension over the last 30 years. Even when it was, analytical information appropriate to the educational process has not always been available. Once the "teachable moment" for policy extension is realized, time is limited for quantitative and qualitative analysis. Thus, micromodeling research activities need to anticipate the policy agenda. Micromodeling can improve understanding of the firm-level impacts of macro policy. Additionally, micro- and macromodels can show how aggregate behavior may yield results different from those expected by individuals at the micro-decision level.

Keywords Extension education, farm policy, micromodels

Policy extension education has been a regular component of agricultural extension only since the late 1940s. At its inception, the primary concern was the agricultural sector, and in some cases the farm firm, under the impact of changing government policies. In those cases where the farm firm was a focus of attention, a definite distinction was made between farm management extension activities and policy education activities that became part of the ethos of policy extension. Over time, the subject matter concerns of policy education in extension have broadened, with the farm firm becoming a less explicit reference point. In part this reflects the broadening mission and clientele of USDA and the land grant institutions, and in part it is a reflection of the broadening definition of farm firm concerns.

The author is a professor of agricultural economics at Purdue University, Lafayette, Ind.

In the years since the 1940s, a variety of modeling activities with a policy label have been undertaken. It is not clear, however, that there was successful interaction between those doing the modeling and those involved in policy extension. At times, those undertaking the analytical modeling little understood what the policy education activity was all about. On occasion policy educators may have had limited understanding of the problems being tackled. This reduced the effective utilization of the analytical products of modeling efforts for policy education.

This paper focuses on broad agricultural policy and the extent to which it has a farm focus—leaving aside the other focus points of extension like rural development. It is also limited to the policy education process, leaving aside other roles extension educators play, such as advising policymakers. The paper is divided into two parts, a review of policy education in the past and prospects for the 1980s. The first section also describes public policy education as it has evolved over the last 30 years. Both parts give attention to the analytical base available to support public policy education with special attention to microinformation and models.

Policy Education in the 1950s and 1960s

During this period the lore and the methodology of policy education was developed. In brief, the goal of such policy education was to develop in individuals:

1. An active interest in policy problems;
2. An understanding of the facts, issues, alternatives, principles, and values;
3. The ability to make judgments on the basis of critical examination of the evidence and logical thinking; and
4. The desire and ability to participate effectively in the solution of public problems (Dunbar).

The methodology developed for such education was a problem-alternatives-consequences approach that involved:

1. Identifying the problem and its causes,
2. Setting forth the issues in a positive manner,
3. Presenting all known proposed alternative solutions and gaining an understanding of how each alternative would work if put into effect,
4. Analyzing the consequences of following each alternative in terms of criteria important to the audience, and

5. Leaving to the audience the appraisal of the various alternative solutions in the light of its own values and goals.

Some other basic notions have been important in the development and practice of policy education in extension. One of these is that, by definition, policy involves group decisions. Individual firm decisions relating primarily to that firm were not part of this arena. An example would be the Soil Bank legislation and its subsequent implementation. Prior to the passage of the May 1956 act, information was made available to farm people on a policy education basis to enable them to form informed judgments about the wisdom of the act. Once the act was passed, the primary extension activity was to assist individual farm operators in deciding how to apply the provisions of the program to their own farms. This was a farm management activity (Penn, Atkinson, and others). The distinction between a policy education activity and a farm management or other educational activity is seldom clearly defined, but it can be critical in ascertaining the usefulness of firm-oriented analysis in policy education.

It is also important to distinguish between (1) policy problems that surround a new program such as the initiation of a soil bank, and (2) policy problems that involve relatively small adjustments in mechanism—such as support levels—from year to year. Extension policy education tends to focus on the former and not the latter. The distinction between these two types of policy-related problems illustrates an issue for modelers and those who allocate resources for their activities. How broadly based and flexible should macro- or micro-models be in order to cover the analytical needs of policy questions?

Policy education in the 1950s tended to focus on macro issues. They involved very real conflicts of values held by the public. Publications like *Turning the Searchlight on Farm Policy* (Nourse and others) posed basic questions about the future direction of farm policy and the relationship between the farm and the general economy. The analysis was more conceptual than empirical, and this tended to be the style of policy educators when dealing with such issues.

Once some general direction was given to farm policy a period of implementation followed, but basic issues like the extent of public involvement in the farm firm surfaced again along with other issues of content, interpretation, and form of implementation. The wheat referendum of 1963 is an example. Some USDA officials believed or hoped that the policy issue had been decided and that the referendum was merely an implementation step. Many farmers and extension educators believed that important policy choices were being made. This event raised and ultimately settled a number of questions about the role of policy educators at the land-grant institutions as compared with the role of USDA administrators of commodity programs.

Near the end of the 1960s much farm policy involved tinkering with policies and programs on which there was basic agreement. This is certainly true in terms of the farm clientele for extension education for whom the setting of payment levels might be the critical issue. Basic policy issues also were likely to be raised by urban or other interests concerned with the incidence of costs and benefits for commodity programs as well as the absolute levels of costs and benefits. It is on this basis that basic agricultural policies, thought by some to be settled, were questioned by nonfarm groups.

The Analytical Component

Analytical responses to farm policy and farm firm questions of the 1950s and 1960s were both macro and micro. On the macro side, studies were made to estimate the costs of various farm program components (Bottum, Dunbar, and others). Modeling efforts within universities and USDA variously estimated aggregate responses to different farm programs, assessing their program costs and demand and supply impacts (Tweeten; Heady and Mayer; Whittlesey and Heady; and Brandow).

Such analysis was invaluable insofar as it allowed some assessment of the magnitude and incidence of program costs and benefits under different program alternatives. There were exceptions, but much of the analysis was increasingly based on existing program structure with less interest in a broader range of alternative solutions, and it became more useful to administrators than to policy educators. To some extent this reflected limitations of scope and breadth in the existing technology of analysis. Another technical limitation was the difficulty of capturing the dynamic effects of farm programs over time. This limited the usefulness of the models—especially the linear programming models—to shortrun consequences, if the results were to be credible.

Much of the related microanalysis focused on farm management choices by individual firms among different farm program alternatives. As linear programming techniques improved, optimizing models were developed to examine the impact on firms of a wide variety of program alternatives. This served not only farm firms, but also provided clues to policy educators about likely aggregate response.

The style and content of policy extension education reflected the availability and especially the validity of micro- and macro-analytical techniques. When the basic issues of farm policy were discussed in the early 1950s, few powerful and validated analytical tools were available to answer the questions about the economic consequences of various alternatives. The analytical tools that were developed in the late 1950s and the 1960s were limited technically and thus used to answer static questions that often related as much or more to implementation as to policy formulation. When policy educators were asked to address broad-ranging policy questions beyond the scope of existing policies and models, they based their responses on judgments from basic economic principles, as they had done in the 1950s. Existing models could only fill in some pieces within a broader comparative analytical framework.

The 1970s and the Changing Analytical Base

World food considerations dominated much of the agricultural policy in the 1970s with the farm firm of only secondary concern. Insofar as expanded exports supported commodity prices, farm income took care of itself and there was less concern about the general public costs of traditional farm programs and the income transfer to the farm sector.

Policy education programs in extension tended to look beyond the farm as well. The national "Who Will Control Agriculture" and "Your Food" educational programs brought attention to the whole food chain. This approach was partially in response to an expanded audience outside the food and fiber sector that was interested in broader food issues. It also responded to anxieties of some farm people about changes in the ways that farm resources were being owned, organized, and managed.

The impact of nonfarm programs on the farm firm and the food and fiber sector became a major area of interest. Environment, work safety, and resource use and availability became focal points for both farm firm and broader policy analysis. New analytical tools (or the revamping of old ones) and new data sources were often required. Farm adjustment models entered their cocoons and emerged reformed as water assessment models fundable by the Environmental Protection Agency or energy models fundable by the Department of Energy (Meister, Heady and others; and Dvoskin and Heady). A wider range of techniques was applied to the farm firm-level analysis in order to capture interactions among production systems, resource use, and the environment (Doering; Muller, Peart, and others). In analytical terms there was much more concern with trying to capture firm-production system interactions and firm-sector interactions.

In the late 1970s, interest in the farm firm was renewed. Concern with the total food system came full circle. Attention was given to the structure of U.S. farming and to the type and kind of farm firms in the sector, how they were changing, and how they were being affected by developments in the overall food and fiber system and in the general economy as a result of government policies (Bergland).

This has stimulated a variety of new approaches. Existing models are being reworked to address the farm firm in this new context (Drabenstott and Heady). Farm records are being used differently and in a most useful way to give an array of the impact of different factors on different kinds of farm firms with an eye to things like the growth or survival dynamics of these firms (Jensen, Hatch, and Harrington). In addition, simulation techniques, which have not been widely applied to farm firm analysis, are being used to represent the dynamics of firm behavior and national policy interaction (Richardson and Nixon). The nature of these approaches reflects an increasing awareness of the limitations of static analysis.

The Timing Dilemma of Policy Education in the 1980s

Some basic characteristics of policy education are critical to the way it is carried out and to the impact it might have. Policy education seldom has a direct short-term impact on the profitability of a farm. Policy education, by design (because it deals with values such as equity and freedom of choice), does not provide technically correct answers or prescriptions that can be followed by farmers as can fertilizer recommendations from agronomists. Little, if any, demand exists for public policy education on the basis of personal short-term self interest.

As a result, educators in policy extension have come to depend upon using the "teachable moment"—the moment when educational clientele, for selfish or other reasons, are interested in a policy issue but have not yet taken a firm stand on it. Policy education can be both difficult and frustrating if attempted before interest develops on an issue or after people's opinions have become fixed. The teachable moment is usually too short in time span to allow the planning, development, and delivery of an educational program within its duration. A policy educator must often risk planning ahead for such a moment or depend on existing broad-based expertise. This short time frame has important implications for research, including micromodeling, which generates policy-relevant information.

One either takes a risk in specifying the content and form of policy analysis in advance of the emergence of a policy issue or depends on a general research program to have the information available on the issue when required. The same kinds of constraints and risks exist with respect to staff analysis for policymakers. The problem of meshing critical inputs and having them sufficiently developed at the right moment is a major concern in evaluating the usefulness of analytical tools for policy extension education, just as it is for providing information to policy makers.

Policy education cannot always be limited to topics at their teachable moment. At times some aspect of conditions is sufficiently important that the policy educator must address it even where client interest has not yet been stimulated or even where the ideal moment has passed and opinions have already been formed. In such cases the policy educator must be prepared, in the one instance, to demonstrate why people should be interested in the topic, and in the other instance, to attempt education even though many clients have formed a definite opinion about preferred policies for such conditions.

Policy Education and Analytical Resources

As our focus here is on micromodels, the first concern is the extent to which broad policy issues in the 1980s will involve farms. The second question is the extent to which the financial or other conditions of the farm will have an impact on these issues; that is, will the status of farms matter to anyone other than those associated with them?

I would expect a number of broad basic issues to affect farms in the 1980s. Resource use questions, such as the availability and pricing of water, energy, and land, will be of great public interest and of critical importance to the operation of farms. It is more difficult to speculate whether or not the status of the farm itself will be a policy issue. My guess is that this will not be so in the first half of the 1980s as the definition of appropriate public involvement narrows. It is likely to be a matter of public interest in the last half of the decade as further resource constraints make the public impacts of private acts more apparent. Then we can expect interest in public activity to influence industry structure and equity among firms.

Even if they have clairvoyance, to what extent can public policy educators skew the research investment decisions of micro- or macro-economic analysts? In the past it has often been a matter of circumstance when a thorough analysis of alternative courses of action has been available at the teachable moment for a given issue. In many cases, policy educators have continued to practice the art of most political economists in adapting what information was already available and then subjecting it to some economic logic and simple applied analysis. This analysis has not been bad from a communications standpoint, because many policy educators have made their own basic economic logic more understandable to the general public than an equally lucid interpretation of someone else's complex quantitative analysis. Although the teaching function may have been better served, it may have been purchased at the cost of improved accuracy and perspective had more thorough analysis been available.

We are unlikely to get either farm firm or macroanalysis that is easily used for policy education without some give and take between policy educators and analysts. The timing demands for research results as well as basic issue and problem definitions can be different for the policy educator as compared with the analyst. The general questions of public choice which might be addressed by a policy educator may well be addressed in a similar way by a macroanalyst. However, the approach of microanalysts may be less compatible with the needs of policy educators. Microanalysts do not often address firm-level questions in a way that provides the most relevant information needed for an analysis by the policy educator.

This raises a number of questions. How closely should the policy educator and the firm analyst work together? Should micro-level analysis be specified from the start to answer the policy questions believed to be of future importance to the policy educator? The first question is usually answered on the basis of individual preferences and accidents of location and personality. As to the second question, it seems logical to structure microanalysis so that it meets the requirements of some specific policy questions. However, this may not be the most useful approach. A better approach might be to construct microanalytical models along a more general framework with a great deal more flexibility than we have done in the past. The general reason for this is that the focus of policy issues can change radically and rapidly. A generalized approach to firm analysis that allows a wide range of analytical options would be adaptable to changing circumstances.

There is a specific technical rationale for constructing more generalized micromodels. Low-cost, high-performance computer hardware and software will result in much of the specific firm analysis being carried out by the farmer at the farm site. While we will design much of that software for the farmer, the main thrust of our own activity should be along lines that farmers cannot accomplish themselves (even with the help of the private software vendors). In addition, large amounts of core memory capacity have become so cheap and readily available that we should be using this new technical advantage to assemble highly flexible and generalizable software—along the lines that have already been achieved by private systems such as EXPRESS. The sound technical and economic reasons for not doing this in the past no longer exist, and new technical capacity has been created. We need to recognize the hardware and software revolution and take advantage of it.

One need of public educators relates to both micro- and aggregate macromodels. Clients are most responsive when they can easily relate policy, program rules, and provisions to their particular situations. Easily understood micromodels facilitate this process. For example, the initial direct microeffects of a reduction in federal estate taxes on a specific family and related farm activity can be quickly understood. Policy education might well proceed on such a basis of firm-level analysis. However, what is not clear is how the same families will be affected by the secondary and feedback effects of estate tax reduction which might impact land prices, the availability of land to purchase for farm expansion or for a new enterprise, local and state taxes, inflation, and the system of family farms.

Accurate anticipation of feedback effects is essential to a satisfactory policy education program. Clients understand that their own actions in response to these policies do not by themselves significantly affect the broader community. However, it is much more difficult for them to understand and anticipate how others will respond to policies and, therefore, how the combined behavior of all individuals will affect aggregate variables which in turn may affect the individual. Micromodels have an important role in educating the public about these types of relationships. Macromodels do, too. But such models need to be related and each type constructed with the realization that they can be helpful in public education on identical public issues.

I hope the modelers will relate to these needs. This hope is not based only on the importance of public education needs, although they are substantial. It is based on the realization that policymakers also need to accurately anticipate aggregate behavior and its impacts. Neither the public nor policymakers should be surprised when a policy stimulates such things as changes in behavior that lead to large increases in program participation by individuals who would never qualify if their behavior were not adjusted in response to the policy.

My hunches are that micro behavioral models are important to the accurate anticipation of aggregate behavior; that macromodels can give a broader perspective to managers of micro units such as households and farms; and that policy educators and policymakers and the public welfare will gain if analysts are able to devise systems of micro- and macromodels that are in some way related and support each other for use in policy analysis and education.

References

Bergland, Bob. *A Time to Choose: Summary Report on the Structure of Agriculture.* Washington, D.C.: U.S. Dept. Agr., 1981.

Bottum, J. C., J. O. Dunbar, and others. *Land Retirement and Farm Policy.* Research Bulletin 704. Purdue Univ., Agr. Expt. Sta., 1961.

Brandow, G. E. *Interrelations Among Demands for Farm Products and Implications for Control of Market Supply*. Bull. No. 680. Pennsylvania State Univ., Agr. Expt. Sta., Aug. 1961.

Brewster, J. M. "The Role of Rural Beliefs and Values in Agricultural Policy," *Increasing Understanding of Public Problems and Policies, 1959*. Chicago: The Farm Foundation, 1959.

Doering, O. C. *An Energy Analysis of Alternative Production Methods and Cropping Systems in the Corn Belt*. National Science Foundation Report NSF/RA-770125. Purdue Univ., Agr. Expt. Sta., 1977.

Drabenstott, M. R., and E. O. Heady. *Small Family Farms in American Agriculture: Projected Incomes for North Central Iowa Small Farms*. CARD Report 96. Ames, Iowa: Center for Agricultural and Rural Development, 1980.

Dunbar, J. O. "Development and Execution of a Public Policy Education Program," *Increasing Understanding of Public Problems and Policies, 1963*. Chicago: The Farm Foundation, 1963.

Dvoskin, D., and E. O. Heady. *U.S. Agricultural Production Under Limited Energy Supplies, High Prices, and Expanding Agricultural Exports*. CARD Report 69. Ames, Iowa: Center for Agricultural and Rural Development, 1976.

Jensen, H. R., T. C. Hatch, and D. H. Harrington. *Economic Well-Being of Farms: Third Annual Report to Congress on the Status of Family Farms*. AER-469. U.S. Dept. Agr., Econ. Stat. Coop. Serv., 1981.

Meister, A. D., E. O. Heady, and others. *U.S. Agricultural Production in Relation to Alternative Water, Environmental and Export Policies*. Ames, Iowa: Center for Agricultural and Rural Development, 1976.

Muller Jr., R. E., R. M. Peart, and others. *Developing an Energy Input-Output Simulator (AGNRG) for Analysis of Alternative Cropping Systems in the Corn Belt*. National Science Foundation Report NSF/RA-770126. Purdue Univ., Agr. Expt. Sta., 1977.

Nourse, E. G., and others. *Turning the Searchlight on Farm Policy*. Chicago: The Farm Foundation, 1952.

Penn, R. J., T. E. Atkinson, and others. "Extension Programs on the Soil Bank" in *Increasing Understanding of Public Problems and Policies, 1957*. Chicago: The Farm Foundation, 1957.

Richardson, J. W., and C. J. Nixon. *The Farm Level Income and Policy Simulation Model: FLIPSIM*. Departmental Technical Report No. 81-2. Dept. of Agricultural Economics, Texas A&M Univ., 1981.

Tweeten, L. G., E. O. Heady, and L. V. Mayer. *Farm Program Alternatives, Farm Incomes and Public Costs Under Alternative Commodity Programs for Feedgrains and Wheat*. CAED Report 18. Ames, Iowa: Center for Agricultural and Economic Development, 1963.

Whittlesey, N. K., and E. O. Heady. *Aggregate Economic Effects of Alternative Land Retirement Programs; a Linear Programming Analysis*. TB-1351. U.S. Dept. Agr., Econ. Res. Serv., 1966.

Discussion of Part 8

A Conceptual Micromodel for Extension Farm Management and Policy

Harold W. Walker

Research Lag and Extension's Dilemma

Nelson presents a very good picture of the current situation of farm growth and planning models in farm management extension education. Research is far behind extension's needs in this area. Yet those engaged in such research long ago realized the limited payoff for this type of research and oriented their efforts towards professional and administrative requirements for recognition—basic research and journal articles. Applied research of this nature receives a kindly nod and an admonishment to do something productive. I see no changes in evaluation systems to correct this in the near future.

I must agree with Nelson that extension does not have the personnel or equipment to train farmers to operate personal computers and teach the successful use of required techniques. However, I disagree with his solution of passing this function to high schools and community colleges.

Education is faced with a major decision that is long overdue. Extension must either train or hire people and obtain the equipment to do the job or step aside while others take over this function. Commercial farmers are tired of waiting for extension to catch up. We are losing commercial farmers to other information sources, some of which are even less capable of doing the job than extension. This decade will find extension serving the needs of commercial farmers or losing them entirely as a clientele.

Nelson's Opinion

Nelson's opinion that we are still some distance from solving the rapid response turnaround time with computers is not quite right. Immediate response has been available for several years. We have farm management agents with portable terminals visiting farmers and leaving

The author is an associate professor of agricultural economics and an extension economist in farm management at Virginia Polytechnic Institute and State University.

computer results with them. They can use the farmer's telephone to enter the farmer's record data, produce a record analysis, create an aggregate of farms similar to the farmer's, and produce a progress report of the farm along with the aggregate in a couple of hours. Turnaround time is not a problem with properly designed programs placed on reliable computer systems.

Microsolution or Microproblem?

Microcomputers are with us, but at a cost. A reasonably adequate system costs $5,000 to $8,000 today with a life expectancy of 3 to 5 years. Obsolescence reduces this time to about 2 years. The obvious lack of good software restricts usage after the hardware is purchased. Considering all costs of micros, the cost per job run is quite high. The next year or two should bring some improved units onto the market along with a better supply of useful software for farmers. I am not recommending the purchase of micros at this time except for games and learning experience that must be acquired.

Action or Reaction?

Doering presented a good paper on the development of policy extension activities. I think he said that past policy education has been in the area of reaction and not of formative action. I suspect this is partially due to the unwillingness of some extension policy specialists to risk being wrong, especially on major issues.

Micromodels have tended to be static, as Doering states. I embrace the notion that models should incorporate the dynamics of human behavior. Particular attention needs to be given to taking account of how behavior in response to policies will affect production of individual farms and the use of resources and in turn the supply of food and fiber.

As Doering states, "We need to recognize the hardware and software computer revolution and take advantage of it."

Conceptual Micromodel

Neither paper proposes a complete and comprehensive model for the firm. I would like to complete the picture by proposing a conceptual model of what is needed for extension farm management and policy decisions (see the figure).

The farm firm is the center of this model, but the concept applies to other firms. Resources, enterprise selection criteria, and institutional

CONCEPTUAL MICROMODEL

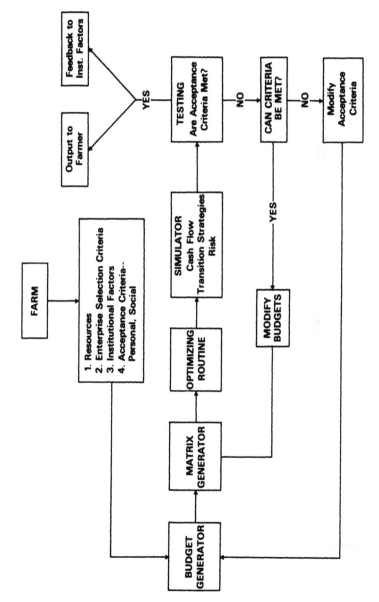

factors are already built into present models. Personal and social acceptance criteria need to be added. We have budget generators, matrix generators, and optimizing routines already linked so that the output from one phase becomes input into the succeeding phase. Some of these models are doing a good job. Some work is being done with simulation to examine cash flow consequences, potential transition strategies, risk, and other relevant factors in resource use adjustment. I know of no successful linkage between optimizing routines and simulators for a good evaluation of various optimizing strategies.

The testing phase is now largely a subjective evaluation. Some type of program can be developed for this testing to be done as another link in the total model. If the acceptance criteria are met, a report is produced for the farmer and results are fed back to those engaged in establishing institutional factors so the effects of institutional factors on the firm may be evaluated. These factors may not be producing the desired results or decisions by the firm.

If acceptance criteria are not met, some type of analysis is needed to determine if they can be met. The original budgets can be modified until an acceptable strategy is derived. If not, acceptance criteria will have to be modified.

Social-Personal Conflicts

Conflicts between personal and social acceptance criteria may arise that will require concessions on the part of either or both. Suppose a social goal relating to some environmental objective prohibits the firm from surviving or places severe restrictions on survival. Obviously, something has to give. In such a situation, it may be unfair to insist that the firm bear the total cost of achieving social goals. At the same time in other situations, it is not reasonable for high profits of a firm to rely on externalized costs. Thus, achievement of social goals such as the assurance of an adequate food supply at reasonable prices may require society's bearing some or all of the costs incurred in achieving environmental objectives.

Ex Nihilo Nihil Fit

It will take years of diligent work by many people before this model is operational. We do not know how to develop some of the components today. Yet, I have confidence that it can be done. Until then, we will continue to evaluate adjustments in agricultural production on a fragmental basis and express puzzlement over unexpected results.

Discussion of Part 8

Computer Software Needs in Farm Management Extension Activities

Gerald Perkins

I have worked with the development of computer software that has been used at the individual farm level for the last 8 years. I have found some evidence of the first problem that Dr. Nelson has identified when farm-level computer models are used—the problem of incorporating individual farm data into a given model. In most cases, this problem has been solved by simply involving the farm operator in the design of the model itself. This involvement takes several forms. First, a group of farmers are interviewed on a given subject. Several from this group are later selected to help test the software developed after their objective and needs have been discussed. By involving the people you are trying to help, you eliminate 3 problem situations. First, you know that the software will be accepted, because it was designed around the end user's needs. Second, you assure yourself that the model can be operated successfully by the end users. Finally, you have created a situation where data gathering and entry has been made easier, because the end user's peers support the product.

The second problem, response time, is almost nonexistent now, as a result of portable terminals. The farmer can use several nationwide time-sharing systems by merely using a telephone and a $1,600 terminal.

The biggest shortcoming I have encountered is that the end users are not consulted before a model is designed. If the people whom you deem necessary to operate your software are not involved with its development, your project will certainly fail.

Financial Analysis

As Dr. Nelson has pointed out, the most pressing demand of a farm operator is finding a way to keep track of financial data. Micro-, mini-, and mainframe vendors have often attempted to produce this type of software for the farm. Their biggest problem is trying to modify the

The author is president of World-Wide Computing, Inc.

old double-entry system used in many businesses. This will not work at the farm level.

Financial analysis software is available in certain formats on several different computer systems. The Computerized Management Network system at Virginia Polytechnic Institute has a farm-record analysis program; the Agnet system at Nebraska has a package called BusPak. Several universities also have this type of software.

If farmers want to invest in only a terminal, then they must use the software of that particular system. If farmers decide to buy their own system, the problem is twofold. First, they quickly discover that the software they want is almost nonexistent. Second, they have to either find someone to write the desired software or create the software themselves.

In the past, technological advances were usually passed on to the farmers by their local extension agents. Since the introduction of the microcomputer and its ever-decreasing price tag, farmers have bypassed these agents. The land-grant universities and the extension service had to scramble to catch up to their clients.

Why do farmers buy and use microcomputers? There are several reasons, such as:

1. Keeping tax records and farm books;
2. Making their farm more competitive;
3. Benefiting from the organizational ability of a computer when making day-to-day decisions;
4. Saving time by getting better, quicker, and more detailed reports on all farm enterprises—a time savings that means saving money; and
5. Alerting themselves to a "red flag" situation involving one of their enterprises.

None of these decisions enhances or improves the local extension agency.

User-Friendly Software

With these facts in mind, how should a given model be constructed? Dr. Nelson has pointed out the data requirements. I would like to expand on the subject of the software being user friendly.

Once the decision is made to work with the farm operator, the problem of software development is raised.

Contrary to the belief that only Ph.D.'s can operate a computer terminal, most people with a sixth-grade education can use a terminal.

This finding holds true in all the countries that I have worked in. The software must have 2 necessary commodities: It must be "user friendly," and it must be written in the user's terminology. User-friendly software has one overriding rule—be patient. If the user makes a mistake, don't print a group of error messages and stop the program. Instead, notify the user of the error, and give the individual another chance to input the necessary data.

During the development of software, the intended users should be consulted on content, accuracy of input and output, and terminology. This appraisal can be done by working with people selected from the user group. After the software has been developed and tested in the laboratory, it must be field tested. Here it will be modified to meet the users' needs. This is also the first time that training of uninformed users must take place. If the software is designed correctly, this training can be as short as 10 minutes to simply show the basics of terminal operation. The terminology of the software can now be tested. If a user who has never used a computer terminal before can sit down and run the program after only 10 minutes, the program is well on its way to being accepted.

The last important aspect of software development is that the output be easily understood, short, and arranged in a logical order. The user should not have to skip all around the output to find the critical input results. If possible, farmers—or at least their input—should be identified on the output. For example, a crop rotation program should identify the field location. A recordkeeping system should identify the record dates and enterprise.

Experiences in Foreign Countries

In the foreign countries in which I have worked, the use of computers to analyze conditions at the farm level is minimal. In most lower-income countries, the computer is located at a central point. Because of the poor communication system, remote terminals do not exist. Development of software to produce nationwide data bases or networks that are used by many different government agencies has been emphasized. In a few cases, computer software is being used in working with farm operators. In Costa Rica, the main crop is coffee. The agricultural institute (CATIE) is setting up a data base to fight coffee rust and other plant diseases within Central America. In Trinidad, the Caribbean Agricultural Research Institute (CARDI) is using a computer network to survey farms on 8 islands to find out the best placement for markets, shipping routes, and warehousing for their farm products. In Egypt, the government is using a computer to analyze data related to livestock populations and feed resources. This analysis is then used to determine crop placements, slaughterhouses, warehousing, and market outlets. In all 3 of these situations, the government is working with the farm operators to provide them with a needed resource.

Conclusion

I feel the emphasis placed on modeling is misdirected if the real objective is to help the farmer. The construction of an inexpensive, nationwide computer network that provides data used in management decisions would be much more beneficial to farm operators. This network could be used to store crop and livestock prices, simple matrix-inversion models, recordkeeping systems, market prices, and so forth. By providing this information, extension personnel could have all the input data they require and the farmers would have their needed resource.

I travel and work in countries that do not usually have much in the line of computer systems or software packages. In this situation, the basic questions asked by government are: "What resources are available?" and "How can we improve the farmers' standard of living?"

Discussion of Part 8

Use of Farm Models in Extension Activities: An Old Concept with New Techniques

B. H. Robinson

Many macromodels provide guidance for policymakers on the impacts of alternative programs and policies on the agricultural sector, and some of them provide insights into intersectoral adjustments. However, they frequently fall short of providing information useful in microdecisions or in assessing the impacts of policy alternatives on farm adjustments.

After a broad and varied discussion of micromodeling efforts and firm decision analysis, including the implications for policy research, the conference organizers provided an opportunity for extension educators to respond to the use of farm models in extension.

Although Dr. Nelson and Dr. Doering took different approaches in their presentations, each discussed some of the current models employed in extension programs, responded to real and imagined problems of linking research objectives and output to extension objectives and needs, and commented on the current techniques for building micromodels. Both addressed the need for more integrated efforts in micromodeling and policy analysis among macro-policy analysts, microresearchers, and extension educators, and they arrived at the same conclusion: More cooperation and coordination is both feasible and desirable. They also identified the complexities and problems of the task at hand and called for increased flexibility in modeling as a possible remedy.

I find little in either paper with which to disagree, so I will take some of their arguments an additional step and discuss an approach to the integration of efforts, information, and analysis among specialized areas in our profession. As presumptuous as this may seem and with the pitfalls involved, it might provide a basis for discussion and a framework for improved communication and cooperation.

During the last decade, the shift to more macroanalysis of the impacts of alternative policies and programs and the reduction in ERS field staff have both contributed to reduced links between ERS and the land-grant

The author is an extension economist and professor of agricultural economics and rural sociology at Clemson University, Clemson, S.C.

research and extension education complex. The old micro-based, policy research production adjustment models and the ERS organization of the 1960s provided a direct professional link as well as a firm-based link in policy research.

Useful information flows between macroanalysts and microanalysts have been interrupted, and the extension user of research results has been left with a dry well. Current research products require significant interpretation to be of use to extension economists in programs oriented to either firm decisionmaking or policy education, and they provide little basis for the macroanalyst to advise policymakers on microadjustments. The historical link that allied macropolicy analysts and microanalysts has undergone as dramatic a change in the last decade as has the industry they serve. One problem is to find a common ground and a way to reestablish cooperation. Our respective clientele have already identified the need through their demands for information.

We should strive to identify possible ways that researchers can make input into extension education activities and ways that extension educators can better communicate needs and problems to researchers. A brief descriptive categorization of extension modeling efforts and the current and potential information flows may be helpful in identifying areas of mutual concern and benefit.

Modeling and Extension

Modeling itself is not a new concept for extension. The establishment of the "Porter Demonstration Farm" by Seaman A. Knapp in 1904 is representative of the basic education model adopted by extension at its inception in 1914. The basic model has been refined and expanded over time. Also, extension clientele has broadened, and techniques, information, and relevant theoretical concepts have expanded to address an ever-widening range of issues. Yet the basic approach remains intact; providing a flow of relevant information to address public issues and aid in better decisions remains the paramount objective.

Categories of Models

Although it is somewhat naive to broadly categorize the models used in policy analysis and education, they fall roughly into 3 broad groups. Admittedly, categories overlap and some ideas and models will fall through the rather broad gaps in the skeletal structure.

Explanatory and Guidance Models

This group includes the broad range of micro- and macro research models, many of which were discussed at this conference. It includes both the mathematical micro- and macromodeling effort in which researchers often communicate with researchers and the efforts of many researchers to analyze the broad sectoral and intersectoral impacts and adjustments to policy variables. Many of these models are geared to addressing either narrow and specific questions in a rather sophisticated way, or they focus on broad adjustment issues in a highly aggregated manner. In either case, considerable interpretation and modification are required to render the results useful to extension educational programs.

Many of the macromodels currently in use for policy analysis do not depend on a micro base as did the old micro-aggregation models. Rather they are independent of a micro linkage, utilize macro variables, and ignore intrasectoral adjustments. Timeliness is another problem in the information flow. The latter problem was addressed by Doering in his discussion of the "teachable moment" theory. Yet, improved cooperation and increased model flexibility could permit each group to analyze the variables and issues critical to the other.

User Models

Included in this category are the innovative new firm decision and information models designed to provide "quick turnaround" answers to specific firm problems. These models are evolving from a high input requirement to a low input by extension economists. Technology, private software development, and a concentrated effort in many land-grant universities on development of efficient, user-easy software have rendered many of these models directly available to clientele.

Extension and research activities in developing firm decision guides and models no longer require the operational and interpretative input of only a few years ago. In fact, users are currently doing for themselves what professional economists did for them only a few years ago, and extension economists seem to be doing much of what microresearchers did only a few years ago. As a result, the professional can devote more time to the development of new models for users and tie into a much broader data base than ever before. These models may offer a viable opportunity for feedback between the macroanalyst (concerned with prices, input, costs, and so forth) and the microanalyst (concerned with firm adjustments, cash flow problems, investment and tax strategies, and so forth).

Education and Intuitive Models

My final category includes the myriad of models employed by extension educators and others concerned with providing information and decision aids directly to users. The models range from well-defined educational tools such as the policy education models described by Doering and others to "on-the-spot" response models to specific firm decision questions such as, "Should I participate in the feed grain program?"

In some cases, educators have weeks to gather information, prepare their approach, select the appropriate educational model or theoretical framework, and conduct the applied analyses to address the problem. More frequently, educators are faced with only a few days to address the available information and techniques to the task at hand.

In the case of policy education activities, educators confront the difficult task of anticipating issues in order to provide relevant data and analysis to decisionmakers at the "teachable moment." Doering discussed the inherent problems of using rigorous analytical information in the policy education process. Instead, the policy educator frequently relies on intuitive models, economic logic, and what analysis is available. Even if macroanalysis is available, it is often so complex that the interpretation required to make it useful to the public is prohibitive. Often the policy educator must also address a wider range of alternatives than has been addressed by the analyst. Yet, macroanalysis has great potential for improving the effectiveness of policy education if the problem of timeliness can be overcome and if the models are sufficiently flexible to permit analysis of changing issues and parameters.

The analysis of firm adjustment to changing economic and institutional environments has been considerably enhanced by the micromodeling of university researchers and farm management specialists and the revolutionary new hardware and software of the past decade. Many models have also proven effective in analyzing alternatives at the local government level.

Linkage Models

Macroanalysts have not adequately addressed the farm firm adjustment problems of the past few years. Macromodels of entire production sectors evaluate aggregate impacts, but by their nature ignore intrasectoral adjustments. Outcries from subsectors of the agricultural complex, such as the American Agricultural Movement, have acquainted policymakers with serious problems within the agricultural sector. Policymakers have begun to ask questions about the differential impacts of policy alternatives.

Many of the needed micromodels and the hardware and software to use them are in place or will be here soon. Micro decision models (cash flow, investment decisions, storage decisions, linear programming, and so forth) and user-oriented hardware are quickly evolving. The major problem is to reestablish the links among the research and education professionals that existed only a decade ago. This is not to argue for a return to the previous ERS organizational arrangement or the old micro/macro production response models typical of the 1960s. Rather, I am suggesting the continued use of powerful macromodels for assessing the impacts of alternative policies on both sectoral and intersectoral adjustments, but supplemented with microanalysis of impacts on different types, sizes, and groups of farms.

Improved communication between the policy analyst (at the national level) and the extension educator and land-grant researcher could facilitate a much needed flow of information in both directions at the critical stage of analysis, but would require a commitment by all parties.

The critical link that needs to be reestablished is tripartite and requires a continual flow of data and information among macropolicy analyst, policy educator, and university farm management specialist/microresearcher. Cooperative efforts could permit the macroanalyst to link with the microanalyst to assess microadjustments to policy alternatives. The link between policy educator and macroanalyst would provide much-needed data and analysis for improved policy education programs. Often, the intuitive models of the policy educator can only indicate the direction in which institutional and economic variables change in response to policies. Input from macroanalysts could provide information on the magnitude of change.

The 3 parties would have to maintain communication throughout the planning and execution stages of the work. Objectives of each member may change as work progresses. Thus, the macropolicy analyst, policy educator, and university farm management specialist/microresearcher must have model and procedural flexibilities that permit each of them to contribute needed insights and to effectively utilize new information as work progresses. The potential payoff from the suggested linkages and flexibilities are large for they involve the quality of information supplied to policymakers about current adjustments by farmers and the impacts of policies on farmers in differing situations.

PART 9

Needs of the Future

18. Micromodeling in Perspective
B. R. Eddleman

The complexities introduced into micromodeling activities since the 1960s, when some of us worked on firm growth problems, have been impressive. Many of the papers presented were cast in the framework of firm growth models. I was left with the impression from these papers that the primary objective was to develop models that could be used by farm decisionmakers under conditions of risk and uncertainty to make incremental improvements through time in ongoing farm operations. This focus contrasts with what I thought was to be the major purpose of the conference—understanding how micromodels can be used for analyzing the effects of policy changes on a farm firm or groups of farm firms.

As with any research endeavor, it is important for micromodelers to ask and answer 3 questions:

- What are we trying to do?
- For whom?
- In what time frame?

My comments will focus primarily on the first of these 3 questions. R. J. Hildreth's comments deal with the question, "For whom?"—that is, for peers within the agricultural economic discipline, for farm decisionmakers, or for policymakers.

If we are to use micromodels to measure impacts of policy change on farm firms, it is important to ask: "What do we know about producer behavior under conditions of uncertain prices, yields, and institutional

---

The author is a professor of agricultural economics at Mississippi State University.

structures?" The empirical models presented by the conference speakers would seem to imply that we know quite a bit about farmers' behavior relative to these variables. Many of the models imply that the yield distribution for a particular farm is readily available and known. My experience has shown that these data are difficult to discover and that variations in county-average yields are often used in micromodels as if the data were applicable to an individual farm. But are they? And does it make any difference if they aren't? The point is that our data for specific farm situations are often meager and that results of models based on such data should be presented with great care. Too often, the claims for the results imply empiricism that does not exist.

Much of the discussion at this conference was focused on slight modifications and small incremental improvements in techniques and approaches. These improvements enrich the accomplishments of firm-growth research that was done in the 1950s and 1960s. Less emphasis, with the possible exception of Richard Day's paper, was given to searching out ways to advance micromodeling activity by quantum amounts. These recent improvements have been realized at a cost—the opportunity to have focused more diligently on farmer behavior. The papers presented seem to depend excessively on previous econometric studies as the source of the behavioral relationships included in the micromodels. I encourage you to also give emphasis to obtaining knowledge of the decisionmaking processes of farmers from: (1) direct contacts and observations of experiences of farmers themselves; (2) experiences of people who work closely with farmers—merchants such as buyers, feed dealers, implement dealers, as well as extension workers; and (3) theoretical models of human behavior—not entirely limited to economic behavior. Interdependencies among the sources are evident.

Most of the models imply that financial gain is the only incentive shaping farm decisionmakers' behavior. Other attributes of behavior such as status in the community, leisure, aesthetic enjoyments, and religious values also may be important. Additionally, the goal of satisficing[1] or adaptive economic behavior may be much more relevant in describing the precise strategies producers follow to achieve multiple goals. The econometric approach—with its emphasis on maximizing and minimizing behavior—is an appropriate paradigm in dealing with such questions as: "Do I go into farming and how do I organize the farming activities if I choose farming?" In contrast, the econometric approach is often deficient in dealing with behavior that reflects satisficing and rank-order goals. Furthermore, even though financial considerations probably dominate most of the year-to-year production and marketing decisions so long as the basic organization of the farm is not reconsidered, farmers are largely reactors to situations; they react to soil conditions, to insects, to diseases, to weather, to the current market, and even to exogenous policy changes. These decisions—that one might think of as operational versus long-term organizational decisions—have not been adequately captured in the econometric models presented in this conference.

What do we conclude from this shortcoming with regard to micromodeling the impacts of policy actions on farm firms? The late 1950s marked the emergence of a new frontier of empirical analyses as computer technology and new econometric techniques were developed. With the introduction of this technology, micromodeling became a possibility. Economists began to make fewer restrictive assumptions that characterized the partial budgeting techniques of the 1940s and early 1950s. Greater use of empirical data produced numerous estimates of statistically derived economic parameters and relationships. Empirical analysis of microeconomic farm firm relationships involved detailed budgets and statistical estimates of production functions. These data and relationships were incorporated into optimizing procedures—for example, linear programming analyses. Factor-factor, factor-product, and product-product economic relationships were investigated within the context of economies of scale and intertemporal analyses.

Advances in computer technology have continued to reduce the cost of making a given number of computations, and the size of the micromodels in terms of the number of relationships included have increased accordingly. However, relatively few new empirical tools were developed for microeconomic analysis in the 1970s. The result has been a lack of detail concerning interrelationships between the farm sector and nonagricultural sectors of the U.S. and world economies. The profession of micromodeling appears to have continued to move along the plateau reached in the late 1960s in dealing with these more aggregate international and national markets for products and inputs and with associated policy changes in the United States and abroad.

I anticipate that micromodeling and firm analyses will:

- Continue to be a highly respected activity of our profession,
- Stay basically people-oriented,
- Be limited in precision by behavioral variability of decisionmakers, and
- Become increasingly more quantitative.

In carrying out our micromodeling activities, I encourage each of us to:

- Develop better relationships with other professional fields,
- Undertake more multidisciplinary efforts,

- Recognize the changing problem setting in which we deal, and
- Remember the pragmatic origins of our efforts and the pragmatic orientations of our past successes.

We must temper our fascination with quantitative techniques with a willingness to maintain an orientation to reality, and we must recognize that economic concepts as well as techniques can become obsolete. We have a responsibility to revise both. Furthermore, in our efforts to do all of these things, we need to constantly concern ourselves with the questions: "For whom are we doing it, and in what time frame are we trying to accomplish it?"

Notes

1. Satisficing means that decisionmakers choose alternatives that provide a level of goal achievement that is "good enough" rather than making an all-out effort to maximize the goal.

19. The Use of Micromodels

R. J. Hildreth

The objectives of the conference were the following:

• To improve our understanding of advantages and limitations of using models of farm growth and adjustment behavior to analyze policies and regulations and to conduct extension activities.

• To foster current and future professional interactions among economists at USDA, universities, and other institutions interested in models of farm growth and adjustment.

The second objective was well achieved. There was vigorous interaction among the participants. People left the sessions mumbling to each other and to themselves. The discussion in free time was lively. I personally gained considerable consumer surplus from the conference and especially appreciated the opportunity to review new developments in micromodeling.

The first objective of the conference was not as well achieved, especially the development and improvement of understanding of the advantages and limitations of micromodels in analyzing policies and regulations. Why? There were no poor papers. Thus, the failure to fully achieve the first objective must lie elsewhere. I think the reason was a lack of sufficient perspective on problems encountered in using micromodels. It is this topic I now intend to explore.

What are the uses of micromodeling? At least 3 can be identified. They are not mutually exclusive. Micromodeling can be used as a means of academic scholarship, a means of improving farmer decisionmaking, and a means of policy analysis.

The author is managing director of the Farm Foundation.

Academic Scholarship

The building of micromodels can facilitate obtaining tenure and promotion in our universities. A focus on building models for this purpose leads to different behavior among economists than would focus on the other 2 purposes. At the risk of being misunderstood, I like to speak of model building as an art form, as did Reichelderfer, to illustrate the focus on academic scholarship.

Art forms have value and can communicate reality. Focusing on the "art form" enables the profession to be more productive and useful in the long run. One can explain and predict some of the behavior of model builders by viewing them as artists. Artists view reality and communicate (at least send messages) in creative and innovative ways. Artists learn by doing; that is, they develop their human capital. Artists place emphasis on interesting new forms; that is, they do not place high value on using a model a second time. Technicians and artisans do that. Artists dispute among themselves about style, methods, and material, but not about how the results of their efforts are viewed by nonartists. Artists also feel they are misunderstood and underpaid.

Similarly, model builders pursuing academic scholarship contribute to the productivity of the profession. They are creative and innovative. They emphasize new forms. But, they tend not to be especially interested in tasks such as adapting models for use in extension or policy analyses. They tend to have limited interest in the usefulness of their theoretical advances to these applied endeavors.

Aid to Farmer Decisionmaking

Micromodels are increasingly used as an aid to farmer decisionmaking. The microcomputer enables more farmers to adapt micromodels to help them evaluate alternative courses of action or develop new courses of action for their specific situation. The paper by Nelson and the discussion by Perkins and Walker explored this aspect of micromodels.

The value of the model for this use has to be judged from the farmer's perspective. The probable criterion of the farmer is relevance, not logical validity of the assumptions. Farmers are interested in quantitative and qualitative results. That is, they are interested in point estimates or predictions of consequences as well as an understanding of the structure of relationships involving their farming activities. Farmers are also interested in developing their own human capital and can find models to use for that purpose.

As Perkins brought out clearly in his discussion, building models for this use involves the clear definition of the client. It is obvious that if modelers see "clients" as their peers, they will develop a different model than if they consider their clients to be farmers.

Policy Analysis

Clearly, micromodels cannot replace macromodels in policy analysis. However, they can provide perspectives on the consequences of a policy alternative to individual firms and on the distribution of consequences of policies among farm firms, and thereby contribute richness to analyses of food or agricultural policies. A number of issues in the use of micromodels as a policy analysis tool arise:

1. Can we have multipurpose micromodels; is it possible to serve the 3 clients with a single model? Will a client in the academic community, a farmer decisionmaker, and a policy decisionmaker each be happy with the analysis provided from one model? I suspect that the answer is "no." The strategic detail requirements may vary significantly among the 3 purposes. Thus, it may be necessary to respecify models for different kinds of clients or perhaps develop different models for the different purposes.

2. What kind of policies should be evaluated? Among those that could be examined are commodity policy, tax policy, regulation and administration of policy, marketing policy, and general economic policies. Is it possible to specify one model that will determine the consequences of each kind? I suspect not. Again, the strategic detail requirements may vary significantly.

3. What consequence is to be measured? Is it net income, acres of specified crops planted, distributional impacts (of what), or resource ownership patterns? Is the modeling to encompass consequences to all types of factor owners, or is it to concentrate on farm operators? Even if it is to focus on farm operators, should it concentrate on farming activities or should it focus on households and thereby encompass farm and nonfarm activities? If nonfarm activities are not included, how useful are the models in the 1980s where most farm-related households have nonfarm activities as well? The decision of which consequences are to be measured by the model is not distinct from decisions about which policies are to be analyzed. They must be given joint consideration. There was little discussion in the papers as to what consequences should be measured.

4. Should the model builder ask the policy decisionmaker what consequences are to be evaluated? Just as it is important in building models for farmers to observe them and be involved in the building process, it may be useful to involve policymakers in building models for policy evaluation.

5. Can models be built to do more than analyze the consequences of a policy already passed by the Congress? Why not evaluate proposed policy alternatives? The evaluation of proposed policy alternatives may be more useful than analyzing consequences of approved policies. Regardless, major attention should be given to using the modeling insights to conceptualize other policy alternatives.

6. Who are the policy decisionmakers? The papers did not reveal a consensus about who they are. Are they people in the administration who propose a policy? Are they members of the Congress or their staffs who vote on policy? Are they officers or organizations which try to influence the administration or the Congress? They are all the above plus, on occasion, the public at large.

7. How can micromodeling be linked with macromodeling? It is, of course, possible to use micromodels to formulate hypotheses to be tested by macromodels. However, although microeconomic units individually behave as if they do not have an effect on markets, they, of course, do. Thus, the aggregation problem remains a fundamental issue in using micromodels in this way. At the same time, macromodels can be used to generate variable estimates that are exogenous and crucial to individual models of farm firms.

Consistent specification of related macro- and micromodels would be a requirement for effective joint use of such models. Thus, combinations of micro- and macromodels could be complementary and enhance the quality of estimates of aggregative, as well as distributional, effects of alternative policies.

In the near term, at least, the aggregation issue will have to be finessed. Thus, results of microanalysis of typical or proxy farms will have to be presented as case studies. Their potential contribution should not be discounted. Information from case studies is often the driving force in the political process. A clear statement of the consequences of the policy on people with differing economic conditions could contribute substantially to enlightened policymaking. There is always the question of how many and what kind of case farms or households are needed. This problem will have to be partially dealt with as a political process problem as well as one with economic dimensions. Political scientists could perhaps help us deal with this issue.

8. How complex a model should be used? My answer to this question is a pragmatic one. We should use as simple a model as we can to deal with the consequences to be measured. The question to be answered or the problem to be dealt with should determine the complexity of the model.

9. How can the human organizational linkages be arranged among policymakers, macromodel builders, micromodel builders, extension educators, and data suppliers? Bob Robinson points out clearly in his discussion the gaps that now exist between ERS and the research and extension economists at universities. Each of us can and should work toward closing these gaps.

10. How can adequate data sources be developed and maintained? Micromodels draw from USDA series, census data, farm records, and surveys. The coefficients for specific farms are difficult to obtain. I found it interesting that many of the models that were used were for typical farms in Illinois because of the availability of farm records from Illinois. Few of the papers, except for Ken Farrell's, mentioned data as a problem. I see the data problem as one of the major limitations in the development of micromodels and their potential use.

There is also the issue of reality checks in the process of model building. It is difficult for ERS personnel to have close contact with farmers. Even model builders from universities do not take advantage of the opportunity of "reality checking" their models with farmers. I was encouraged that Rob King plans to use his model in workshops with Colorado farmers and applaud his plans to use the farmers in the workshop as panelists for the further development of his model.

In conclusion, I focus on the financial support needed for micromodeling activities. Bobby Eddleman's points—to be clear as to what we are trying to do and for whom we are doing it—are crucial to obtaining the necessary support, whether it be adjustments in the allocation of currently available funds or additional appropriations. Skill in explaining the rationale for this research to people who control budgets and appropriations is of enormous importance. But, in the longer run, success in applied micromodeling to support policymaking will depend on our creativity in supporting policymaking with objective analyses and on the intrinsic usefulness of micromodels. This conference is an important step in stimulating that creativity. But it is only one step. As my list of issues suggests, many are yet to be taken.

Index

AAM. *See* American Agricultural Movement
Adaptive control theories, 30, 32–33, 52–53, 115
Adaptive economics, 23, 24–33, 35–40, 41–48, 50–55, 57–58, 59
Adaptive strategies, 359, 365
"Add factors," 97
Administrative Behavior (Simon), 30
Administrative function, 23–24, 25, 35, 40
After-tax income. *See* Income, net
Age, 264
Aggregate feedback, 39
Aggregate forecasts. *See under* Agricultural forecasts
Aggregate supply response function, 3, 5, 8, 153
  error, 3, 7
Aggregate variables, 63
Aggregation bias, 35, 69, 70, 92, 93. *See also* Macromodel, aggregation error; Micromodel, aggregation error
Agnet system, 390
Agricultural change, 18–19, 34, 39, 44–46, 51–52, 55, 159, 379–380. *See also* Adaptive economics
Agricultural economics, 1, 19, 106–107, 148, 150. *See also* Agricultural change; Economic Research Service; Farm management; Micromodel
Agricultural forecasts, 73–79, 84–85, 97, 102, 196
  aggregate, 75, 79–84, 85, 96–97
  aggregate, defined, 79
  evaluation, 80–81
  weighting, 80
Agricultural Stabilization and Conservation Service (ASCS), 340, 341, 344, 345
Agricultural variables, 73, 102, 103
Agriculture, U. S. Department of (USDA), 3, 4, 6, 11, 73, 74, 80, 81, 82, 84, 93, 102, 103, 153, 177, 298, 372, 375, 377, 378, 407. *See also* Economic Research Service
Agriculture and Food Act (1981), 103, 104–105
Akerlof, G. A., 58
Alfalfa, 319, 320, 321
Almon lag models, 237–238(figs.), 240–241, 244
American Agricultural Movement (AAM), 102, 396

Anderson, J., 162, 170
Anderson, Jock R., 335, 336
Animal unit months (AUMs), 319
ASCS. *See* Agricultural Stabilization and Conservation Service
Assets, 263, 264–265, 266–270, 271(table), 272–273(figs.), 274(table), 275, 316, 321
  fixity, 217. *See also* Johnson, G. L., model
  pricing model, 127, 128, 129(table)
  returns, 129, 132(n4)
  transfer, 136, 141, 142
  value, 127–128, 130, 165, 224, 225. *See also* Land, value
  See also Farming related households model
AUMs. *See* Animal unit months
Autofeedback, 23
Automobiles, 143
Autoregression, weighted, 167

Baker, C. B., 120, 123, 124, 290, 297, 304
Barley, 222
Barry, Peter J., 123, 146, 148, 149, 150, 153, 154, 155, 156, 157, 167, 290, 293
Basis contract, 186–187, 188–189, 192(n5), 193(n12), 202
Basis risk, 185, 186
Baum, Kenneth H., 40, 279, 288
Bayes Rule, 32
Behavioral economics, 30–31, 32, 58. *See also* Adaptive economics
Bell, D. E., 162
Bellman, R., 32
Bermuda pasture, 319, 321
Beta distribution, 336
Binswanger, Hans P., 169
Bioeconomics, 53
Black, J. Roy, 96, 97, 239
Boehlje, Michael D., 125, 213, 236, 290, 293
Bounded rationality, 22, 24–25, 26, 57
Boussard, Jean-Marc, 291, 292, 295
Bradford, G. L. 294, 295
Brigham, Eugene F., 293
Brink, L., 170
Brown, Robert A., 185
Budgeting. *See under* Farm management
BusPak, 390

Calvin, Linda, 282, 284
Capital accumulation, 115, 254, 257

Capital gains, 113, 126, 129, 130, 131, 151,
   156, 165, 254
Capital goods, 113
Capital rationing, 293, 294, 304
Captick, Daniel F., 245
CARDI. *See* Caribbean Agricultural Research
   Institute
Caribbean Agricultural Research Institute
   (CARDI), 391
Cash flow. *See* Income, and financial flow;
   Production, cash flow
CATIE (Costa Rican agricultural institute), 391
Cattle
   prices, 183–184, 185, 205, 320(table), 321
   production, 319, 322, 323, 324(table)
CATTLE FAX, 74
Cattle replacement decisions, 78(table)
CCC. *See* Commodity Credit Corporation
CDF. *See* Cumulative distribution functions
CE. *See* Certainty equivalent
Census of Agriculture, 147, 223
   1974, 13
Certainty equivalent (CE), 168
Chase Econometrics, 74, 81
Chavas, J. P., 359
Chien, Y. I., 294, 295
Choice criteria, 45, 169–171, 359
Choice variables, 338, 339
Chow, G. C., 335
Clements, Alvin M., Jr., 223, 321
Cobweb model, 39, 41
Cochrane, W. W., model, 57, 59
Cognitive dissonance, 52
Colorado, 334, 340, 341, 345, 407
Commodity brokerage services, 74
Commodity Credit Corporation (CCC), 222,
   225, 279
Commodity Economics Division, 5
Commodity News Service, 74
Commodity programs, 6, 103, 104, 107, 219,
   223, 224–225, 378. *See also* Agricultural
   Stabilization and Conservation Service
Commodity services, 107
Commodity supply, 3, 4, 24
Common good, 20
Competition, 20, 21, 51, 57. *See also*
   Multiple-farm opportunity set simulation
   model; Resource availability
Complementary models. *See* Midromodel, and
   macromodel
Computerized Management Network system,
   390
Computers, 11, 12, 31, 33, 54, 59, 74, 84, 85,
   147, 191, 382, 385–386, 389, 391–392,
   401
   data and program system, 373–374. *See also*
      Dynamo computer language
   onfarm, 75, 97, 98, 368–369, 370–372, 385,
      389, 390
   programs. *See* FLIPRIP; FLIPSIM; TELPLAN
   user-friendly, 390–391
Condra, Gary D., 181
Conservation policies, 17, 104, 107
Conservation selection, principle of, 30

Constraint function, 25, 28, 31, 32, 34, 35,
   36, 37(fig.), 38, 45, 57, 151, 160, 163,
   167, 176–177, 202, 213, 217–218, 219,
   222, 275, 317, 319, 335, 338. *See also*
   Hogs, production model
Consumer Price Index, 129, 242(table)
Consumption. *See under* Household
Consumption-labor goal, 322, 325, 327(figs.),
   328
Consumption-profit goal, 325, 326–327(figs.)
Continuation clause, 137
Controlling. *See* Administrative function
Cooperative Farm Credit System, 112
Corn belt, 68, 170
Corn prices, 184, 205, 222, 224, 236, 239,
   240(table), 241, 243(table), 248–249(table),
   270, 296, 298, 340. *See also* Planting
   decisions
Costa Rica, 391
Cost minimization, 2, 3, 11
Cotton, 220, 224–225, 229, 230
Council of Economic Advisors, 4
Cramer, Robert, 74
Creative morphogenesis, 46
Credit, 6, 11, 24, 25, 35, 112–115, 116, 125,
   151, 155–156, 176, 182, 254, 264,
   272–275, 304
   status, 140, 143
   See also Income, taxation; Loan rates; Risk,
   credit
CROPMX, 222
Crop Reporting Districts, 225
Cross-commodity issues, 5
Cumulative distribution functions (CDF), 173,
   174
Cyert, Richard, 31

Danok, Abdulla B., 176, 177
Dartmouth College Computing Center, 276
Darwin, Charles, 33
Data and program system, 373–374
Data availability problem, 146, 147–148
Data banks, 75
Data Resources, Inc., 73–74, 81
Day, Richard H., 35, 38, 39, 41, 42, 50, 51,
   52, 53, 54, 56, 57, 59, 61, 62, 91, 92, 94,
   217, 400
Day-Singh model, 35–39
Dean, G., 168
Decision making, 400–401, 404–405
   authority, 136, 139, 143
   flexible, 335, 336, 338, 347. *See also* Risk
   linear, 335
   strategic, 75, 76(table), 78(table), 85, 97
   tactical, 75, 76(table), 85, 97
   See also Adaptive economics; Agricultural
   forecasts; Farm sector, decision making;
   Goals; Household, -firm decision making
Decision trees, 84, 85
Deficiency payments, 225, 345
DeHaan, H., 40
Delta states, 68, 224
Demand shifts, 5
Depreciation, 223, 254–257, 302–303

Depression, 11
Deterministic models, 351, 352
Dillon, John L., 162, 170
Discounted cash inflows, 292
Disequilibrium, 22-24, 31, 33, 40, 44, 50, 51-52, 54, 56, 57, 197, 298
Distribution of outcomes, 161, 162, 356
Distributive effects, 7, 8, 10, 12, 59, 63, 65, 91, 261, 275
Doane's Agricultural Service, 74
Dobbins, Craig L., 353, 358, 359, 361, 364
Doering, Otto C., III, 386, 393, 395, 396
Dual control theory. *See* Adaptive control theories
Duncan, Marvin, 282
Durable inputs, 217
Dynamical systems theory. *See* System dynamics
Dynamic economics, 21, 33, 40-41(figs.), 42, 43(fig.). *See also* System dynamics
Dynamo computer language, 262, 263, 275, 276, 283

East-Central states, 224
Econometric models, 8, 35, 46, 53, 66, 67, 68, 69, 73, 74, 75, 81, 106, 185, 189, 198, 207, 281, 400-401. *See also* Macromodel; Micromodel
Economic Recovery Tax Act (1981), 105, 136, 138, 140, 141
Economic Research Service (ERS), 1, 4, 5, 6, 7, 8, 15-17, 94, 212, 223, 285, 298, 407
baseline price data, 224
forecasting program, 4
priority changes, 7, 8, 10
reorganization (1973), 5
research, 107-108, 109, 288, 393-394
See also Commodity Economics Division; National Economic Analysis Division; Natural Resource Economics Division
Economics, 53. *See also* Agricultural economics; Behavioral economics; Bioeconomics; Dynamic economics; Neoclassical economic theory
Economist-lawyer conflict, 146-147, 148
ECROP. *See* Expected size of crop
Eddleman, B. H., 407
Education, 57, 93, 107. *See also under* Extension programs; Risk
Egypt, 391-392
EHP. *See* Expected harvest price
Eidman, Vernon R., 195, 196, 200, 201, 202, 204, 205, 206, 223, 298
Eisgruber, Ludwig, 312. *See also* Patrick-Eisgruber model
Elasticities of demand, 57
EMV. *See* Expected monetary value
E(MVP). *See* Expected marginal value product
Endogenous structural change, 45
Endogenous variables, 31, 32, 51, 52, 53, 106
Energy, 5
Energy, Department of, 379
Environmental problems, 5, 34
Environmental Protection Agency, 379

Equilibrium, 31, 33, 39, 44, 50, 51, 52, 53-54, 56, 57, 61, 124. *See also* Market equilibrium; Prices, competition equilibrium; Social equilibrium
Equity, 63, 157, 163, 246
capital, 112, 118-119, 161, 163, 164
markets, 127
terminal, 292, 293
-to-asset ratio, 129-130
See also Farm debt, -to-equity ratios; Hogs, production model
Erosion, 107
ERS. *See* Economic Research Service
Estate planning, 136, 137, 154, 155, 288
corporate, 141-142, 144
EV. *See* Mean-variance; Net returns; Portfolio model, mean-variance
Evaluation. *See* Model efficacy
Evolutionary model, 30, 33
Exogenous variables, 51, 106, 223, 232(n8), 264, 269, 270, 275, 279, 283, 316, 373
Expectations model, 167, 206
Expected harvest price (EHP), 342, 343
Expected marginal value product (E(MVP)), 170
Expected monetary value (EMV), 168, 311
Expected size of crop (ECROP), 342
Expected utility hypothesis, 160, 170, 172-173, 174, 196, 218, 311, 333, 334, 340, 341
Experiment stations, 11, 107
Exponential smoothing function, 321
EXPRESS, 382
Extension programs, 73, 74, 75, 84, 287, 367-368, 369, 385, 386, 390, 393, 394, 407
education, 375-383, 394, 396

Factor markets, 276
FAHM 1981. *See* Farming-related household model
Fan, Y.-K., 42
Farm Bureau (Ill.), 74
Farm Business Farm Management Association (FBFM), Illinois, 128
Farm Credit System, 155, 167
Farm debt, 112, 113, 116, 121, 229-230, 246, 247-252, 267, 271(table), 296
-to-asset ratios, 113, 114(fig.), 129, 229, 247, 248-249(table), 252, 254
-to-equity ratios, 119(table), 129, 182, 218, 229-230, 264, 265-266, 267(fig.), 269, 274, 275
See also Farm firm, and finance
Farmer-owner reserve (FOR) program, 222, 225, 226-228(table), 279
Farmers Home Administration, 115, 153
Farm firm, 14, 63, 279-280
conceptual micromodel, 386-388
defined, 13
and finance, 115-116, 118, 125, 149-150, 151, 155-157, 160, 161, 219, 239, 254-257, 389-390. *See also* Credit;

Income, and financial flow; Portfolio model; Risk, financial
    and inheritance, 136, 139, 141, 144, 147.
    See also Intergenerational transfer plan
    organization, 135–145, 149, 154, 155. *See also* Farm management; Farm sector; Farm trusts; Household; Incorporated farms; Partnerships
    and taxation, 136, 138–139, 140, 239, 254, 257, 264. *See also* Income, taxation
    See also Computers; Extension programs
Farming-related household model (FAHM 1981), 262–277, 281, 282, 283, 284
Farm-level continuous optimization models for integrated policy analysis (FLIPCOM), 211, 214–224, 230, 279–280, 283
Farm management, 2, 5–6, 11, 104, 296, 301
    aggregate, 7
    budgeting, 2, 10–12, 14
    and growth and adjustment, 6, 11, 12, 13, 14, 153, 224–230, 367–368
    and universities, 7
    See also Extension programs; Government policies
Farm operator families, 13, 14. *See also* Household
Farm sector, 1–2, 5, 74, 105, 379
    aggregate response, 7, 8, 63
    change, 4, 5, 6, 7
    decision making, 6, 12, 19, 33–34, 54, 73, 96–98, 104, 144–145, 151–152. *See also* Decision making, authority
    diversity, 6
    growth, 112, 115, 154, 155, 290, 292
    needs, 368–369
    structure, 106
    See also Government policies; Urban-regional sectors
Farm size, 6, 11, 104, 106, 112, 213–214, 236, 248–249(table), 252, 253(table), 254–257
Farm trusts, 144
Farm type, 6, 104, 106, 154
Farrell, Kenneth R., 407
FBFM. *See* Farm Business Farm Management Association
FCIC. *See* Federal Crop Insurance Corporation
Feasible region, 25–26, 27–30, 38, 45
Federal Crop Insurance Corporation (FCIC) Program, 125, 222, 223
Federal Land Bank, 143, 242–243(tables), 264
FEDS. *See* Firm Enterprise Data System
Feedback, 34, 35, 38, 39, 40, 41, 42, 44, 52, 56, 61, 71, 96, 115, 206, 341, 359, 383
    nonlinear, 45, 50, 51, 52, 263
    See also Aggregate feedback; Autofeedback; Interactive feedback; Macrofeedback
Feeder cattle prices, 183, 184, 185
Fel'dbaum, A. A., 32
Feldstein, Martin, 126
Ferguson, C. E., 311
Financial institutions, 148, 182
Financing. *See* Credit; Farm firm, and finance; Household, financial behavior;

Micromodel, of accounting; Portfolio model
Firch, R. S., 206
Firm Enterprise Data System (FEDS), 223
Fisher Effect, 127
Fisher separation theorem, 292
Fitzsimmons, Cleo, 245
Fixed factor utilization, 14
Fixed strategies, 359
Flexible efficiency criterion. *See* Stochastic techniques
FLIPCOM. *See* Farm-level continuous optimization models for integrated policy analysis
FLIPRIP, 211–212, 224, 226–228(table), 279, 288
FLIPSIM, 211–212, 279, 288
Flood, D., 183
FLOSSIM. *See* Multiple-farm opportunity set simulation model
Food exports, 5, 379
FOR. *See* Farmer-owner reserve program
Forecasts, 207–208, 209, 297
    one-step-ahead, 167
    See also Agricultural forecasts; Price, forecasts
Forrester, Jay W., 31, 32, 40
FORTRAN computer programs, 263. *See also* FLIPRIP; FLIPSIM
Fox, Karl, 4
Fox report (1974), 4–5
Franzmann, John R., 209
Frey, T., 224
Friedman, Milton, 20, 361
Fundamental analysis, 189–190, 194(n10)
Futures markets, 156, 182–183, 184–185, 186, 187, 188, 190–191, 198, 199, 204, 205, 206, 340, 344

Gabriel, S. C., 120, 124, 163
Game theory approach, 295, 333, 334
Gardner, Bruce L., 106, 184
Generalized risk-efficient Monte Carlo programming (GREMP) model, 336–338, 339–341, 343, 344–347, 352, 353(fig.), 355(table), 356, 359
Generating techniques (GT), 352(fig.), 353–354, 355(table), 356
Georgia, 290, 295, 298
Georgia State University, 81
Goal programming (GP), 313–314, 315(fig.), 318(table), 328, 352(fig.), 353, 354, 355(table), 356, 359. *See also* Goals, model
Goals, 28, 29–30, 35, 37(fig.), 38, 45, 53, 61, 202, 245
    model, 310–328
    priority, 317, 329(n4)
    safety, 35, 36, 37(fig.), 38
    social, 388
    subsistence, 35, 36, 37(fig.), 38, 39
    weights, 312–313, 316, 317, 319(table), 321, 328
Goicoechea, A., 353

Goodsell, Wylie, 5
Gordon growth model, 127
Government policies, 2, 3, 5, 362
  alternatives, 5, 16
  analytical responses to, 378–380
  and decision making, 13, 34, 159
  impact, 6–7, 10, 16, 19, 104, 106–107, 109,
    125, 306, 345, 347, 381–382, 386, 401.
    See also Agricultural economics; Economic
    Research Service; Regulatory impact
    studies
  impact, nonfarm, 379
  and models, 70–71, 102, 152, 154, 258,
    276, 279, 286–288, 361, 365, 372, 395,
    401, 407. *See also* Farming-related
    household model; FLIPRIP; FLIPSIM
  and programs, 103–105. *See also* Credit;
    Extension programs
  See also Goal programming; Macroeconomic
    policies; Micromodel, and policy analysis;
    Price, supports; Tax policy
GP. *See* Goal programming
Grain market, 73, 77, 82, 184
Grain reserves, 103
Grain storage decisions, 81–84, 186–187, 189,
  239, 246
Great Plains, 102–103
GREMP. *See* Generalized risk-efficient Monte
  Carlo programming model
Griffin, S., 125, 213, 236
Growth and adjustment. *See under* Farm
  management
GT. *See* Generating techniques

Halter-Dean model, 214, 215–216(table)
Hardaker, J. Brian, 162, 170
Hardin, D. C., 213
Harl, N., 224
Harmon, Wyatte L., 245
Hauschew, Larry D., 240
Havlicek, Joseph, Jr., 154
Heady, Earl O., 40, 290
Hedging, 183, 185–186, 187, 188, 190, 191,
  195, 196, 198–199, 208, 340–341, 342,
  343, 345, 346(table), 347
Heifner, Richard G., 185
Helmers, Glenn A., 187
Herr, William McD., 240
High Plains states, 224
Hildreth, R. J., 399
Hinman, H. R., 213. *See also* Hutton-Hinman
  model
Hirshleifer, Jack, 292
Hochman, E., 85
Hogs, 40–41(figs.), 205, 298, 299(table)
  markets, 184, 185
  production model, 290–292, 295–305
Holland, D., 214
Holmes, E. G., 245
Holt, C. C., 335
Homeostasis, 44
Homogeneous regions. *See* Land-grant
  regional adjustment studies
Hoskin, Roger L., 239

House Committee on Small Business, 182
Household
  consumption, 12, 35, 36, 37(fig.), 38,
    161–162, 163, 164, 165, 246, 252, 267,
    276, 282, 297, 306
  financial behavior, 262, 266–277
  -firm decision making, 35, 54, 264, 282
  needs, 112, 262
  See also Farming-related household model
Households and Institutions Sector, 13
Human capital output, 93–95, 287
Hunting and food-collecting activities, 18
Hutton, R. F., 213
Hutton-Hinman model, 214, 215–216(table)
Hybrid models, 213, 214. *See also* Farm-level
  continuous optimization models for
  integrated policy analysis

Ikerd, John E., 185
Illinois, 128–129, 224, 225, 226–228(table),
  229, 230, 236, 239, 244, 270, 407
Illinois Farm Business Records, 236
Illinois Land Price Index, 242(table)
Illinois State Statistical Office (Statistical
  Reporting Service), 243
Income
  adequate, 2, 3, 4, 5, 106
  and borrowing, 123
  data, 102–103
  depressed, 2, 11–12
  and financial flow, 45, 130, 151, 152, 156,
    163–166, 200, 223–224, 245, 246,
    250–251(table), 264, 265, 266–269, 343,
    346(table). *See also* Goals, model; Linear
    programming, polyperiodic
  fluctuations, 105, 202
  growth, 128–129
  level, 91, 131
  maintenance, 112
  mean, 229
  net, 6–7, 73, 161, 163, 164, 167–168,
    175–176, 178, 197, 199, 223, 239, 246,
    292
  nonfarm, 13, 14, 105, 113, 262, 264,
    271(table), 272–273(figs.), 274(table)
  operating, 113, 165–166, 246,
    250–251(table)
  perpetual, 127
  production, 143
  supplements, 24
  support programs, 160
  taxation, 126, 137, 140–141, 142, 143, 151,
    160, 165, 167–168, 199, 200, 224, 239,
    246, 250–251(table), 252, 297, 300–304
  See also Assets; Capital gains; Inflation;
    Insurance, income; Partnerships
Incorporated farms, 140–145, 147
India, 169
Indiana, 176
Individual equilibrium, 21, 22
Inflation, 81, 105, 112, 113, 123, 126–131,
  132, 156, 197, 219, 241, 243(table), 254,
  297–298
Innovation, 34, 35

Institutional variables. *See* Farm firm, and finance
Insurance, 11, 24, 58
  crop, 160, 177, 222, 225, 226–228(table), 340, 341, 342–343, 345, 346(table), 347
  disaster, 125, 222, 223
  income, 104
Interactive feedback, 23, 25, 27, 42
Interest rates, 105, 112, 113, 114, 115, 116, 121, 123, 125, 129, 130, 151, 156, 165, 167, 197, 219, 242–243(tables), 270, 298, 343
Intergenerational transfer plan, 310, 314, 316, 322
Intuitive models, 396
Investment, 25, 104, 105–106, 115, 116, 117, 125, 130, 201, 218, 290. *See also* Land, value; Linear programming, polyperiodic; Partnerships
Iowa, 214
Iowa Grain Livestock Hedging Corporation, 74
Iowa State University, 6
  Extension Service, 287–288
Irwin, George D., 154, 290
Iterative approach, 317

Jevons, William Stanley, 20
Johnson, Bruce B., 243
Johnson, G. L., model, 57, 59
Johnson, Sherman R., 5
Johnson, Stanley R., 98
Just, Richard E., 59, 80, 81, 98

Kaiser, Eddie, 293
Kansas City contract, 341
Kansas Transitional Planning Model, 368
Keeney, R. L., 162
Kellogg Foundation, 74
King, Robert P., 84, 171, 175, 223, 239, 336, 338, 341, 353, 354, 358, 359, 361, 364, 365, 407
KISS (keep it simple, stupid) philosophy, 92
Knapp, Seaman A., 394
Knowles, G. J., 169, 172
Koyck distributed lag model, 222
Kuhn, H. W., 353
Kuhn, Thomas S., 53

Labor
  costs, 219
  family, 264
  markets, 58, 222, 223, 224, 225, 246, 247
Ladd, George W., 53
Lagged dependenet variable, 185, 241
Land
  crop, 222, 223, 224–225, 229, 230, 269–270
  diversion, 103, 104, 222
  foreign investment in, 147
  indivisibility, 136
  prices, 128, 154, 156, 240–241, 242–243(tables), 244, 252, 267(fig.)
  set asides, 103, 104, 222
  supply, 241–245, 254, 264, 266(fig.), 286

  value, 105, 113, 126–129, 136, 213, 223, 224, 298
  See also Farming-related household model; Multiple-farm opportunity set simulation model
Land-grant regional adjustment studies, 3, 4
Lee, John E., Jr., 10, 15, 16, 91
Lee-Rask land price model, 244, 252
Legal variables. *See* Incorporated farms; Partnerships; Tax policy
Leontief, Wassily, 101
Lexicographic approach, 311, 312
Limitation coefficient, 25
Limited partnerships, 139
Lin, William, 168
Linear programming (LP), 2–3, 6, 15, 69–70, 115, 153, 213, 214, 219, 239, 263, 279, 334, 352(fig.), 378, 401
  algorithms, 334
  decision process, 2
  and multiple goals, 312, 359, 362. *See also* Goal programming
  polyperiodic (PLP), 289–307
  See also FLIPSIM; MOTAD model
Lins, David A., 224
Livestock commodities, 182. *See also specific types*
Loan rates, 103, 112, 116, 121, 130–131, 167, 225
Loftsgard, Laurel D., 290
Looney, J. W., 146, 147, 149, 150, 151, 153, 154, 155
Lorenz, E., 45
LP. *See* Linear programming
Lutgen, Lynn H., 187
Lybecker, Donald W., 341, 359

McCarl, Bruce A., 170, 176, 177
Machinery, 152, 177, 290, 347
McRae, E. C., 335
Macroeconomic policies, 105–106, 286, 377, 378
Macrofeedback, 39, 40, 41–42
Macromodel, 4, 5–6, 7, 15, 17, 54, 65–66, 98, 102, 103, 212, 393
  aggregation error, 6, 16, 64, 65–67, 68, 69
  flexibility, 64
  of nonfarm economy, 34–35
  use, 382, 383, 395, 396, 405
  variables, 16, 212, 261, 275
  See also Micromodel, and macromodel; Normative models; Positive models
Malthus, Thomas, 33
Manetsch, T. S., 39
Mapp, Harry P., Jr., 223, 295, 353, 358, 359, 361, 364
March, James G., 31
Marginal factor cost (MFC), 170
Margins, 182, 193(n3), 343
Market conditions, 2, 4, 5, 6, 19, 22, 36, 41, 50, 57, 71–72(n1), 112, 116, 185, 206
  and financial deregulation, 113–114
  international, 5, 73, 182, 195, 202, 304
  nonfarm, 39

See also Distributive effects; Grain market
Market equilibrium, 21, 22
Market-groping, 20
Marketing, 117, 161, 203, 206
Marketing firms, 2
Marschak, J., 32
Marshall, Alfred, 20, 33
Martin, J. Rod, 290
Mass, Nathaniel, 51, 52
Maximizing behavior, 311, 312, 313, 314,
    334, 336, 351-352, 353
MCP. *See* Monte Carlo programming
Mean-variance (EV), 160, 334. *See also under*
    Portfolio model
Mechanization, 112
Melichar, Emanuel O., 128
Meyer, J., 174-175
Meyers, J., 84
MFC. *See* Marginal factor cost
Michigan, 340
Michigan State University (MSU), 75, 84
    Department of Agricultural Economics, 74
*Michigan State University Agriculture Model
    Project*, 74, 240, 258(n3)
Microeconomic behavior, 28, 263
Micromodel, 1, 3, 4, 5, 6-8, 10-13, 15, 16,
    39, 47, 54, 231, 341, 378, 379, 386-388,
    393-394, 399-402
    of accounting, 130-131
    adaptive, 29(fig.), 35
    and aggregate behavior, 8, 10, 15, 16,
        63-64, 132, 153-154, 190, 212, 245-246
    aggregation error, 15, 69
    analytic framework, 57-58
    cost-effectiveness, 17
    data, 3, 10, 14, 16, 19, 34, 54, 59, 98, 102,
        103, 104, 107-108, 153, 154, 368, 370,
        372, 373, 397
    dynamic, 107
    formal, 12, 14
    informal, 12-13
    and macromodel, 8, 15, 16, 34, 42, 51,
        63-65, 70-71, 88-90, 91, 92-93, 94, 96,
        97, 109, 154, 212-213, 258, 279, 283,
        382, 394, 395, 397, 406
    and policy analysis, 115, 161, 177
    specification error, 15
    specifications, 154
    use, 35, 56, 57, 59, 70-71, 98, 105, 106,
        109, 148, 159, 213-214, 215-216(table),
        286-288, 365-366, 380, 382, 383, 397,
        403-408
    variables, 151
    See also Adaptive economics; Day-Singh
        model; Decision making; Farm firm;
        Goals, model; Multiple-farm opportunity
        set simulation model; Portfolio model;
        Risk
Microsoft, 370, 371(table)
Minimization process, 174
Minnesota, 172
Mississippi, 224, 225, 226-228(table), 229,
    230
Mitchell, Donald O., 96, 97

Mixed integer programming, 304
MOA. *See* Multiple objective analysis
Model development, 89, 94-95, 108-109,
    153-154, 306-307, 362, 366, 368, 404,
    407. *See also* Economist-lawyer conflict
Model efficacy, 358-361
Modiglioni, F., 335
Monte Carlo programming (MCP), 352(fig.),
    354, 355(table), 356
Monte Carlo simulation model, 168, 240, 247,
    250-251(table), 254, 257, 359. *See also*
    Generalized risk-efficient Monte Carlo
    programming model; Multiple-farm
    opportunity set simulation model; Risk,
    -efficient programming; Risk-efficient
    Monte Carlo programming model
Moore, C., 168
Model referrant control, 33
Model specification, 65-71, 88, 92, 93
MOTAD model, 170, 176, 214, 245, 333, 334
MSU. *See* Michigan State University
Muller, G. P., 41
Multiperiod models, 153, 162, 177, 197,
    212-213, 217, 219, 224, 347
Multiple-farm opportunity set simulation
    model (FLOSSIM), 235-241, 244-258,
    283, 287
Multiple objective analysis (MOA), 353
Multisector model, 40, 42
Multivariate Beta process generator, 239,
    240(table)
Multivariate distribution, 338, 344
Multivariate models, 53
Musser, Wesley N., 296, 365
Muth, J. F., 335

National agricultural policy. *See* Economic
    Research Service; Government policies
National Economic Analysis Division, 5
National Income and Product Accounting
    System, 13
Natural Resource Economics Division, 6
Natural resource issues, 16-17
NCON. *See* Number of contracts already sold
NDES. *See* Number of contracts desired to be
    sold
Nelson, Richard, 33
Nelson, Ted R., 385, 389, 390, 393, 404
Neoclassical economic theory, 20-22, 38, 51,
    53, 56, 57, 58, 292, 296, 305-306
Nerlovian lagged price, 182
Net returns, 185, 293-294, 311, 316, 336, 340
Net worth. *See* Assets
New industrial organization models, 57, 58,
    69(n1)
Nondurable inputs, 217, 219
Nonfarm economy, 34-35, 155. *See also*
    Income, nonfarm; Macromodel, of
    nonfarm economy
Nonlinear relationships, 276
Normative models, 66, 69-70, 96, 159, 160,
    166, 196, 204, 205, 306, 339, 345
NSOLD. *See* Number of contracts sold

Number of contracts already sold (NCON),
342
Number of contracts desired to be sold
(NDES), 342
Number of contracts sold (NSOLD), 342

Oamek, George E., 175, 354, 358, 359, 361,
364, 365
Oats, 222
Oilseed market, 73
Oklahoma, 322
Oklahoma State University, 209, 214
Optimal control theory, 115, 175, 209, 335
Optimization, 20, 21, 27, 28, 31, 42, 54, 57,
123, 124, 131, 190, 196, 209, 213, 245,
353, 360(fig.), 369, 378, 388, 401
myopic, 59, 217
subroutines, 222
See also Farm-level continuous optimization
models for integrated policy analysis;
Linear programming, polyperiodic
Overproduction, 2

Pareto optimality, 20
Paris, Q., 334, 335
Partnerships, 135, 136–139, 144
Patrick, George F., 245, 312
Patrick-Eisgruber model, 214, 215–216(table)
Payment limitations, 103, 106, 131, 156, 222
Peanuts, 222
Peck, Anne E., 182, 183, 184, 187
Penrose, E. T., 296
Performance indicators, 161
Perkins, Gerald, 404
Perlack, R. D., 354
Pesticide use, 202
Pest management. See Bioeconomics
Petit, M., 295
Pinto beans, 340
Planning function, 23, 24, 25, 35, 40, 247,
279–290, 291–292, 316, See also Goals,
model; Risk, -efficient programming;
Target planning
Planting decisions, 184, 339
Plaxico, J. S., 183, 290
Policy education. See Extension programs,
education
Porter Demonstration Farm, 394
Portfolio model, 111, 116–124, 131, 147, 150,
154, 201
adjustments, 124–126
mean-variance (EV), 116, 117, 118, 150,
157, 201
Positive economics, 20
Positive models, 66–69, 70, 96, 166, 204,
281–285, 306, 345
Predictions, 52, 54, 56, 159, 160
Prescriptive objective function, 209
Price, 5, 39, 73, 124, 153, 198, 218, 222, 254,
263, 304–305, 344
aggregate, 8, 63, 91, 155, 160
competition equilibrium, 20, 22, 23, 298
-cost relationship, 104
cross-, 68

decrease, 57
forecasts, 77, 79, 80, 189, 207, 208, 297,
298, 299(table), 342
forward, 182, 183, 184, 188, 192(n6), 340,
342, 343
high, 11, 113
increase, 57
marginal-cost, 30
risk, 164, 181–182, 183, 188, 189, 191, 199,
202, 205
supports, 3, 19, 91, 107, 125, 181, 213,
222, 225–229, 279
target, 103, 187–188, 189, 191, 198, 222,
223, 225, 226–228(table), 279, 344
trend, 297–298
variability, 181, 182, 185, 186, 195, 198,
205, 206, 230
and yield, 239–240, 241, 245, 247,
248–249(table), 270, 320–321
See also Expected harvest price; Futures
markets
Priority programming, 30, 35, 37(fig.)
Private sector, 7–8, 19, 74
Probability. See Stochastic techniques
Problem-alternatives-consequences approach,
376–377
Production, 73, 104, 112, 117, 151, 159, 160,
161, 219, 236, 239, 379
cash flow, 264, 267, 294
controls, 19, 107, 151–152, 181
costs, 106, 125, 218, 219, 223, 241
decisions, 187–188, 201, 206, 222–224, 290
risk, 164, 185, 202
See also Farming-related household model;
Income, production; Risk, -efficient
programming
Production Credit Association, 143
Product supply and factor demand, 67–68
Pro Farmer, 74
Profit, 152
maximization, 2, 33, 35, 36, 39, 54, 55, 57,
67, 68, 80, 170, 195, 282
non-maximization, 202
Public policy. See Government policies
Purcell, Wayne D., 183, 185, 195, 198, 200,
202, 204, 205, 206, 207, 208
Putty-clay model, 217

Quadratic programming, 176, 236, 334, 335
Qualitative change, 46–47, 50, 53, 54, 64, 71,
92
Quantitative features, 47, 50, 53, 54–55, 58,
59, 63, 64, 71, 88, 91, 92, 107, 147, 209

Raiffa, H., 162
RAP. See Recursive adaptive programming
models
Ratajczak, Donald, 81
Rausser, Gordon C., 53, 59, 80, 81, 85, 98,
217
Ray, Daryll E., 205
Recursive adaptive programming models
(RAP), 214

Recursive interactive programming models
(RIP), 214, 279
Recursive programming (RP) techniques, 3,
12, 29(fig.), 30, 31–32, 33, 40–41(figs.),
42, 43(fig.), 45, 52–53, 214, 362. *See also*
FLIPRIP; FLIPSIM
Regional adjustment studies. *See* Land-grant
regional adjustment studies
Regulatory impact studies, 5, 107
Reichelderfer, Katharine H., 404
REMP. *See* Risk-efficient Monte Carlo
programming model
Resource allocations, 20, 24, 54, 80, 151, 214,
231, 252, 377
Resource availability, 163, 235, 257, 316, 319.
See also Land, supply; Multiple-farm
opportunity set simulation model
Resource flexibility, 187, 188
Resource levels, 2, 3, 5, 25, 104, 321
Retirement planning, 136, 139, 142
Rice, 222
Richardson, James W., 181, 185, 279, 288
Richmond, Barry, 91, 92
RIP. *See* Recursive interactive programming
models
Risk, 12, 14, 26, 58, 59, 104, 112, 113, 114,
115, 116, 147, 150, 154–155, 160, 178,
201–202, 306
aversion, 115, 116, 123, 124, 125–126, 131,
168, 169, 170, 171, 172, 173, 174, 175,
176, 177, 195, 196, 201, 311, 339,
344–345, 346(table)
business (BR), 119–120, 121, 124, 125, 126,
131, 163, 164, 165, 166, 167, 176, 177,
178, 295
credit, 121–123, 155, 156, 167, 168
and education, 172
-efficient programming, 332–345
-efficient set, 117
financial (FR), 119, 120, 121, 124, 125, 131,
161, 163, 164, 165, 166, 167–168,
177–178, 195, 197, 252, 295
modeling, 195–209
neutral, 168, 171, 172, 173, 293, 311, 344
nonfarm, 155
preferences, 84, 160, 161, 168–177, 185,
197, 201, 311, 333, 338, 340, 343, 345,
347, 352
premium, 168
quantification, 162–167
total (TR), 119, 120, 121, 123, 124, 163,
164, 166
See also Basis risk; Portfolio model; Price,
risk
Risk-efficient Monte Carlo programming
(REMP) model, 335–336, 352(fig.), 353,
354, 355(table), 356–357
Robinson, Bob H., 407
Robison, L. J., 84, 171, 175, 338
RP. *See* Recursive programming techniques

Safety principle, 162, 196, 333, 334
Salination, 34
Salter, W. E., 217

Samuelson, Paul A., 20
Sanint, L. R., 123
Satisficing behavior, 27, 28, 29, 30, 50, 51,
57, 196, 198, 311, 312, 313, 314, 317,
400, 402(n1)
Scalar objective function, 354
Scale farm program benefits, 222
Schaller, Neill, 91
Schertz, Lyle P., 282, 284, 288
Schultz, T. W., model, 57, 59
Self-interest, 20
Self-management, 35
Sensitivity analysis, 197, 252–254, 263, 272,
275, 276, 291, 356
Shields, Mike E., 209
Short crop yields, 84
SIC. *See* Standard Industrial Classification
Sigmoid pattern, 42, 43(fig.)
Simon, Herbert A., 22, 30, 32, 311, 335
SIMPLEX, 222
Simulation models, 39, 115, 164, 165, 167,
176, 209, 213, 306, 380
and mathematical programming, 294–295
See also specific models
Singh, I., 35, 38. *See also* Day-Singh model
Situation and Outlook reports (USDA), 73, 74
Skees, Jerry R., 286, 287, 288
Slutsky equation, 72(n2)
Smale, S., 45
Smith, Adam, 20, 33
Smith, Donnie A., 245
Social equilibrium, 22
Socioeconomic variables, 42, 263
Soil Bank act (1956), 377
Soil depletion, 34, 107
Soil management, 11
Sonka, Steven T., 295
Sorghum, 222, 224–225, 230, 319, 320(table),
321, 322, 323–324(table), 325
Soviet grain sale, 4, 181, 202
Soybeans, 181, 184, 205, 222, 224, 236, 239,
240(table), 241, 243(table),
248–251(tables), 296, 298, 340. *See also*
Planting decisions
SP. *See* Stochastic techniques, programming
Sparks Commodities, 74
Spatial analysis, 3
Spence, M., 57
Spillman, W. J., 11
Sposito, V., 40
SRS. *See* Statistical Reporting Service
Standard Industrial Classification (SIC), 13
State Cooperative Extension Service, 74, 75
Statistical Reporting Service (SRS), 17
Stiglitz, J., 58
Stochastic techniques, 12, 45, 65, 69, 116,
209, 213, 241, 294, 334–335, 338, 339,
340, 343, 351, 352
dominance, 84–85, 161, 162, 171, 172, 176,
339, 340, 345
dominance, first degree (FSD), 173, 174,
175, 339
dominance, second degree (SSD), 173–174,
175, 336, 339

dominance, third degree (TSD), 173, 174, 175
flexibility, 359
iteration loop, 358, 360(fig.)
programming (SP), 352(fig.), 354, 355(table), 356
Storage, 3, 103
Storage decisions. *See* Grain storage decisions
Storage file codes, 273–274
Subsidies, 115
Supply response, 184, 188, 189, 190
Surpluses, 3, 68
Swanson, Earl R., 290
Swine. *See* Hogs
Symmetry, 68, 72(n2)
System dynamics, 30, 31, 32, 45, 61, 262–263, 275, 276

Target planning, 26–28, 51
Tatônement, 20
Tax policy, 6, 19, 105–106, 149, 151, 152, 247, 254
legislation, 136, 138, 140, 141, 148
and valuation, 137
See also Farm firm, and taxation; Income, taxation; Incorporated farms; Partnerships
Tax Reform Act (1976), 136
Taylor, C. Robert, 88, 91, 92, 93, 96, 359
Tchebychev inequality, 163
Teachable moment, 380–381, 395
Technical analysis, 186, 189, 190, 191, 192(n9), 206–207, 208, 209, 352
"Technological treadmill." *See* Cochrane, W. W., model
Technology, 2, 11, 18, 19, 39, 42, 44, 57, 104, 223
and change, 159, 160
constraint, 25, 217–218
TELPLAN programming, 239
Tenant farming, 106, 244–245, 252
Texas, 224–225, 226–228(table), 229, 230
Texas A & M University, 212
Department of Agricultural Economics, 223
Thunen, von H.,20
Top Farmer, 74
Tractorcade, 102
Transportation, 5, 148
Trapp, James N., 205
Trinidad, 391
Tucker, A. W., 353
Turning the Searchlight on Farm Policy (Nourse), 377

Uncertainty, 26, 41, 57, 75, 81, 84–85, 104, 159, 163, 178(n1), 311. *See also* Risk, -efficient programming
Unemployment compensation, 24
Uniform Partnership Act (1914), 137
Unincorporated farms, 13
Universities, 7, 73, 74, 94, 153, 390, 403–404

land-grant, 11, 93, 95, 153, 375, 378, 390
research, 107–108, 109, 407
University of Georgia, 296
University of Missouri, 74, 98
Unviability, 45, 46
Urban-regional sectors, 42, 43(fig.)
USDA. *See* Agriculture, U. S. Department of
User models, 395
Utility function, 26, 28, 36, 84, 85, 115, 116, 117, 124, 131, 150–151, 155, 161, 162, 168, 170, 172, 173, 174, 201, 245, 311–312. *See also* Expected utility hypothesis
Utility models, 38, 196
Utility-of-money function, 168

Vandeputte, J. M., 290, 297
Van Horn profit index values, 229
Variable amortization plan, 116
Variability, 201. *See also under* Price
Virginia Polytechnic Institute, 390
VISICALC (computer program), 370, 371(table)
Vitosh, Maurice, 239

Walker, Harold W., 404
Walras, Leon, 20
Washington, 214
Water assessment models, 379
Water management, 11, 17
Water pollution, 34
Weed management, 340
Weeks, Eldon E., 285
Weld County (Colo.), 341, 345
West Germany, 40(fig.), 41–42
Wharton Econometric Forecasting Associates, 73–74, 81
Wheat, 222, 319, 320(table), 321, 322, 323–324(table)
risk management, 334, 340–345
Wheat referendum (1963), 377–378
White, T. Kelley, 176, 177, 290, 296
Whole farm models, 213–214, 215–216(table)
Willett, Lois Schertz, 282, 284
Williamson, Oliver, 31
Willis, C. E., 354
Willman, David R., 293
Winter, Sidney G., 33
Womak, Abner, 98

XMP, 222

Yale Economic Essays, 33
Young, Douglas L., 166, 171, 172, 214, 295

Zellner, A., 335
ZFR. *See* Zone of flexible response
Zilberman, David, 59
Zone of flexible response (ZFR), 26, 27–28, 38